Herschel at the Cape

DIARIES AND CORRESPONDENCE
OF SIR JOHN HERSCHEL, 1834–1838

NUMBER ONE

History of Science Series

HUMANITIES RESEARCH CENTER

THE UNIVERSITY OF TEXAS

The "admirable likeness," Pickersgill's portrait of Sir John Herschel.
(By permission of the Master and Fellows of St. John's College, Cambridge)

Herschel
at the Cape

DIARIES AND CORRESPONDENCE OF SIR JOHN HERSCHEL, 1834–1838

EDITED WITH AN INTRODUCTION BY

David S. Evans, Terence J. Deeming, Betty Hall Evans, Stephen Goldfarb

UNIVERSITY OF TEXAS PRESS · AUSTIN AND LONDON

Standard Book Number 292–78387–6
Library of Congress Catalog Card No. 69–63013
Copyright © 1969 by The University of Texas
All Rights Reserved

Printed by The University of Texas Printing Division, Austin
Bound by Universal Bookbindery, Inc., San Antonio

To Our Parents

FOREWORD

With the publication of *Herschel at the Cape,* the Humanities Research Center in The University of Texas initiates a series on the history of science.

The proposed publications will range over a wide variety of subjects. Each should be significant in itself. The significance of sponsorship by the Humanities Research Center is manifest: This series bears evidence to the fact that the history of science provides essential threads in the broader history of man's experience.

Harry Ransom

ACKNOWLEDGMENTS

This book begins in the bar of the Grand Hotel, Saltsjöbaden, in Sweden. Lowering our voices to match the subdued and expensive decor, Terry Deeming and I discussed the fact that the Manuscript Library of The University of Texas at Austin had acquired a considerable collection of the papers of Sir John Herschel. One thing led to another, and in 1965 I found myself at the University as a National Science Foundation Senior Visiting Scientist, mainly to do some astronomy. I am more than grateful for this award, and to the chairman of the Astronomy Department, Professor Harlan J. Smith, for permitting me to spend a corner of my time on the rather less immediate topic of astronomical history. My own thanks are due to Dr. Deeming, who introduced me to the possibility and worked with me on the production of this book; to my wife, Betty Hall Evans, with her knowledge of South African history and her talent for deciphering even the worst specimens of Herschelian handwriting; and to Mr. Stephen Goldfarb, who has worked with me ferreting out knotty points and, by his cataloguing of the collection, made it easier for us to find our way through it. The thanks of all of us are due to Dr. George Basalla of the Department of History, to Dr. James H. Leech, special consultant for the History of Science Collection, both of whom greatly smoothed our path, and to the Manuscript Committee, who authorized our work on the documents we selected. Special words of thanks go to Mrs. Mary Virginia Daily of the Humanities Research Center for turning our transcripts, only one degree more legible than Sir John's original, into a typescript appropriate for the attention of an editor, and to Miss Wanda Wolf of the Department of Astronomy for her valuable secretarial help in the final stages of preparation of the manuscript for the press.

In the editorial process we have had much help, always generously and gladly given. Those who aided us have included Dr. A. M. Lewin Robinson, chief librarian of the South African Public Library, Cape Town; the chief archivist, South African Archives, Cape Town; Dr. W. S. Rapson, scientific adviser to the Chamber of Mines, Johannesburg; Dr. Anthony V. Hall of the Department of Botany in the University of Cape Town; my colleagues at the Royal Observatory at the Cape; Mr. Heath of the Naval Library of the Ministry of Defence, London; Dr. George Doig and Dr. William H. Hess, both of the Department of Classical Languages at The University of Texas; Dr. Andrew T. Young of the Department of Astronomy, The University of Texas; Professor John H. Day, head of the Department of Zoology in the University of Cape Town; Mrs. M. K. Rowan of the Percy Fitzpatrick Institute of Ornithology, University of Cape Town; Mr. Frank Bradlow of Cape Town, the author, with his wife, of the standard work on Thomas Bowler and an expert on many of the individuals whose names appear in the diaries; and Herr Günther Buttmann, author of the recently published biography, *John Herschel, Lebensbild eines Naturforschers.*

The aid given by these experts has often involved them in many hours of arduous research. Our acknowledgment of their generosity and skill in rescuing us from errors which might have been committed, cannot be too warmly expressed.

Illustrative material has been drawn, not only from the diaries themselves, but also, by permission, from material in the possession of the South African Archives Depot; the South African Library; the Master and Fellows of St. John's College, Cambridge, England; the National Portrait Gallery, London, England; and members of the Herschel family. To all of them we express our grateful thanks. Also to be acknowledged are Professor J. M. Winterbottom and Messrs. Maskew Miller, Ltd., for permission to quote from E. Leonard Gill's *A First Guide to South African Birds*; Graaff's Trust Limited, Cape Town, for permission to quote from G. R. McLachlan and R. Liversidge's revision of Austin Roberts' *Birds of South Africa*; and Random House, Inc., for permission to quote from Robert N. Linscott's edition of *Complete Poems and Selected Letters of Michelangelo.*

To conclude this list I would like to add the name of the late Mr. Donald McIntyre, of Cape Town, who first introduced me to the Herschel documents at the South African Archives and communicated to me some of his own fascination with the engaging personality of John Frederick William Herschel.

D.S.E.

Austin, Texas, U.S.A.
Observatory, Cape Province, South Africa

INTRODUCTION

Those who know the name of Herschel will know its possessor as the discoverer of the planet Uranus, an incident immortalized in the lines of Keats:

> Then felt I like some watcher of the skies
> When a new planet swims into his ken

They may also know that he made his own telescopes, aided by his devoted sister, who fed him with her own hands as he worked uninterruptedly at grinding and polishing his mirrors.

This indeed was Herschel—but not the one we are interested in. The celebrated Herschel, to whose discovery of Uranus in 1781 the quoted lines apply, was Frederick William Herschel (1738–1822), usually known simply as William Herschel. His devoted sister was Caroline Lucretia Herschel (1750–1848), and it is the son of the one, and the nephew of the other, John Frederick William Herschel (1792–1871), who is the author of these diaries. To concentrate attention on the son is not to denigrate the father; rather, perhaps, we shall find the son comparable in distinction with his father, almost certainly exceeding him in versatility, and portrayed in far more fascinating human detail through the medium of his own words.

Our first task then, is properly to distinguish our dramatis personae. William, the father, is the subject of a number of excellent biographies, and we need do no more than recapitulate those incidents of his life which bear upon our own topic.

He came of a musical family resident in Hanover, at a time when the Hanoverian house were the sovereigns of Britain. His first employment was as an oboist in the Hanoverian Guards at the age of fourteen. At an early age he became interested in philosophy, but after the defeat of Hastenbeck in 1757 his health began to fail, and his parents sent him off to England. He landed there with no more than a French crown piece in his pocket, some say, technically a deserter from his regiment, though this has been contested.

William's life was hard but he eventually made a musical career for himself, working his way up in 1766 to the "agreeable and lucrative" situation of organist at the Octagon Chapel in Bath. Bath was then a fashionable spa, the resort of the nobility and gentry, who occupied their leisure with routs and entertainments in which instrumental and vocal music played an important part. There he directed concerts and oratorios, gave music lessons, and composed a number of not very distinguished works.

His study of harmony led him on to mathematics, that of mathematics to astronomy and other branches of what was then called "Natural Philosophy." After using a rented small reflecting telescope, he bought the tools and equipment of an optician and began in 1773 to construct his own instruments. By this time he had with him in Bath his brother Alexander, who played in the Bath orchestra, and his sister, Caroline Lucretia, eleven years younger than himself. He began a life of unremitting toil, teaching or conducting for many hours, and then hurrying in the intervals to work on the manufacture of his mirrors. Gradually his observing programs extended until he had observed the whole of the accessible heavens. This he did by a technique carried on in the south by his son. This was called "sweeping" and consisted in holding the telescope fixed at a certain inclination on the meridian. In this way, as the earth rotated, all the objects at a certain declination would be carried successively into the field and could be noted. Their positions could be derived either from the inclination of the telescope and the time shown on a sidereal clock, or from a knowledge of the positional coordinates of a number of standard or "zero" stars of which the positions were known by other means. Because of the frequency with which this subject is mentioned in the diary it is of some importance to have an appreciation of this simple technique.

It was while employing this technique that William discovered Uranus, the first addition to the classical list of planets. By the perfection of his telescope he was able to distinguish the planetary disc, and, although he first took it for a comet, he soon realized its true

nature. He was at first, through a devotion to his sovereign, impelled
to call it Georgium Sidus, but the Greek name took precedence, and
the planetary symbol is based on the letter *H* in honor of its discoverer. Two consequences flowed from this: One was an interview
with the King (George III) in 1782, when the monarch handed him
a pardon for his desertion, if desertion it was, from the Hanoverian
forces. The other was his appointment, unique to him, as Royal Astronomer, with a salary of £200 per year. It was at this time that he
moved, accompanied by his sister, first to Datchet, later to Windsor,
and finally to Slough, all fairly close to the royal residence at Windsor Castle.

The only way to summarize the vast extent of his observational
researches is to say that he was the founder of stellar astronomy. By
this we mean that he surveyed and catalogued the whole of the
northern sky, in the course of which he discovered 2,500 nebulae and
studied their distribution. He undertook a vast program of star
gauging, that is, counting the numbers of stars of various magnitudes
in selected areas of the sky. If one assumes that all stars are of equal
intrinsic brightness and that they differ in apparent brightness solely
because of their different distances from us, one can use the results
of such surveys to make a scale model of the distribution of the stars
in space. The assumption is wildly incorrect, but the results are less
at variance with the truth than might have been expected. In this
way William gave astronomy a metrical picture of our own galaxy,
seen by us as the Milky Way, which bears a remarkably close resemblance to the correct one. The two aspects in which it is most
deficient are the placing of the Sun near the center of the system,
and the interpretation of the dark rifts in the Milky Way, in fact
due to obscuring matter in interstellar space, as dark lanes void of
stars, which have the remarkable, to us, property of all pointing
toward the observer in William's model.

When William flourished, nothing was known of the real distances
of stars, except that they must all, even of the nearest, be very great. If
this were not so, then during the course of a year, as the Earth moves
in its orbit round the Sun, the nearer stars would be seen to shift their
positions with respect to the more remote ones. This is in fact the
method by which a rather small sample of stars have had their distances measured, but the shifts are all minute, and in only one or two
cases are they detectable by visual measures made at the telescope.

These cases, such as Thomas Henderson's measurement of α Centauri and Friedrich Bessel's of 61 Cygni, came only in the next century, and are mentioned in the notes to the diary. William's hope was that if he could find pairs of stars, apparently very close together in the sky, and of unequal brightness, this inequality of brightness would be caused by the fainter star being by far the more distant, the near perfect alignment being accidental. In such a case the fainter stars would provide a distant reference point, unchanged by the shift of the earth's position in its orbital motion, while the nearer, brighter star would show this shift in enhanced measure. This expectation was not realized. Most of William's pairs are physically connected pairs, relatively close to each other in space and of intrinsically different luminosities. They will show a relative shift, but this is due to the fact that such binary stars are necessarily in motion in orbits about their common center of gravity because of their mutual gravitational attraction. Thus William failed to measure stellar parallax; instead, he demonstrated orbital motion in binary stars and went a long way toward proving that gravitation, the force which we know as controlling the behavior of the solar system, operates in the same manner in the remotest stellar systems then available for detailed study. We shall see his son continuing these investigations with an especial interest in the binary star γ Virginis.

William also devoted attention to the estimation of the apparent brightness, or apparent magnitudes, of stars, and devised the scheme of a "sequence" still current in modern stellar photometry, that is, a group of stars of diverse magnitudes in a given area of sky giving a standard set into which the magnitude of an unknown star may be fitted by interpolation.

In addition to Uranus, William also discovered two of its satellites and two of the satellites of Saturn, the latter a topic which much exercised his son.

William remained a bachelor until he was fifty, and his sister was much distressed when he married. It seems certain that she was not jealous in the ordinary sense, for she was too noble and amiable a lady for that. However, she had worked with him in his rise from obscurity, she had been his official assistant and housekeeper, and she had watched over him with the utmost devotion. And now she had suddenly lost her job and hardly knew what to do with herself. Caroline's history follows a little later. Now we are concerned with

William's marriage to a widow lady. Born Mary Baldwin, she had been married to a London merchant, Mr. John Pitt, by whom she had had a son who died. It is often said that the portion brought by Mary Pitt to the Herschel household was the means of relieving the family from financial stress, and this may in part be true. It is also the case that William's business in the manufacture of telescopes was a large one.

William perfected his polishing machine in 1788. Before 1795 he had made 200 seven-foot, 150 ten-foot, and 80 twenty-foot mirrors, as well as a multitude of smaller ones. These mirrors were specified by their focal lengths and not by their diameters as they would be today; John Herschel's own twenty-foot was 1½ feet in diameter, and in all probability this ratio of focal length to diameter obtained for almost all the mirrors. The money paid for these instruments was considerable. The King paid William 600 guineas apiece for four ten-foot telescopes. William received £3,150 for a twenty-five–foot telescope for the Madrid Observatory. He also embarked on the world's first large telescope. This was the forty-foot, with a mirror of 49½ inches diameter weighing 2,118 lbs. (made, of course, of speculum metal). This is the instrument shown on the seal of the Royal Astronomical Society. George III gave a grant-in-aid of £4,000 for this instrument and a sum of £200 per year for maintenance. The instrument remained at Slough until 1839, and was then dismantled. On this occasion son John composed a "Requiem of the forty feet reflector at Slough/Sung on New Year's Eve, 1839–40," inspired by the fact that it had become a fashionable amusement to get inside the telescope tube. Now it lay horizontal, and they sang:

> In the old Telescope's tube we sit,
> And the shades of the past around us flit,
> His requiem sing we with shout and din,
> While the old year goes out and the new comes in.
> Merrily, merrily, let us all sing,
> And make the old telescope rattle and ring.

To return to the Herschel family finances: the subject is fortunately well summarized for us in a manuscript list in William's handwriting detailing all the larger telescopes which he made and the prices he obtained for them. This document, which is in the Texas collection of Herschel papers, shows a total of nearly £16,000. This enormous sum was, certainly, not all profit, but it probably ought to be multiplied some thirty times to get the modern equivalent in dollars, giving a considerable turnover for a one-man part-time business. In reading the diaries one should keep in mind the finances of the family, for they are a little surprising. Perhaps this is the correct point for a digression on this topic. In 1757 William's assets were one crown piece. His income thereafter consisted of his musical fees, his stipend of £200 a year from the King, the profits on his telescopes, and the jointure from Mary Pitt. He was prosperous enough to send his son to Cambridge, but John does not seem to have had a regular income except his fellowship stipend from St. John's College, Cambridge, until his appointment quite late in life as Master of the Mint. When he went to South Africa he (John) could afford to pay fares which, for example, for the return voyage amounted to £500. His rent at Feldhausen was £225 per annum, and Lady Herschel's account book (also in the Texas collection) shows that they spent something like £300 per quarter. John Herschel then bought the property for £3,000—in itself rather a mystery. Why should he tie up so large a sum in a distant land when it would have brought in at interest the major part of the rent? His letters show that he was prepared to be extremely generous financially to his wife's family. One also gets the impression, however, that on returning to England he was somewhat short of money. In other words, the son was able to live like a landed gentleman, dispensing sums considerably greater than the salary of Maclear as H.M. Astronomer at the Cape (£600 per annum). It all remains a little mysterious, as other people's finances so often are.

We return briefly to William Herschel. Of his scientific work we note only that he was the discoverer of infrared radiation and a diligent observer of sunspots. In connection with the latter he made one of his few really bad deductions, to wit, that the spots were the real surface of the Sun, dark and possibly even inhabited, with luminous clouds above. On the other hand, his claim to have observed vulcanism on the Moon was derided, and here, if we are to believe recent observations of vulcanism or some kind of luminescence, he may have been more correct. We shall know the answer to this for certain quite soon. William was loaded with scientific honors of all kinds—honorary degrees, membership in foreign academies, a knighthood of the Royal Hanoverian Guelphic Order in 1816, and the presidency of the Royal Astronomical Society. He died in 1822 in his eighty-fourth year and is buried near Slough. Lady Herschel died in 1832, a date of some significance for this chronicle, for it seems that only after her death was son John willing to embark on his expedition to South Africa.

One of the chief qualities of the Herschels seems to have been their amiability, and certainly we detect this in full measure in the character of the attractive and devoted sister, Caroline Lucretia Herschel. Her father desired to have her well educated; her mother wished to make her a drudge. She began by learning dressmaking, but went to England in 1772 when William offered her a home. She then trained as a singer, but when William turned to astronomy she, too, at first with some reluctance, changed to a scientific career. She noted his observations, did his calculations, kept his house, and sometimes ground his mirrors. She discovered, on her own, eight comets between 1786 and 1797. Then, in 1787, she received an annual salary of £50 as her brother's assistant and cooperated with him till his marriage. Even after that event she continued with an astronomical career and continued to work with him. So overwhelmed was she by grief at his death in 1822 that she rashly determined to spend the rest of her life looking after her relatives in Hanover. She made over all her property to them, and, as we shall see from the correspondence, exhibited an almost pathological reluctance to accept any money. She, too, was widely honored, receiving the Gold Medal of the Astronomical Society in 1828 and being made by that, then all-male, body an honorary member, with Mrs. Somerville to keep her company. All the scientific great went to see her in Germany. For her part she remained as lively as a cricket and of undiminished devotion to astronomy into her nineties. She regularly attended the theater and the concert hall. When she was eighty-three her nephew wrote of her: "She runs about the town with me, and skips up her two flights of stairs. In the morning until eleven or twelve she is dull and weary, but as the day advances she gains life, and is quite 'fresh and funny' at ten o'clock P.M., and sings old rhymes, nay even dances! to the great delight of all who see her." On her ninety-seventh birthday she sang a composition of her brother William's for the Crown Prince and Princess of Prussia. This fascinating and lively little old lady lived long enough to see the record of her nephew's Cape observations in print, passing away in her ninety-eighth year, and being buried in Hanover.

Now at last we come to the diarist, John Frederick William Herschel, only son of William Herschel, who was born into a position in the scientific milieu such as might have been occupied by a son of Goethe or Einstein. This gave him an immense advantage in life, and he possessed the native ability to profit by it. Educated mainly at home by a private tutor, he entered St. John's College, Cambridge, at the age of seventeen, and graduated in 1813 as senior wrangler and first Smith's prizeman, upon which he was immediately elected to a fellowship. He was described as a prodigy in science, interested in poetry, and very unassuming. At Cambridge he made several friendships which were to last all his life, notably with George Peacock, Charles Babbage, and William Whewell. Biographical footnotes on these three are included in the diary text. Suffice it to say here that they were all mathematicians and that, with the former two, the young Herschel founded the "Analytical Society" dedicated to the reform of mathematics in England, which then lagged behind that of continental Europe. The three published some numbers of a journal and wrote a mathematical textbook. Whewell specialized during his career in the study of tides, and we shall continually meet his name as a correspondent on this topic in the diaries. Herschel first distinguished himself in mathematics, being awarded the Copley Medal of the Royal Society in 1821. His father wished him to enter the Church, but John resisted, and briefly turned to the law. Becoming acquainted with Wollaston and James South, he turned to science, and after failing to obtain the chair of chemistry at Cambridge carried out experiments in chemistry and physical optics at Slough. He finally turned to astronomy in 1816, and later worked with South at the latter's observatory in London on the reobservation of his father's double stars. This brought him the Astronomical Society's Gold Medal and the Lalande Prize of the French Academy of Sciences. At this time he was on good terms with South, but the latter incurred the hostility of his colleagues by his dispute, later on, with Troughton. Herschel's changed attitude is reflected in his correspondence from the Cape. Herschel took an active part in the foundation of the Astronomical (a little later the Royal Astronomical) Society, wrote its inaugural address, and was its first Foreign Secretary. He traveled in Italy and Switzerland in 1821 with Charles Babbage and visited Holland in 1822 with James Grahame. Later he again traveled with Babbage and made a barometrical determination of the height of

Mount Etna in 1824, visiting other countries on his way home. The diaries of these journeys are in the Texas collection, and only their illegibility and their allusive style prevent their being made available to a wider audience. The ascent of Etna also indicates that when Herschel holds forth in his South African diaries on the resemblance between mountain fires and volcanic eruptions, he may be speaking from experience. On these journeys he also made the acquaintance of almost all the contemporary scientists of note in the countries he visited. As he was an indefatigable correspondent, his mere lists of addresses are of absorbing interest, for they include practically all men of mark in the world.

Working at Slough, Herschel reobserved all the heavens accessible to him, cataloguing and drawing nebulae, clusters, and planetary nebulae. He produced catalogues of these objects and of double stars, and a method for the investigation of binary star orbits.

His interests were not confined to astronomy, but embraced optics and chemistry as well. His suggestion of a transparent glass cover to protect the cornea of the eye in cases of trachoma led him to propose what we now know as the contact lens, but the idea could not be put into practice because of lack of means for local anesthesia of the eye. He wrote, as the first volume of Lardner's *Cabinet Cyclopaedia*, "A Preliminary Discourse on the Study of Natural Philosophy," which was an immense success.

In 1829 he married Margaret Brodie Stewart, second daughter (by his second marriage) of the Reverend Dr. Alexander Stewart, of Dingwall, Ross-shire, a well-known Gaelic scholar. At the time of the marriage, on March 3, John was thirty-seven years old, all but six days, while Margaret, who was born on August 16, 1810, was still under nineteen years old, and a fascinatingly good-looking creature. Their marriage was one of unclouded happiness; they had twelve children, three before they went to South Africa, three in South Africa—altogether, three sons and nine daughters. One cannot help being staggered by the sheer physical endurance of Lady Herschel. She withstood two long ocean voyages in spite of being a ready victim to seasickness, in vessels of much less than a thousand tons displacement. In South Africa she was pregnant more than half the time. The home party comprised five adults and three children on arrival, and she had to run an establishment which employed something like ten servants: Clara and Candassa, who did washing; Somai, the (Malay) coachman; David (? David Brown), the groom; Minto, the houseboy; Dawes and Jack, the laborers; Andreas; Jephtha and his wife, who cooked; Hannah; Leah; Catherine Quinn; Nancy, the needlewoman; the cowboy—and, as Sir John would say, &c &c &c. Evidently this menagerie was not all on hand simultaneously, but there can be little doubt that poor Margaret (or Marg^t as Sir John would have it) was manageress to a small hotel, usually with about twenty people in it. She clearly doted on her husband and would follow him anywhere, even for a couple of hundred miles across rough country in an ox wagon. She has left her own diary, which is in the South African Archives, but so far its minute "crossed" writing has defeated all attempts at deciphering.

This is getting us a little ahead of ourselves. We need now to envisage Herschel some time in 1832: He is one of the most celebrated scientists in Europe, a knight, honored by numerous scientific bodies, happily married, acquainted with all the most distinguished scientists of Europe, a polyglot, with German, French, Italian, Latin, and Greek at his command. He has made a name for himself in mathematics, astronomy, chemistry, and several other fields. Now he thinks of observing the southern sky just as he has done the northern, and he seeks for a place to carry out this ambition.

The choice is very limited: at the thirtieth parallel of south latitude only 20 per cent of the length of a parallel of latitude is solid ground. All the rest is ocean. His possible choices, then as now, were South America, South Africa, and Australia. He chose the second for obvious reasons: South Africa, or rather the colony of the Cape of Good Hope, originally a way station founded by the Dutch East India Company for the refreshment of their ships bound for the Far East, had passed, after various vicissitudes caused by the Napoleonic Wars, into British hands in 1814, having been occupied by them, for the second time, from 1806. There were plenty of traditions of the Dutch rule still very much alive, but the British had provided an active economic ferment to act in a previously rather static situation. In 1820, concerned about the safety of their own naval and merchant ships, they founded the Royal Observatory at the Cape of Good Hope, for the "improvement of astronomy and navigation." In this choice they were much influenced by previous visits of astronomers to the Cape, most important of whom was the Abbé de la Caille, now more usually called Lacaille, who went there and made observations from

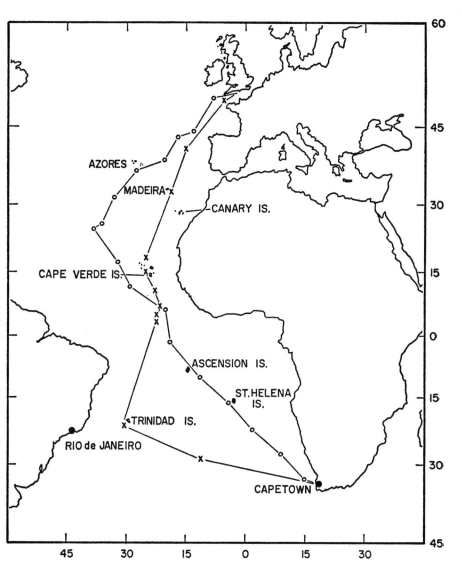

Map 1. Track chart of voyages of the *Mountstuart Elphinstone*, 1833–1834 (x–x–x), and the *Windsor*, 1838 (o–o–o).

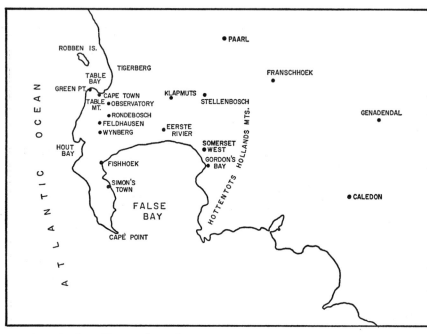

Map 2.   Cape Town and vicinity.

1751 to 1753. He had determined the positions of some ten thousand stars, devised a scheme of southern constellations, and made geodetic observations for the determination of the shape of the Earth. These had led to an anomalous result which much exercised both Lacaille and those who came after him. In addition, Lacaille made use of the fact that Cape Town is in the same longitude as Eastern Europe to organize a scheme of observations for the determination of the solar parallax (equivalent to the distance of the Earth from the Sun). Thus Herschel had, ready made, a stable long-established civilization— dating from 1652, so that we are nearer in time to Sir John than he was to the first settlement—an astronomical tradition and an active observatory, a healthy climate, and a convenient longitude.

Nevertheless, it was a hazardous voyage, to be compared with a trip to the South Pole in modern times, and might be dangerous for the children's health and that of his wife. The Observatory existed, but it was not the honored institution it has now become. Its first director was the Reverend Fearon Fallows, who built it, but died in 1831 following an attack of scarlet fever; its second director was Thomas Henderson, who hated the place and only stayed two years, though he brought it scientific fame during this brief time. The third director, Thomas Maclear, was about to take up his post almost simultaneously with Herschel's departure from England. He had been born in Ireland in 1794 and, after the age of fifteen, had been brought up by two medical uncles.

After completing his studies, Maclear became house surgeon at the Bedford Infirmary, where he came into contact with the astronomically minded Smyth family. Their history is given in footnotes; suffice it to say that, although eccentric, they were also evidently able and persuasive, and quite soon Maclear became interested in astronomy. In 1823 he moved to Biggleswade, where he practiced in association with one of his uncles, and soon after married Mary Pearse. In 1833 he was appointed to the post of H.M. Astronomer at the Cape in succession to Henderson. It seems clear that there was not much competition for the job, for Maclear was really an amateur, though a very assiduous one. Indeed one wonders how he ever found time for his patients, so engrossed was he with observations of the planets, transit observations of stars, and a little geodetic triangulation on the side. When he left for the Cape, his patients thought enough of him to present him with a piece of silver plate. His uncles thought him insane and refused any help with the onerous matter of the fares, which, as we have seen in the case of Sir John, could amount to an enormous sum. The Maclear family eventually traveled unhappily on the *Tam O'Shanter*, embarking on October 10, 1833, and arriving on January 5, 1834, just ten days before the Herschels. Theirs was not a large party, but it did include one member, Thomas Bowler, mentioned at greater length in a footnote, who eventually gained considerable fame as an artist and a portrayer of the contemporary Cape scene.

From this point on, the diaries may be allowed to speak for themselves. They do need a word of introduction concerning their style and format. In the first place, Sir John was not writing for publication. In his flyleaf note he says that the "Memoranda are Purposely Made Unintelligible," and so they are, or very nearly. Not, however, for the reasons envisaged by Sir John. They are not in code; they are simply in his handwriting, which has the following characteristics:

In excitement it is written at a quick galloping scrawl, with numerous abbreviations, the long *s* for the first of a pair, occasional use of the thorn, and other idiosyncrasies. He would divide words at the end of a line with no hyphen, capitalize or not as he felt inclined, omit nearly all punctuation except dividing dashes, overwrite errors rather than rubbing them out, and vary the spelling of proper names in a manner likely to drive the transcriber into a tizzy. Some of this is natural. Rare is the diarist who can sit down in comfort at the day's end and in a rounded literary style describe the day's events. If one were Herschel, round whom events happened all the time, each line is by itself an explosive record of something, and when the day was over, work at his sweeping was only just beginning for Sir John. He is at his most connected and legible when writing to his aunt, whom he loved deeply and whose age demanded a legible hand. To James Calder Stewart, his brother in law, who perhaps filled the place left empty by the dead half-brother, he wrote warmly, sometimes with a sententious degree of philosophizing which half repels modern taste, and sometimes, when in haste, in a difficult steady scrawl. Lying in bed ill, and watching a mountain fire, he writes legibly and connectedly. On horseback, half-way up Table Mountain, blunt pencil in hand, he is at his exasperating worst.

But what is fascinating about him is the breadth of his interests, the never failing ingenuity, the close observation of the world about him. Sometimes he reminds one of the White Knight with a "little contrivance of my own," sometimes of a ham actor with his condescensions and prejudices, sometimes of the corniest of comics with his hideous puns. Even so, this was a man worth knowing. He wrote his diary for himself and it is at once a piece of social history, a slice of life, and the most rollicking adventure story. There are shipwrecks and wars, political intrigues and treachery, duels and masquerades, murders and assassinations. There are betrothals and marriages, confirmations and funerals. Less surprisingly, in the diaries of a scientist, there are records of long hours spent at the telescope, and of new discoveries. There are, too, accounts of experiments in many other fields: botany, zoology, ornithology, and chemistry. In meteorology there is every indication that Sir John knew a front when he saw one; a piece of paper asymmetrically stained with graduated pigment which is in the Texas collection, hints that he may, with his strong interest in flower colors, have been on the verge of discovering chromatography. He had his quirks as well, one of which was a half-formed idea that the phases of the Moon were correlated with the weather.

The world beat a path to his door. As he was at an important way station on the route to India, many came on this account alone: admirals and governors, soldiers and legislators all came, were sometimes shown pet astronomical objects, and went their ways. Others came for other reasons: Charles Darwin came when the *Beagle* touched at the Cape. Franklin, of Northwest Passage fame, came on the way to his governorship. For South Africans, other names will touch chords. Herschel saw Philip often (but rarely spelled him right), and Michell (and never got him right); D'Urban and Napier called and were called upon; even Kok and Waterboer paid visits.

Not all the questions raised by the diaries are answered, a fact which should be part of the value of this record, which would bulk out enormously if we tried to do this ourselves. Each will have his own questions, to which the answers may be more or less readily accessible. One of mine is "Was the meteor, seen to fall north of the Orange River, the great meteorite of Hoba West?" A friend who once, quite legitimately, took a sample of the latter for metallurgical analysis discoursed vividly on the difficulty of cutting it with a hack saw. If the identification were correct how could the natives have got pieces off it? At this writing we are still mystified by "Chedar Tigers" (what are they?) and Indian hippopotami (do they exist?).

The reader may be a little surprised—until he embarks on the diaries—that we have not been able to trace the significance of all the allusions. Once he has read a little, he, too, may be fascinated, as we have been, by the opportunity to play detective and solve mysteries. Since we have not uncovered all the answers, our correct scholarly course has obviously been to reproduce the diaries as exactly as we can, with all their eccentricities of punctuation and spelling, so as to make accessible a proper record. We have done our best to present them "warts and all," but the few facsimile reproductions included in the book will show that we have had our difficulties.

For the four years which they cover, the diaries, read along with the notes, should speak with sufficient eloquence for themselves. This should be the more so not only because of the inclusion of the more significant sketches from the body of the diaries themselves, but also because, through the courtesy of the South African Public Library, we have been able to reunite with the journal a number of the drawings executed by Sir John and mentioned by him in the entries.

This is not a biography of Sir John, and until the publication of *John Herschel, Lebensbild eines Naturforschers*, by Günther Buttmann, none seems to have existed. That broader account should be consulted by those for whom the present sketch is insufficient. On the other hand, the diaries give a much more detailed account of their own period than could be included by Herr Buttmann in his book.

To complete the stories of the principal figures in the diaries need not take long. Maclear remained at the Cape until his retirement, but even when old and blind he remembered with affection those days of collaboration with Sir John. He and his wife are buried at the Cape Observatory, which, under his tutelage and that of one of his successors, Sir David Gill, acquired that international reputation it now possesses.

Sir John went back to a rapturous welcome in England, and to a baronetcy conferred on him by the new queen. In spite of the enormous families of himself and some of his descendants, this is now extinct. The preoccupation with the camera lucida carried him on to photography, and he discovered the action of hypo as a fixer and invented the terms *positive* and *negative*, now in general use. His Cape observations were published through the financial aid of the Duke of Northumberland, and dear old Caroline saw them and her nephew before she died. Herschel himself left Slough and bought a house at Collingwood; he became Master of the Mint in 1850 and was deeply engrossed in its problems. He sat on innumerable committees and commissions: for the promotion of magnetic observations, for the reform of the coinage, for the reform of constellation nomenclature, for the improvement of standards of weight and measure, on the Council of the Royal Society, the Board of Trustees of the British Museum—and so on almost ad infinitum. He amused himself with translations of Schiller, Homer, and Dante. He wrote compendia and treatises. In experimental science he worked on the optical and crystallographic properties of quartz, on microscope objectives, and, with Babbage, on the magnetic properties of rotating plates.

He died in 1871 and is buried in Westminster Abbey not far from Newton. Few men have done as much in their lives; that he has received rather little attention from historians is less than just. The flying start which Sir John got by being the son of his father has been paid for by his being overshadowed by him. Perhaps this record will help to give him the place he deserves.

D.S.E.

CONTENTS

1836

1837

CONTENTS

PLATES

FIGURES

MAPS

1833

The Diary: Selected Entries

[The entries in the diaries are rarely continuous, and blank days or weeks often occur. The diary volumes, purchased from stationers in England, contain numbers of printed pages of general, postal, legal, and other information. In the earlier parts are spaces labeled "Monthly Summary" and at the end, pages headed "Occasional Memoranda," "Observations," and the like. Before the voyage to South Africa commenced, Sir John made relatively few entries in the diary for 1833. A few which relate to the African journey have been picked out for inclusion here. Once the journey has begun, the narrative is almost continuous.]

[Flyleaf]

This book belongs to J.F.W. Herschel

Slough Bucks

If lost, whoever may find it will receive five shillings reward on returning it to the above address, or to the care of Messrs. Smith & Elder & Co N° 65 Cornhill, London[1]

---

[1] Agents, bankers, and publishers having a special connection with India. Their publications included Sir Andrew Smith's *Illustrations of the Zoology of South Africa* (aided by a government grant of £1,500), Sir John Herschel's *Results of Astronomical Observations made during the years 1834–38 at the Cape of Good Hope* (for which the Duke of Northumberland contributed more than £1,000 toward production costs), and Charles Darwin's *Zoology of the Voyage of the* Beagle (with a government grant of £1,000). Publishers of Ruskin, Charlotte Brontë, Thackeray, Wilkie Collins, Browning, and others, the firm also produced *The Cornhill Magazine* and the *Dictionary of National Biography*. The business passed to John Murray in 1917. (See Leonard Huxley, *The House of Smith Elder.*) It seems likely that Peter Stewart, one of Lady Herschel's brothers, was employed there.

No more reward will be offered nor will the book
be advertised for, not to encourage dishonest finders

The Memoranda are purposely made unintelligible
and valueless to anybody but the owner

[Monthly Summary: November, 1833]
   November 13. Embarked for the Cape of Good Hope with Margt
& the 3 children & Stone,[2] Rance[3] & Nanson,[4] servants.

[Monthly Summary: December, 1833]
   The whole of this month at sea—

[Selected Entries]

Wednesday, January 9, 1833
   This morning at 6 AM our third child, a son born. M had a fine time
& thank God is doing admirably well

Wednesday, April 10, 1833
   Our son was Christened William James.[5] His Uncle J.C.S.[6] & Sir J.
South[7] stood Godfathers. Dr Jennings performed the ceremony—

   [2] Herschel's mechanic, John Stone, who also assisted with astronomical observations. His pay was about 16s per week.
   [3] James Rance, personal servant to Herschel. His pay was about £1 per week.
   [4] Woman servant, Mrs. Nanson, who, although suspected of hypochondria by
Herschel, died in South Africa in 1834.
   [5] William James Herschel (1833–1917), Bengal Civil Service (1853–1878),
succeeded to the Baronetcy (conferred on Sir John by Queen Victoria at her
Coronation), 1871. He initiated the public use of finger prints for purposes of
identification.
   [6] James Calder Stewart (J.C.S.), a brother of Lady Herschel. He was a busi-
nessman and merchant who traveled in Europe, and later spent some time in
China, being compelled by ill-health to return to England. He was a frequent
correspondent of Sir John's, who, an only child himself, held J.C.S. in warm and
almost brotherly affection.
   [7] Sir James South (1785–1867), a founder, and, in 1829, president of the
(Royal) Astronomical Society. He was assisted by Sir John Herschel (1821–
1823) in his reobservation of Sir William Herschel's double stars. Through a dispute with Troughton, the instrument maker, he became alienated from the British
scientific community, including Sir John Herschel.

Thursday, May 30, 1833
Pd JS 17£. 12:7d   for Claret to Cape

Friday, August 2, 1833
Finished packing the 20 feet[8] all but finally screwing down the cases & some small items, and cleared the observatory & dismissed the carpenters (pro tempore)

Stone 2 days. 6[s]

Saturday, October 5, 1833
Slough-London. Breakfasted with M[t] at 20 Charlotte Street— then went down to the Dock to arrange the Cabin furniture—Joined M in Cornhill & called with her on Capt Beaufort[9] who shewed us the original drawings (by Webster) of Capt[n] Cooks Voyages of Discovery chiefly of Otaheitian[10] Scenery.—Made some calls—Dined at home then went to McLear[11] at Dr Lee's[12]

---

[8] The reflecting telescope of eighteen inches aperture and twenty feet focal length.

[9] Sir Francis Beaufort (1774–1857), at this time Hydrographer to the Royal Navy, and, as such, closely connected with the work of the Royal Observatories. He was noted for his marine survey work in the estuary of the River Plate and in the Mediterranean, and for his invention of the Beaufort scale of wind strengths. He was a fellow both of the Royal Society and of the Royal Astronomical Society. Promoted to rear admiral on retirement, he was made K.C.B. in 1848. A man of diminutive stature, he is said always to have insisted that he be portrayed sitting down.

[10] Otaheiti, an obsolete rendering of Tahiti, scene of the observations of the transit of Venus made by Captain Cook's expedition of 1769. John Webber (not Webster) was the artist on the last expedition, which left England in 1776, and returned after Cook's death. Webber, the son of a Swiss sculptor settled in England, was noted for the accuracy of his drawings. (See Roderick Cameron, *The Golden Haze.*)

[11] Illustrative of Sir John's erratic way with names. He passes through several variants before settling down to the correct version *Maclear.* Other names are consistently misspelled, even those of intimate associates.

[12] Dr. John Lee (1783–1866), a wealthy lawyer and amateur astronomer who bought Admiral Smyth's astronomical instruments and set them up at Hartwell House, near Aylesbury, Bucks. One of the founders, and, in 1862, president of the Royal Astronomical Society, he financed the publication of the eccentric *Speculum Hartwellianum,* compiled by Admiral Smyth (see note 84, May 11, 1834). Thomas Bowler (see note 110, June 21, 1835), was a grandson of a housekeeper of Dr. Lee. Lee married another of his housekeepers.

Monday, October 7, 1833

Went to Deville & at Dr Lee's request had a cast taken in plaster of me.—NB puts One in mind of being buried alive

Tuesday, October 22, 1833

Left Slough with Margt & the children for London

Monday, October 28, 1833

J Stone and his wife came. Advanced J.S. 12:10:0 and arranged to pay Mrs S. 16^s a week through P.S.

The Diary: November 13 to December 27

Wednesday, November 13, 1833[13]

Portsmouth. 57 High Street

11^h 20^m. AM. took boat and in about ½ hour embarked on board the Mount Stewart Elphinstone [Figure 1] in which we had engaged & prepared the 3 Stern Cabins in what is called the Round House on the larboard side and one below, forward, for the two men servants, James Rance and John Stone. We (Margaret & Self, the 3 children, C. I. & W,[14] and Mrs. Nanson our Nurse) were accompanied on board by Jones, Ja^s Stewart, Thos and M^rs and Maj. Baldwin and Mrs Moorsom & our late nursery maid Sarah who took leave of us when the ship got under weigh & we then (about 1½ PM. stood out along the Solent through the Needles Passage with a fair wind, the

[13] In the early hours of this morning in the eastern part of the United States, there occurred the largest meteor shower ever witnessed (Leonids). The event was approximately simultaneous with Sir John's embarkation, but was invisible in England because the shower radiant was below the horizon.

[14] Caroline Emilia Mary Herschel, eldest of Sir John's twelve children, married, in 1852, Sir Alexander Hamilton-Gordon. She was a bedchamber woman to Queen Victoria, 1855–1891. She is usually called "Carry" in the diary.

Isabella Herschel, usually referred to as "Bella," was the second daughter of Sir John. She died unmarried in 1893, aged sixty-one.

W. is William James Herschel (see note 5, April 10, 1833), then a baby of ten months.

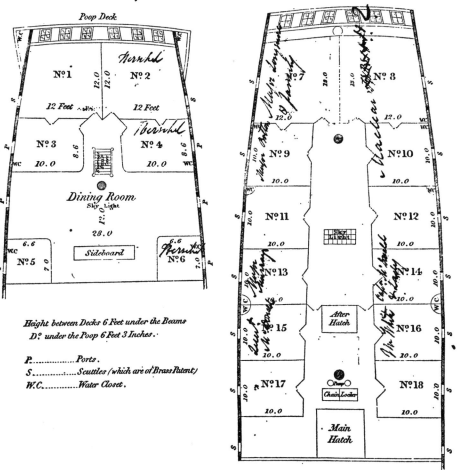

Figure 1. Maclear had intended to travel to Cape Town on this ship; the copy reproduced here was used by him for writing a note on the back to his wife, and therefore has the Herschel cabins correctly marked, as well as those on the lower deck originally reserved for the Maclear party. (By permission of the South African Archives, Cape Town)

Isle of Wight looking very beautiful & the view of the Needles and Scratchells Bay very fine. Sir R. and Lady D'Urban[15] &c &c &c.—

Dined about 4, and evening soon closing in got an uncomfortable night's rest.

Thursday, November 14, 1833

Out of Sight of Land somewhere about the Start Point. Wind still fair and the sea moderate.

Little motion & altogether pleasant sailing.

A good beginning. M. a little ill but able to come on Deck—but . . . M. occupied þᵉ Couch. I tried the Swing cot which seems to me ill contrived and no better than a fixed bed—indeed not so good. It has too much motion of its own, vibrating long after the ship is still

Friday, November 15, 1833

Sea still subsiding as well as the wind and Ship making slow way. At daybreak said to be about 30 miles S of the Lizard Point, and standing out nearly W. or a little S of W, in the direction of the Sole Bank.

Saturday, November 16, 1833

Sea almost glassy as to surface, but still the greater waves not subsided or inclined to subside giving the effect of a heaving surface of oiled skin, intersected with a sort of meshed or net-worked texture of smaller low rounded waves—In the afternoon, a *cats-paw* or rather

---

[15] Sir Benjamin D'Urban (1777–1849). After a distinguished military career, he was successively Lieutenant Governor of British Guiana and of Barbados. Appointed Governor and commander-in-chief of the Cape, 1833, he was overwhelmed by numerous difficulties with native tribes and the white settlers. The slaves were liberated in 1834, and the terms of compensation and other circumstances helped to precipitate the Great Trek (withdrawal of inhabitants of Dutch and Huguenot origin—the Afrikaners—to empty lands beyond the Cape boundaries and outside British jurisdiction), which began in 1835. Using Andrew Smith (see note 21, January 24, 1834) as his agent, he negotiated various treaties with native tribes, but after a frontier war he annexed all the land up to the Kei River. This aroused the hostility of the missionary Dr. John Philip (see note 22, January 24, 1834), who campaigned against D'Urban in England. D'Urban was relieved of his post in 1837, and succeeded as Governor by Sir George Napier (see note 1, January 31, 1838). The settlers of Port Natal renamed their city Durban. In 1847 D'Urban took command of the military forces in Canada, dying two years later in Montreal.

sharper and smaller crispation began to come on & the glassy smooth-
ness vanished. As soon as dark noticed the great quantity of luminous
masses like stars, lamps, or fire-balls in the wake of the ship some
more some less luminous & durable. The small ones only flashed for a
moment the great ones shone for several seconds.

The wind Rising and a considerable ripple coming on the water,
both increased till at length it blew hard & the Cap^n was obliged to
take in Top Gallants and put 2 or 3 reefs in main topsails an operation
accomplished slowly & with difficulty, by reason of the extreme in-
efficiency or insu*fficiency* or both of the crew, who are declared as
bad a set of lubbers as ever worked a ship.—Sea grew at last very
heavy & we passed a dreadful night. M^t very ill indeed.

Sunday, November 17, 1833

The stiff breeze and heavy sea continue and are very distressing.
The Stern Cabins have much more motion than those nearer mid-
ships, nothing can be kept safe that is not on Swing trays, and it is
difficult to keep a footing.—M very ill with sickness and headache—
No possible relief to be had.—Sea sickness is *not* an affection of the
stomach so much as of the whole nervous system. I should call it a
general giddiness which the stomach in particular resents by a spasm.
It is therefore vain to attempt to allay it by swallowing things such as
burnt brandy, or sal volatile &c. Custom only & patient endurance
can overcome it. A wretched day. Towards evening the sea & wind
rather less.

Monday, November 18, 1833

After a comfortless night* in my own little awning cabin found M.
still extremely ill with sickness & headache but able to eat some meat.
The Wind abated and more favorable, and our course which had
been nearly due W or even a little N. began to tend to the SW. It was
some comfort to hear that we had got on 50 or 60 miles in the night
towards the Azores.

Took Merid. Alt ☉[16] whence concluded Lat. 46° 4′

* filled with strange, connected, tragical dreams.

[16] The sign denotes the Sun. A measurement of the angular elevation (altitude) of the Sun, at apparent noon, when it is due south or north (on the
meridian) gives the latitude of the observer's position.

Tuesday, November 19, 1833

5 of the sheep died since yesterday—probably from overdrinking as they have been allowed 2 quarts of water a day each (Sheep drink little & seldom) On the other hand the ducks of which there is a great collection on board get little or no water!

M. rose & passed the day chiefly in my Cabin & on deck & seems to have got over the worst of the sickness. Cap^n Jones—an Engineer? officer—a very sensible well informed person

Wednesday, November 20, 1833

Mt Stewart Elphinstone.

A fine sunny morning—Ship going right before the wind 7 or 8 knots an hour and with little motion—

Therm.[17] Lowest during the night $= 52°.0 + x$

Highest in the day—$64.3 + y$

Day generally cloudy

Lat. at noon by Capt^{ns} obs. $= 41°37'$. Long 15° W.

A most agreeable temperature. M. decidedly much better   passed the greater part of the day on Deck & dined in the Cuddy.[18]—

NB. Have seen no fish nor bird today.

Thursday, November 21, 1833

Therm. Minim. at Night of Nov. 20–21 $= 60.1 + x$

Maxim. in day $= 63.5 + y$

A heavy day among the Crew. All sail set and wind right aft scudding away 7 or 8 knots & very little sea or motion—All hands at work sail making, pulling to bits old rope, cleaning up compasses &c   The Passengers Male & female all about on Deck and betaking themselves to their avocations—M. quite free from sickness & occupied on Deck like the rest. In evening all assembled, a fiddler struck up & the Middies set to dancing horn-pipes, one of them dressed in a cloak & straw hat bent down into the resemblance of a bonnet so as to look like a Billingsgate[19] fishwife—&c &c.—Came on to blow pretty hard & the roll proportionally increased.

---

[17] Maximum and minimum temperature readings, with as yet undetermined zero corrections, x and y, to be applied to the two thermometers.

[18] The galley or pantry of a ship.

[19] The London fish market, traditionally notorious for foul language.

Friday, November 22, 1833

Minimum Therm. last night 62.5 + x

A Cloudy & windy night with a little rain followed by a very fine day   plenty of Sun which we have not had for a great while & great areas of clear blue sky.—The blue of the Sea is now decided, & little tinged with green.

Saturday, November 23, 1833

Off Madiera, but too far to get sight of it. Sea now very blue.

Sunday, November 24, 1833

The Captn Read Prayers on the Quarter Deck   All þe crew and Passengers assembled, an Ensign spread over the Capstan for an Altar and one over a table for a Reading Desk.

Thursday, November 28, 1833

Place at Noon. Lat 18 51 N. Lon. 25 28 W.

A fine strong breeze & cross swell consisting of two very distinct lines of waves intersecting, from the N and East or NE which make a very irregular sea. The greatest estimated wave is certainly not 15 feet from summit to lowest depression. In the night we broke two of the Studding sail Booms on which the Captn had all the Studd [in] g sails taken in.—For the 1st time saw a large covey of flying fish, beautiful silvery creatures, darting like white short arrows very near the surface and making pretty long & swift flights. Stone says he saw some on Sunday last. The Cuddy Servant broke the Mercurial therm. of my Register Therm for *Maxima*. By good luck I had an old one by Ronketti of which the *minimum* one was broken. Two halves made a whole & thus I am still enabled to keep the series going. Commenced observing the temp. of the sea water.

Friday, November 29, 1833

Place at Noon. Lat. 15 57 N Lon. 25 44 W. Passed St Antony (Cape Verde Islands) in þe night without seeing it   all the morning on the lookout but no Sign—Here altered our course & steered more to the East, nearer the wind, the effect being to incline the ship to the starboard, making it difficult to walk and laborious even to sit still. The sea flushes up occasionally into the Awning Cabin Window

The climate is now Summer.—More flying fish seen today—but their flights are not frequent. I only saw one covey. At night the

Phosphoric sparks frequent & of long duration. Some lay floating like bright lamps for at least 5 sec. NB. This before the Moon Rose. In the Moonlight nights lately I have seen none though the light is too strong to be *quenched* by the moon. Tonight they ceased when the moon Rose.—Today they got up a main top gallant[20] mast, a formidable process yet very prettily managed.

Saturday, November 30, 1833

At Daybreak found the sea smooth & little wind   Also the water emerald green in place of blue & a large white bird sailing about. All indicate land near. The Capt⁹ was all night on the look out for Brava (The SW of the Cape Verde) but got no sight of it—yet the Ship he says was almost becalmed under its lee—About 9 am the colour of the sea was again dark blue (but a dull colour) & the wind & sea as strong as ever. We have therefore now passed þᵉ Cape Verds.[21]

At night being fine tried a set of Lunars[22] but it seems a most impracticable matter to get a good contact. However at length I did satisfy myself that one or two minutes may be discerned one way or other by a beginner. Then gave in.—NB. Jupiter nearly vertical. Saw Canopus. It is much inferior to Sirius which is superb.[23]

Sunday, December 1, 1833

Splendid morning after a superb Starlight night. In Lat 11°13′ Long. 22°51′. The wind wʰ. has hitherto been steady now began to

---

[20] Topgallant sails are above the topsails and below the royals. Studding sails are light sails set at the side of a principal square sail in free winds.

[21] The Cape Verde Islands off the African coast.

[22] A very interesting reference. Before the invention of the marine chronometer in the previous century, it was impossible to determine longitude at sea with any precision. One method proposed was equivalent to the determination of time from observations of the position of the Moon relative to certain stars in the zodiac. The observations were difficult, and to make use of them required an accurate predicted position for the Moon at all future times. The preparation of accurate lunar ephemerides became possible only in relatively recent times, though approximations adequate for navigation, of a rather crude kind, were probably available fairly early in the eighteenth century. At all events, even with chronometers available, the captain of the *Mountstuart Elphinstone* seemed to find the method useful, and difficult, and was probably somewhat embarrassed by the presence aboard of the most famous astronomer of the day.

[23] Sirius and Canopus are the two brightest stars in the sky, the latter, the fainter and much more southerly, not visible from England.

flag a little yet we still make good way. The temperature is steadily increasing & it is now decidedly warmer (See Register)   Flying fish pretty numerous—no dolphin or shark yet seen. M. very ill today with Headaches, Pains in limbs & Sore throat.

Monday, December 2, 1833

M. extremely ill—towards noon began to feel, myself great debility—with extremely low pulse.—Today exposed thermom. in wet linen. The effect of evaporation in a fresh breeze was as follows.— Dry thermom^r = 80º0 Wet = 74º4—Evap. = 5º6—

Saw a Sheerwater[24] (a large brown backed bird) Dist. from land = 300 miles. The Captn tells me he once caught a Vulture 7 feet wide from wing to wing at a thousand miles from land, between the Azores and West Indies. Maximum of my own Pulse = 53; Minimum = 47, beats per minute.

Tuesday, December 3, 1833

Maximum of Pulse = 53—Minimum 46. Accompanied with excessive langour and debility—The Crew today sent a Petition or Round Robin to the Captn to be allowed to perform Neptune on Crossing the Line.[25] M. Better & able to come on deck in the Evening but the day has been one of mere exhaustion to both of us—lounging on Sofa and Matrasses—and reading or trying to read. We are now fast losing the trade in Latit 7° 7′ Long. 21 ±. Squalls & rain but the squall helps us on.—Vivid lightning in Evening. too far for thunder but saw one vivid *flash* (the real spark) *vertical,* from the Western edge of the blackest cloud down to the sea. Got all prepared for a bad squall but it did not come.

Wednesday, December 4, 1833

A dead calm the whole day with exception of ½ hour when they tried to get up a squall but did not succeed.

Threw over board 1st a lump of Pewter wrapped in white paper— it remained visible 88 beats of my watch = 22^s.—last visible tint a rich bluish green. A Cube of Coal 3^{in} in the side in white paper was

[24] This might be any of several species of shearwater.

[25] Traditional ceremonies and horseplay inflicted on those first crossing the equator by sea. King Neptune and his Queen hold court; the victims are shaved with huge wooden razors and ducked under water in punishment for the commission of various invented "crimes."

visible 150^b = 37^{s}½ In Evening lowered a piece of Iron by a string (just after Sunset) it was visible 67 feet below the Surface, & the last visible colour was greenish blue—but the daylight was rapidly failing.—M. much better & about as usual—myself less languid and more alive, though the depressed pulse still continues. Threw over Bottle No. 2 with a Note inside. Pulse from 46 to 53

Thursday, December 5, 1833

In the night a short squall from the East which lasted about half an hour. After Breakfast a steady breeze rose which carried us on to the S (by compass). Wind about E.—Can we *now* (Lat. 6° [erased] or 5° 30′) be already in the S.E. trade?[26]—Last night false alarms of a shark. This morning Lady D'Urban saw one from the Stern window. Heavy rain in the night— NB. In these Equatorial Squalls the rain & wind go together—or rather the rain seems to lag a trifle behind the wind.

Friday, December 6, 1833

Last night a brisk squall, and the "Phosphoric Sea"[27] in high perfection running a train out behind the ship for several ship's lengths. It is an assemblage of shining individuals which when seen on the surface are like stars, when turned down deep under water and mixed with air, look nebulous.—But there are 2 sorts—for when the ships speed relaxed as the squall died off there appeared *well defined masses* of irregular forms, (like lumps of boiled rice in a glow) some of considerable size floating in the clear transparent nonluminous water, when the eddy under the stern left a smooth surface. These though violently driven about & many feet deep yet did not *break*,

[26] The strong steady geostrophic prevailing winds of the southern equatorial, tropical, and subtropical regions. For a ship which might have spent many weeks in the equatorial calms of the doldrums, the southeast trades were of immense commercial importance. During the southern summer (northern winter) the Cape area is in the region of these winds. Known as the southeasters, they blow strongly even on land (40 m.p.h or more) and usually bring fair weather, with only scattered cloud. Exceptionally, when they do bring rain, they are known locally as "Black Southeasters." The rain-bearing wind of the Cape winter is the northwester.

[27] Regrettably, these phenomena are much less obvious from a modern ship because the beams from deck and porthole lights illuminate the sea for a considerable distance.

but preserved a sharp outline like lumps of white hot metal intensely luminous#

[Written vertically] # Altogether this was one of the most magnificent sights I ever saw. Regretted extremely having no means of fishing up some.—

Saturday, December 7, 1833

Last night saw for the first time the Greater Magellanic Cloud.[28] It is brighter & larger than I expected to see it & a very odd looking object.—

Much rain fell today, temperature of the stream which ran from the awning very copious 74º5. This however had no sensible effect on the temperature of the surface water of the Sea.—A Dolphin caught. His back fin in dying grew a very dark blue and his skin varied from blue-grey to silver white but no rich or vivid tints appeared. Got his eyes. Their optic nerves have a plicated or folded structure like a ribband doubled longitudinally [a small sketch], thus (a cross section)

Sunday, December 8, 1833

Divine Service in the Morning—Two ships in sight proved to be an English Brig (the Pink) from London to Bahia[29] and a Swedish Brigantine from Viana[30] to Rio. Some of our passengers with our Steward went on board & reported that the Englishman had only 8 persons on Board the Swede 5 and a little boy a passenger. Sent letters home to Mrs S.[31] Cousin May & my aunt[32] to take their chance, with an assurance of their being put on board the packet at Bahia wh goes to England.—A boneta or Benito[33] was speared today, a fish about 15in long and very thick & round tapering remarkably to the

---

[28] Named for the Portuguese navigator, these objects are the two nearest external galaxies. In the far southern sky, they are readily visible to the naked eye, and resemble detached portions of the Milky Way several degrees in diameter. Information derived from them has played a crucial role in modern astronomical development.

[29] Presumably Salvador (Bahia), Brazil.

[30] Presumably Viana do Castelo, Portugal.

[31] Probably Mrs. Stewart, Lady Herschel's mother.

[32] Caroline Lucretia Herschel; the letter appears later.

[33] A bonito, any of various medium-sized tunas, for example, *Sarda sarda*, which is obliquely striped (see "Account of Fishes Eyes").

tail which is small & forked [a small sketch of the fish], flesh very dark like beef.—secured its eyes, of which the Crystalline[34] is singularly formed having a kind of annular protruberance in front [a small sketch] thus. The optic nerve less distinctly folded than in dolphin.

Monday, December 9, 1833

A calm or light Southerly airs the whole day, and very hot. Tacked due E then again a little N. of West.—Dined on the Quarter Deck with Mr. and Mrs. Jones, Major Dutton and ourselves.—A Missionary caught by the men ascending the Rigging & they tied him fast to make him *pay his footing* (a Dollar). A fine Night. Got out the night glass and viewed Southern Clusters. One very beautiful $\oplus$ South of α Eridani (5 or 6° Sf. α)[35] One in the Great Magellanic Cloud &c— At Night fished for Phosphoric insects & caught 3 Sorts—one jelly— one a shrimp about 1/3 inch long & red tailed & one a very small oval active insect. Also caught a black longlegged floating insect.[36]

Friday, December 20, 1833

A fine day and fair strong wind (in the strength of the trade). Got several good sets of Actinometer[37] obs^{ns}. The Rad^{n} does not reach its maximum till after the Culmination of the Sun but the *temperature* is highest before noon.

The Captn in a great puzzle about his Lunars.[38]—All hands at work taking obs^{ns} and at last as a final result the Lunars of yesterday and today give -27^{s} for the Chronometer error while all Chronom^{rs} make it $+ 2^{m} 21^{s}$—Bad work this.—But a Lunar at Sea seems rather a bungling business.

[34] The lens of the fish's eye.

[35] South following (i.e., to the south and east of) the star α of the constellation Eridani. The sign $\oplus$ denotes a cluster. The object seen was not, in fact, a cluster, but the extragalactic spiral nebula designated in modern terms as NGC 782. A rather remarkable observation considering the means available.

[36] Probably a true insect blown out to sea off the Cape Verde Islands.

[37] An instrument devised by Herschel for measuring the heat (infrared) radiation of the Sun, scientifically very advanced for the comptemporary state of knowledge. It consisted of a thermometer with a large bulb containing a dark fluid. Measures were made by comparing the rate of rise for exposures in the Sun and in the shade.

[38] (See note 22, November 30, 1833.) This account of the accuracy of the method will be most revealing to students of navigation.

Saturday, December 21, 1833

More Lunars and more puzzle—at last however by plenty of patience & using masses of obs[ns] came to a satisfactory conclusion & got the Lunars & Chronometers to tally within ½ min. of time or about 8 miles of Longitude.

Sunday, December 22, 1833

Divine Service by Sir Benj[n] D'Urban.

Sun today almost exactly vertical,[39] and being clear, obtained with the assistance of Capt Jones and the Ships Surgeon M[r] M[c]Hardy a very interesting series of Actin[r] obs[ns].

Wednesday, December 25, 1833

Blue colour of sea most beautifully intense, hardly any tinge of green—

Thursday, December 26, 1833

Dined on Deck with Lady D'Urban, Miss D. Marg[t] & party.—In Evening Eclipse of Moon attended with most remarkable phaenomena (See obs[ns]). When the ☽[40] recovered her light & got high the print of a newspaper might be read easily by her light. Nothing ever exceeded the magnificence of this night. I watched Jupiter down to setting & after plunging through a small cloud line, saw him appear like a lamp on the very edge of the sea and go down behind the offing like a fireball.—Saw the small ✴[40] of α Crucis. Crux[41] very insignificant constellation.

Friday, December 27, 1833

A Superb day. The Blue colour of the Sea is now at its maximum and is the deep colour of full indigo—When the light from it strikes up a dark vertical pipe the sides of the pipe appear as if the blue of a prismatic spectrum fell on them.

---

[39] From most observers this might be badly incorrect; stated by Sir John on the day of the solstice it implies that the ship was on the Tropic of Capricorn in latitude 23° south.

[40] Signs for moon and star.

[41] The constellation the Southern Cross. The star α Crucis is a binary with two equal components separated by five seconds of arc. It seems unlikely that he could have observed this from shipboard, and he probably refers to a fainter fifth-magnitude companion distant about ninety seconds of arc.

[Occasional Memoranda, November, 1833]

Capt. Jones 46.th Bengal Native Infantry.

Maj[r]. Dutton.[42] Mil.[itary] Sec.[retary] to Sir B. D'Urban

Capt[n] Beresford[42] Aid[e] de Camp to D°

Letters from the Travelers to

Caroline Lucretia Herschel, Hanover

Portsmouth. Nov. 10 1833

Rec Nov[m] 17 : : [In a different hand]

Dear Aunt

Here we are all ready to embark and awaiting only signal from our Ship the Mountstuart Elphinstone which is lying off Spithead—And a fine ship she is—Maggie and the children are all quite well and lively and we have about us Thomas and Mary Baldwin M[rs] Moorsom—Margaret's Brother James and M[r] Jones and John Henry Nelson, (a pretty party) who are come to see us off. All our goods and Chattels are shipped—the telescopes all stowed safe away, and in short Bating accidents, for which there is not the least ground of apprehension, we shall write you word of our safe arrival and comfortable welcome, at Cape Town about the end of January—perhaps sooner and you will get the letter (allowing for delay between this & Hanover) by some time in April.—This is the longest interruption to our correspondence for after that letters will arrive in succession from 2 to 3 months after date. Meanwhile do you write uninterruptedly and tell us all about you and how you like your new quarters.—James Stewart tells me he means to beg to be admitted into the list of your correspondents and means to write often to you.

The last proof sheet of my Nebula paper left my hands the night I left London and yesterday I got 12 copies to take to the Cape.— One will be forwarded to you tomorrow by Lieut. Stratford RN. Su-

---

[42] Major William Holmes Dutton and Lieutenant George de la Poer Beresford (7th Fusiliers), a relative of the third Marquess of Waterford.

perintendent of the Nautical Almanac who will send it to Prof Schumacher,[43] to whom if you do not soon get it pray write.—And I have also directed a duplicate to be sent to you by Mr Hudson Assist^t Secretary of the RS.[44] and Librarian who will henceforward send you all my papers (duplicates) as will also Stratford.—My Observations of the Satellites of ♅[45] which confirm my Father's results, were sent off to be put in course of publication last night.

Remember me kindly to Dr Groskopff[46] and his amiable lady and to M^{rs} H & believe me you affec^t nephew.

J.F.W. Herschel

[In Lady Herschel's hand]

Dec 8th 1833—Lat 3° 18′

N°2 N°1

My dearest Aunt—Although there is no ship in sight homeward bound, yet being so near the line, where an almost continual calm renders it a general rendesvoux of vessels, we may be called upon daily to furnish our quota of letters "for home"—This is now the fourth Sunday we have spent on board, & our weather hitherto has been most glorious—favourable breezes pursued us, & sent us on at the rate of 3 deg. a day for a full fortnight, & nothing of a *storm* was felt—Now in the region of calms our patience is exercised to its full extent, but even here I believe we are getting on unusually well—Of course we have all been sea sick more or less except dearest Herschel who was nurse & doctor to all needing his assistance—

The ship is very full, & the chatter of the passengers & the bustle of

---

[43] Heinrich Christian Schumacher (1780–1850), German-Danish astronomer; director, Mannheim Observatory, 1813–1815; professor at Copenhagen. He directed the geodetic triangulation of Holstein, which led to the foundation of the Altona Observatory, where Schumacher resided until his death. He published ephemerides, 1822–1832. In 1821 he founded the astronomical journal *Astronomische Nachrichten*, which still continues.

[44] The Royal Society of London.

[45] Sign for the planet Uranus, discovered by Sir John's father, Sir William Herschel, and based on the initial letter of his name.

[46] See note 105, letter to Caroline, June 6, 1834.

the sailors are dreadful, but having been the first to engage accommodations, we have the best cabins on board, & in this snug little
deck cabin of Herschel, I spend the whole day by his gracious permission, the envy of all who have to crowd into the dining *cuddy* for
shelter from the rain—The prettiest sight we have seen in the way of
aquatic novelties, is the brilliantly illuminated track of the ship at
night when the disturbance of the water causes innumerable animal-
culae to emit a phosphoric light,—thousands seen at great depths,
when their light is much spread, & others dancing on the surface, like
stars of the first magnitude as if a thousand Jupiters were created at
our approach, instead of the old story of flowers starting up at our
tread—Sometimes flashes of light are seen in the water, like flashes of
sheet lightning, for the accounting of which I must refer you to
Herschel—We tried to fish up some of the phosphoric animals, but
have not yet succeeded.—The other night we were delighted with
the first sight of a Magellanic Cloud near Canopus—but it was too
near the horizon to be seen very well. We have almost if not quite
lost sight of the Polar Star, & Herschel fancies that already the
Heavens are clearer & more tranquil than in good old England—A
Dolphin was caught yesterday & expired on the deck, but there was
nothing to tempt us to finish our voyage on his back—this was a pretty
small you[ng] fish with a very forked tail, &, its colours al[though]
changeable were not very brilliant—Some amusements are to be
enacted when crossing the line, but I have only bargained that *my
Son's* beard & whiskers are to be untouched—

I must leave room to Herschel to add a word & will therefore only
subscribe myself, with most affec^t love, My dearest Aunt's very attached Niece—M.B. Herschel

[In Sir John's hand]

Dearest Aunt—There is now a ship in sight (two indeed) one said
to be Homeward Bound English—There is therefore no time for me
to say more than all well, for after all the chances are sadly against
your ever getting this   So Adieu   Lebe wohl.

Your aff^{te} nephew JFWH

Dec 8. Lat 3° 18′ Long 22° West of Greenwich

Extracts from a Scientific Notebook
Kept by Sir John During the Voyage

Account of Phosphoric Animals

Dec. 9 fished for them & brought up 3 sorts of creature

1. Longish pieces of clean jelly of a somewhat oval, pointed shape which shines vividly on pressure or agitation Some of them had the appearance of bags but I could not satisfy myself [rough sketches]. They have no appearance of locomotion or life.

2. Pink-red active, shrimplike insects with body legs and a flapping tail about 1/3 inch long. These either shine by flashes as they move or excite other smaller invisible insects to shine by striking them [further very rough sketches].

3. Very small brownish insects which fold up into little oval bodies like ants eggs, but they seem to be a kind of shrimp which rolls itself wood-louse-fashion as they dart about from place to place by sudden starts.—I could not be sure that these shine.

These were put by over night in a glass of Sea Water, in morning all were dead and rapidly decomposing producing a most detestable *fishy salt* odour.[47]

Dec. 10. Fished again & brought up quite a different creature in great abundance. The Sea tonight was more vividly luminous & great Clouds of brilliant light rolled out in the wake, consisting of *lumps* of *light* having an evident *size* & *shape* which, when shining from a great depth as they were rolled down under the eddy of the rudder offered the most extraordinary spectacle as if the clear water were full of rapidly moving lumps of white hot iron.[48] When caught these lumps proved to be animals, insects or mollusca, of about an inch to 1½ inch long, in the form of cylinders as thick as the finger—say ½ inch. They are hollow, and consist of transparent, and very solid jelly,

[47] Identifications of these creatures are probably: (1) salps; (2) hyperiid amphipods or amphipods living in salps; and (3) species of crustacea.
[48] Professor Day identifies this as Pyrosoma, a pelagic tunicate famous for its phosphorescence.

which is set all over with what appears like a fur, giving the creature the appearance of a little muff. This fur is however an assemblage of papillae of a less transparent substance, whitish, and each of which has at outer end (which lies *beneath* the exterior gelatinous coat) a small dot of a rich purplish pink colour.—These dots are what shine. When fresh caught and touched or pressed in the dark it is seen to be covered over with dots of vivid greenish or bluish white light which (seen at some distance) give the whole the appearance of a glowing coal. At this time the animal is firm and hard, (Imagine a piece of the ice plant intensely luminous). But when exhausted by agitation or pressure he becomes more flaccid & shines more feebly or ceases entirely. When cut open the papillae appear on the inside, projecting and rough.

On daylight inspection they are not *muffs* but bags one end only being open, at this end there is a kind of lip which forms the mouth, of thinner substance, transparent [a sketch].

The[y] do not sting the fingers but when applied to the lips, irritate, like nettles, only less powerfully.

Handled they emit a faint, aromatic smell, something like honey which adheres to the fingers when dry, mixed with a fishy smell.

Fished again[49] Dec . . . [his omission] Brought up transparent creatures of a rhombic form which have each in them a hard brown-red nucleus like a pea with a short tail [crude sketch] thus & which are perhaps hearts or brains &c. They have a contractile power & seem to be bags with some spiral thready viscera visible in a magnifying glass—Nanson (confound her) threw them away & I could not catch any more. I could not ascertain that *they* caused the *flashes*.

Account of Fishes eyes

N° 1 Dolphin Crystalline greenish. Optic nerve tail large, swallow-shaped   hog backed or rather hog-headed [text interspersed with several rough sketches]

N° 2 Bonita or Bonito or Benita   About 15 . . . 18^{in} long. a thick fish almost round in section but tapering very rapidly to the tail which is small & so slender in its junction with the body as almost to be cut off.—[a sketch]

Along the back & belly from the tail are several very small fins.

---

[49] The catch possibly consisted of pteropod molluscs.

Colour dark dull blue or slate gray not much variety rather striped. Flesh very dark red like dark col^d beef.—Blood very dark. Crystall. colourless. Vitreous humour[50] very dense. Optic nerve less distinctly plicated than in dolphin.

NB. The Crystalline of the Boneta has an annular protruberance thus [sketch] which perhaps is only the insertion of the Capsule into the ciliary membrane.—the B. eats flying fish

N° 3   Flying fish. Crystalline (not quite fresh) (taken from the stomach of the Boneta) very yellow, or greenish yellow.

N° 4[51]   Toad fish ? or ? young pilot-fish a small ill shaped fish about 3^{in} long or 2½, of a grey colour with a large bag-like belly but laterally thin—taken from stomach of Boneta.—Crystalline colourless [rough sketches of fish].

[An Eclipse of the Moon]

Dec. 26. 1833. An Elipse of ☽. ☽ rose at . . . [his omission] Eclipsed—when first seen (in strong twilight before stars app^d) almost total.—Observed the moment of total obscuration by chronom^r (A) could not be in error more than 1^m

A = 7^h 30^m 0^s I believe the shadow to be over the whole edge. the bright segment is reduced to nothing

A = 7   31   0 Certainly the immersion is now complete. Yet though the total obscuration has without doubt taken place yet the upper half of the disc is perfectly well seen.

A = 7^h 40^m The twilight having diminished so that the stars begin to appear, the whole Moon is visible   the upper part being however considerably brighter than the lower

A = 7   53^m —Nobody would now believe, if not informed that the Moon is in a state of total eclipse—had I not seen the Shadow come up to the edge I should think the eclipse had not begun, but that the moon was dimmed by vapours. The upper limb[52]

[50] Clear colorless jelly filling the posterior chamber of the eye.

[51] Professor Day tentatively identifies N° 4 as a hatchet-fish, which lives at fairly deep levels in oceanic waters (*Argyropelecus* sp.).

[52] In astronomical parlance, a limb is the edge of the disc presented by the Sun, the Moon, or a planet.

is very bright along the dotted part   a   b [refers
to a diagram]

A = 8ʰ 15ᵐ   The Center is now less bright than the two opposite limbs   ab   and cd, but ab is brightest.
The light is very red but the Moon is extremely
conspicuous and the whole disc well seen either
with naked eye or telescope—.With night glass
many small stars are seen within a Diameter of
the Moon. It looks much like a planetary nebula[53]
—The red light inside is of a copper-coloured
cast[54] & very singular.

A = 9ʰ 3ᵐ 30ˢ   Began to watch for the reappearance of the sun's
light on the   ☽ .—There is already come on a
paler hue, almost to be called blue-green on the
right hand edge, which contrasts in a very
marked and extraordinary manner with the rich
deep coppery tints in the interior. I feel assured
that this is not a deception, arising from mere
diff[erenc]ᵉˢ of *intensity* of illumination. It is a
real difference of composition in the rays.

A = 9ʰ   7ᵐ   25ˢ—The first strong suspicion of sun-light striking on
the edge

    9   7   55 —The gleam at the edge increasing rapidly (night
glass)

  +9   8   35 —A decided illumination at edge equal (D°). to

[53] An astronomical object consisting of a central star of extremely high temperature, surrounded by a cloud of gas, sometimes forming a disc, superficially
rather like a planet in appearance. There is, in fact, no physical resemblance to
a planet. The scale of the phenomenon is many times greater than that of the
solar system. The name is exceedingly ill-chosen and misleading. However, in
Herschel's day the nature of these objects was not understood, and as the only
son of the only man who had added a planet to the list known since antiquity he
may have had ambitions to repeat the feat. In later parts of the diary there are
numerous references to discoveries of planetary nebulae, not all of which do belong to this class of object.

[54] The only light reaching the surface of the totally eclipsed moon is sunlight
refracted through the atmosphere of the earth, which has absorbed all the blue
rays.

that at 7^h 30^m 0^s when I judged the shadow to be *quite on*

+9	9	45	—No doubt that Emersion *is begun.*
9	10	40	—Certainly long ago begun—a Bright narrow segment distinctly visible
9^h	9^m	0^s	to be taken for End of total Obscuration
7	30	30	— — — — beginning of D^o—
8	19	45	for middle of total obscuration

(Yet at 8 15 the whole disk of ☽ was conspicuously visible) From this time till the segment of light attained some considerable breadth the appearance in the Nightglass was most beautiful—

At the edge of the bright seg^t was a dull uncertain light which seemed to grow a little brighter & of a feeble greenish hue at 3 or 4′ dist from edge of the shadow & then softened away by an extremely beautiful graduation into a dull cloudy copper-coloured light which grew more & more obscure up to the extreme limb of the Moon. The Phaenom^n in this stage was one of very remarkable beauty. It was well seen by Naked-eye but much better in Nightglass.—The following are the results as compared with Nautical [Almanac]

	Beginning of total obscuration	End of total shadow	End of the eclipse by shadow
By Almanac	8^h 42^m 0^s A.G.T.	10 20 30	11 20 0
By obs. chronom.	A 7 30 0	9 8 35	10 8 30
A's Error on G^{ch} Ap.T.	−1 12 0	−1 11 55	−1 11 30

Therefore A's Error on Gree.^{ch} App^t Time $= -1^h$ 11^m 48^s or at 0^h 0^m 0^s App^t Noon at Green^{ch} A would shew 22^h 49^m 12^s

M. T. at App^t Noon G^{ch} 26 Dec. $= 0^h$ 0^m $55^s = 24^h$ 0^m 55^s (by Alm)

Time by A at App^t Noon at G^{ch} $= 22$ 48 12

Time shewn by A *behind* M.G.T. $= 1$ 12 43

or A's error on Mean Green^{ch} Time $= -1^h$ 12 43

[He compares this with "time by account," that is, extrapolated error of the chronometer A, which should have been $- 1^h\ 10^m\ 30^s$, that is, $2^m\ 13^s$ different from above. "This is very strange and incredible."][55]

[55] These calculations may look a little formidable but are really quite simple. The object of the exercise is to determine the error of the chronometer designated A, whether belonging to the ship or to Herschel, we do not know. The preoccupation with lunars might suggest that he distrusted the performance of chronometer A. It should be remarked that it is not required of a chronometer that it show the correct time, or even that it should not gain or lose. What is required is that the error on some date be exactly known and the "rate" be well known, so that the errors of the chronometer at future times will be capable of correction by the use of an extrapolated error.

In the above computations Herschel finds the chronometer error by comparing its readings with *Almanac* predictions for three phases in the eclipse, and takes the mean. This gives him the error in terms of Greenwich apparent time, that is, a time system in which 0^h is defined as the moment when the Sun is on the meridian at Greenwich. Such a time is not uniformly flowing, because the Sun does not advance uniformly round the celestial equator. A correction is required to the uniformly flowing system of "Mean Greenwich Time" (as he calls it), which is the fifty-five seconds of the line marked "by Alm." A uniformly running chronometer would be rated to keep mean time when required for this purpose. He is appalled by the discrepancy of more than two minutes, but one would think that such ill-defined instants as shadow contacts for a very low altitude moon might easily yield much bigger errors than this.

1834

The Diary: January 1 to January 22

[Note on front cover]   P^d 7^s JFWH
[Pencil note in another hand at top of page for January 1]
   A happy New Year to dear Jack

Wednesday, January 1, 1834

Commenced the New Year in Lat. 29° South   Longitude 11° West
on board the Mount Stuart Elphinstone, expecting to arrive at the
Cape of Good Hope in 10 days or a fortnight (which God grant!).
Hitherto all has gone on most smoothly—no adventures—no accidents—no foul weather—no Bay of Biscay—no long fretting calms at
the Equator—but handed over regularly from one fair wind to another with little more interval than enough to make us acknowledge
the powers under whose influence Neptune's Empire stands.—Since
passing the line it has been beautiful tropical weather.—Today a
change comes o'er the shadow of our dream.—An overcast sky, rain,
wind shifted, and the starboard tack *for the first time* exchanged for
for the larboard. This change of tack in a ship is like a change of
hemisphere—all the habitual conventions of things are *bouleverseés*.
Top becomes bottom & Right left.—Of our party all are well but
Nanson who has a low fever & keeps her bed—Caroline is fretful
   Baby teething & Mamma has contracted a habit of beating me at
chess.—Begin to be tired of keeping a Meteorological Register &
wish for sight of land.—Last night at 12 PM being the exit of the old
year two disguised persons perambulated the ship banging a large
bell—the ladies & gent^n sang Life let us cherish &c—and this morning one of the Cuddy servants was found tied in his hammock, with
his face *painted* black, and quite unconscious of his altered hue.
   Towards ev[en]ing wind increased to a stiff breeze. Royals, studding sails & Topgallants taken in, & 2 Reefs in Main Topsail. Sea &
wind rising and at length a high sea & regular bad weather.

Thursday, January 2, 1834

After a tossing night, found Sea & wind much increased, and the wind being SE. drives us much to North, the course being ENE. This is unfortunate as we are already not south enough,—To add to the comforts of the day Marg^t, the Ayah^1 and Caroline & Bella are all sick, & Nanson helpless in bed.

Ran on with 2 reefs in Main topsail till dinner time & then Tacked again holding a course about S.W. the Sea being now very high & ship pitching detestably.

Nothing can be done on board a ship in bad weather but sit still and wait for better (i.e. for Passengers). All power of occupation is at an end & the whole attention is directed towards preserving some degree of personal equilibrium

At night observed in the ships wake, at long intervals, single, broad, bright *sea lamps* which lay glowing on the surface with a steady brilliant white light till lost in the distance. No possibility of taking them owing to the velocity of the ship—Is this a distinct species?

Friday, January 3, 1834

Wind and Sea greatly abated yet still unable to hold a better course than S.S.W.—Thermom about 72 which feels *cold* an odd enough twist of circumstances, our friends in England are perhaps declaring it a warm day at 42 or shivering round a great fire at 22.—M^t and Nanson both in bed   Ayah sick &c &c.—

No events.

In Ev[en]ing occupied an idle hour with Expts. on melting point of cocoa-nut oil. It is liquid at 82º0—and as it congeals the temperature rises to 84º5. This is analogous to the rise in temperature in water cooled below the freezing point, on fixing or to that of Sol. Sal. Lime^2 in Xtallising.^{3} but *here* no *agitation* will determine the fixation which

---

[1] A Hindi word meaning a native nurse or maid in India. Possibly a servant borrowed on board from one of the other passengers, since, as far as can be made out, the Herschel party did not include any such person.

[2] The transcription is very difficult. The reading adopted depends on the fact that calcium chloride, $CaCl_2:6H_2O$, has the high heat of crystallization of 4.56 kilogram-calories per mole at 18°C.

[3] The symbol X is often used as an abbreviation for Christ, as in Xmas (used by Sir John later), but *Xtallising* for *crystallizing* is unusual.

seems to require the presence of a great quantity of chrystals actually formed.

Rifle Practice with Capt. Jones, at bottles floated astern & tied by long lines.

Saturday, January 4, 1834

The wind still unfavorable & still on the same tack S.S.W.—But as the day got on, it fell nearly calm and then began to come round a little and in the night Jan 4–5 had got pretty fair. The sea also less heavy

The Capt⁰ took some Lunars in þᵉ morning which came out well agreeing with the Chronometers. I worked them by Thomson's Rules & Tables which are *practically excellent*—Plain, & excessively short. —By far the shortest & best method I know.

Read "Memoirs of Great Commanders" by . . . [his omission] An ill got up book, a model of bad style—but his statements are distinct & his summaries not amiss. Life of General Monk[4] is the best & has merit. NB. Monk belongs to the "1st class" of great men, such is the impression one rises with. He might have been Protector with every Suffrage (but ultra royalist's) in his favor—He preferred a restoration, from a perception that the Republic *could* not stand, not from blind devotion to arbitrary sway—and when he restored made no terms though he might have commanded *any*, [words erased]   Good sense, temper, moderation were the keystones of his power.—

On the other hand Gonsalves de Cordoba[5]—great tho' he was as a Captain must be held as a Political character the basest of the base. A willing sacrificer, decked out with all the glory that conquest, talent, temper & private virtue & honour can throw round a character, only to immolate himself under the feet of arbitrary power, deliberately lending himself to the most atrocious villainy over

[4] George Monk (1608–1670), first Duke of Albermarle. He was a noted soldier who became a general in the Cromwellian Army in the English Civil War. Later he brought about the bloodless restoration of Charles II and was ennobled and rewarded by him.

[5] Gonzalo Fernández de Córdoba (1453–1515), known as the Great Captain for his warlike exploits and reform of tactics. He commanded a Spanish expedition supporting the Aragonese house of Naples against Charles VIII of France, 1498. He expelled the French troops from Naples, 1503, where he remained as governor until 1507.

[words erased] his better judgment, & with his eyes broadly open, at the mere mandate of his King (at the Capitulation of Naples)

Sunday, January 5, 1834

Rose late barely in time for Divine Service. After Church got a set of Actinometer obs[us] & was then called to see Nanson who had ruptured a small vessel in the lungs & was ejecting blood. Declared a case of danger, i.e. if a relapse should occur in 48 hours fatal. If not recovery & perhaps better health than before—So spoke Dr Mackintosh a Physician on board who came on þ[e] first alarm.

Got all the Cabins rearranged so as to leave the Middle Poop Cabin all to Nanson the Aft D[o] to M[t] & the 2 children C & I, and the Awning Cabin for self and Baby for which purpose struck the 4 legged bedstead & cleared the area of the Aft cabin.

Dined on Deck. Party the same as usual i.e. Lady D'Urban & Marg[t] Capt[n] Beresford, Major Dutton, M[r] Hall and self.—Wind got round fair, sea abated.

In Evening skimmed Sharon Turner's[6] "Sacred History from Creation to Deluge" a vile trash-book, on the principle of "bringing Science to support religion" as it is now called—i e. "proving" everything it is considered desireable to prove by mustering a roll-call of quotations misapplied and misunderstood out of books called scientific (all being held of equal authority) as the work described   He contends that the 6 days of Creation were really & truly 6 times 24 hours of the same length as at present—in which the Geolog[l] work was done (Vide Lyell's[7] 3rd Volume !!)—He considers that the Atmospheric water if precipitated on the Earth *in toto* would re-drown the world whereas it would not raise the Ocean a foot &c &c

Monday, January 6, 1834

Rose early—the scrubbing and washing of Decks being intolerable —Wind right aft.—All sail set   Royal, stun-sails[8] (? Studding sails)

---

[6] Sharon Turner (1768–1847), a devoutly orthodox popular historian.

[7] Sir Charles Lyell (1797–1875), geologist. He was one of the real founders of modern geology through his introduction of a time scale of millions of years. He was the author of *Principles of Geology* (3 vols., 1830, 1832, 1833), *Elements of Geology* (1838), and *The Antiquity of Man* (1863).

[8] Studding sails.

&c &c and slipping along quietly through almost no sea. Nanson better.

Expt^s on the limits of melting & fixing point of Tallow.—It rises 5°8 in the act of fixing!

Actinometer Measures.   Chess with the Captain

Dined on Deck. Lady D'Urban, Marg^t, Major Dutton, Mr. Hall & Capt. Beresford.

Much Painting & Tarring the Ship to make a shew at the Cape, going on—also the gun's ordered to be "sealed" & Cartridges prepared.—NB. Major D. related case of firing a salute from old guns, where from mere neglect of "Stopping the vent"[9] while springing, at the reloading 2 men were blown limb from limb, & the wife of one was presently after seen picking up the scattered legs & arms of her husband (Major D an eye-witness) Also Captn R^n[10] had a mans arm blown off in the self same act. Surely this must be a bad system of going to work.

After dinner ship began to roll (wind exactly aft) and it grew so dreadfully bad (though almost no sea running) that I hardly know whether it is not worse than the pitching motion in a heavy sea. It takes you unawares. After a long still interval the oscillations begin on a sudden to increase with a slow, equable swing till they acquire such an amplitude as to throw everything about off the tables, chairs &c higgledy piggledy, and almost carry you off your legs. It is a terrible trial of temper, *stomach*, and *understanding*

Made out a list of words for a Universal Comparative Vocabulary to compare languages.—NB. Reading [?] matter.

Tuesday, January 7, 1834

Rose early & unrefreshed, from a dis [last phrase erased]

The rolling motion intolerable.—*This* is the state of things where a swing cot would be useful.—We have one but it cannot be got at be-

---

[9] The guns would be muzzle loaders, mounted on wooden trucks which would run back with the recoil of firing until restrained by rope springs. Reloading would require the charge, wad, and ball to be rammed in from the muzzle. Premature ignition of the charge, even for a blank shot, would blow the rammer to pieces.

[10] Possibly Captain Redman, who commanded the *Hindostan* (see May 31, 1837).

ing over Nanson's couch. Found all the Passengers loud in complaint & looking haggard at breakfast like spectres.—The roll lasted till dinner time with a brisk breeze aft.

The night was a dull overcast Scotch-mist & everything on board was moist & clammy

Saw an albatross,[11] but could not get a shot at him. It is a very bulky looking white-bodied bird. Got a set of Actinom^{rs} in a clear interval

Dined on Deck after which came on to rain & blow the wind coming round to the starboard beam. This took off the rolling & sent the ship along 9 or 10 knots   Stunsails taken in—D^o Royals.—NB The smoking on Deck becomes more pertinacious daily and is a most abominable nuisance

Wednesday, January 8, 1834

After getting along pretty well in the early part of the day the wind began to die away & at length grew nearly calm.

Thursday, January 9, 1834

Baby's 1st birth day.[12]—Nearly calm so much so that the ship got sternway from the swell a-head which is long and powerful making the ship labour & roll intolerably.

A gorgeous sunset &c. See overleaf where this is entered by mistake.

Friday, January 10, 1834
(Entered by mistake here—belongs to Jan. 9)

A most gorgeous sunset—After the sun was quite down (perhaps 10^m or ¼ hour)—Broad red diverging rays tinged at their edges with blue, yellow & greenish rainbow like tints were seen diverging from the place of the Sun below the horizon & melting upward, into the purest, deepest, liquid blue that it is possible to imagine. Have often

---

[11] The albatrosses, Diomedeidae, are primarily birds of the southern oceans. The largest, *Diomedea exulans* (quite possibly the species observed by Sir John), has a wingspan of nine to eleven feet. All are gliding birds, making use of updrafts from the faces of large waves. Some species follow ships persistently, for hours or even days, easily keeping up with them while flying a zig-zag pattern. Sir John and his shipmates evidently did not share the superstition of the Ancient Mariner's comrades.

[12] William James Herschel's birthday.

heard of "tropical Sunsets"—but *this* surpassed all description or expectation.

Saturday, January 11, 1834

This morning a whole covey of Albatrosses seen, sitting on the water.—8 or 10 of them.—A great turn-out of Rifles & fowling pieces, but nothing shot. One which came up astern was a very large & fine bird. White body, whitish or white-brown wings and a soaring, skimming flight.—

This morning M^t extremely exhausted with the rolling of the ship, during the night w^h was *once* most furious, though with little wind or sea. One lurch threw everything into confusion, drove the children in their bassinets from one side to other of the cabin and nearly brained Bella with a chess board w^h leaped off the table.—The swing cot proved of use, and secured *me* at least a night's tolerable sleep.

Occupied the day in packing up rattle traps and preparing for land, and in trying a mode of suspension for a cot which promises well.

Towards Evening Nanson who had been regularly mending in health since her last attack, and who, yesterday & today had sate up some time in her armed chair was reported delirious (NB She had talked very loosely & absurdly for 2 days before, but without suspicion of anything more than ordinary weakness of mind. Mr. M^c-Hardy & Dr Mackintosh declared her delirium a very dangerous indication of increasing weakness and prognosticated a fatal termination. Got all on the alert to sit up with her & keep up the restorative system during the night.

Tuesday, January 14, 1834

No land yet in sight—retired in full expectation of finding it in sight tomorrow morning.

Wednesday, January 15, 1834

At early dawn this morning James knocked on our cabin door and called out Land! Rose & hurried on Deck whence the whole range of the Mountains of the Cape from Table Bay to the Cape of Good Hope was distinctly seen, as a thin, blue, but clearly defined vapour. The Lions Head was seen as an Island the base being below the horizon. Called up Marg^t who also came on Deck just before Sunrise. It was a truly magnificent Scene, but as the Sun got up & strong day-

light came on, the Hills grew dim & almost faded away—till as the
distance diminished they again gained sharpness & reality & began to
break into distinct masses of which all who could muster pencils &
paper were from time to time busy sketching.—Sea Birds became
numerous & were much frightened (none hurt) by our shot &
bullets. —Our enjoyment of this scene was much disturbed by reports from Nanson's cabin who was stated to be all but dying (I saw
no symptoms of it for my part). On taking as usual the temperature
of the water at midday, found to my great surprise a sudden fall of 9°
Fahr. to have taken place. At or soon after, the colour of the water
changed from blue to dull green, not bottle green but grass-green.
The shore still nearing, grew bolder & more rugged & broken   "The
wild Pomp of Mountain Majesty." developed itself & certainly nothing finer than this approach of South Africa can be conceived.—
As the day advanced however the breeze died & at Sunset we were
yet far from entering the Bay, being opposite to a high conical peak
in a mountain gorge, some miles South of the "Green Point" on which
a light house stands. The ship at length hardly went ½ a knot an hour
—it darkened, & lights began to glimmer on shore & presently the light
house grew conspicuously luminous—Again we accelerated our pace
& began to near the light when all at once it disappeared & we were
left on deck all at gaze & in some anxiety as had a foul wind risen we
might have got entangled & perhaps driven on the rugged shore to
the South. We had however before exchanged signals with the Station on the "Lions Rump." so that we knew we were expected.—
After near half an hour the lights reappeared from the thick fog w^h
had covered them.—Rounded Green Point & cast anchor in the nearest anchorage, by advice of the Port Captain

Thursday, January 16, 1834

At daybreak, weighed anchor, and got the ship nearer the town
into good anchorage about a mile from the Jetty. The successive aspects of the town & its surrounding & overhanging mountains especially the Table hill, were extremely interesting in this little Trip.
There are many vessels of considerable size in the harbour (say 20):
The houses are snow white, & apparently roofless.[13]

---

[13] Many had flat roofs that were invisible from the sea.

The situation most remarkable, hemmed in on the sides by steep promontories & backed by the Mural precipice of the Table Mountain which rises sharp & sudden behind it. We had hardly cast anchor when a boat came on board with Col[l] Wade[14] to welcome the Governor and (what was more interesting to us) Marg[ts] brother Dr Duncan Stewart to welcome ourselves. Under his guidance we took Boats & left the ship & (the Lord be thanked) set foot on African ground about 10 AM. Landed on the Old Jetty, a crazy structure of very bad access & proceeded across the large & arid, dusty looking Parade to Miss Rabé's (Rābĕ)[15] boarding house where we were ushered into *immense* apartments, each large enough to hold an English house (a small one) & made extremely comfortable.

Friday, January 17, 1834

Rose pretty early & after breakfast (on Grapes Mulberries Pears &c) took a boat and with my last evening's boatman Spolander, landed Nanson, who has got up her strength and does not seem to have much the matter with her. Got her a nurse & put her to bed at Miss Rabe's. Returned with Two boats and landed some part of the Luggage.

Sunday, January 19, 1834

After Breakfast attended Divine Service in the Principal English Church[16] in this Town.—There are few & small pews the great area is occupied by Chairs & benches, in which all sit indiscriminately. This is right—In God's sight there should be no exclusive aristocratic distinctions—Every one should forget Differences & remember only that he is a worm among worms. The New Governor (Sir B. D'-Urban) was there, with no state.

The organ Loud & harsh, but there was one fine musical effect when a long & loud crashing finale sank at once with a soft gentle air *quite different* in melody from the subject of the noisy piece which went before.—A long loud & swift sermon from D[r] Hough[17] in con-

---

[14] Lieutenant Colonel Thomas Francis Wade, Acting Governor.

[15] Actually Mrs. Rabe, widow of Johannes Christian Rabe, of 2 Keizersgracht.

[16] St. George's Church was only finished for the first service on December 21, 1834, so this must have been the Dutch Reformed Groote Kerk, which the Anglican congregation was permitted to use for its own services.

[17] Conceivably the subject had been chosen for Sir John's special benefit.

demnation of reason and exaltation of faith followed, but he preached so quick it was difficult to *follow him* & I could only perceive him to be a clever man using words in very terse & compact combinations as if he *had* a meaning & wanted his hearers to perceive it.

In the Evening walked out to a quarry on the side of Lion Hill just out of Town. It is Slate (or at least blue hard stone) in nearly vertical very well defined Strata, with a cross stratification extremely distinct, & breaks into very regular blocks almost exactly like the Rhomboid of Carbonate of Lime. Saw no traces of organic remains in any part of this quarry.—It rained & blew from the N.W. and as I got into the Town, the rain grew really violent, and a stiff gale.

Wednesday, January 22, 1834
In the Evening drove out to Baron Ludwig's[18] Botanical Garden Wrote to my aunt No 2.

---

[18] Carl Ferdinand Heinrich von Ludwig (1784–1847), was born in Württemberg in humble circumstances, trained as an apothecary, and then went to Amsterdam. While there he saw an advertisement by Dr. Friedrich Ludwig Liesching for an assistant at the Cape, whither he went in 1805. In 1816 he married Alida Maria Altenstaedt, a widow, who brought him her late husband's snuff and tobacco business. In due course Ludwig became affluent, and was able to indulge his undoubted talents in science. Becoming interested in botany, he made a collection of plants and insects which he presented, in 1826, to the Royal Cabinet for Natural History in Stuttgart, as well as more than a thousand specimens of birds. For this he was made a Knight of the Royal Order of the Crown of Württemberg, and thus acquired his baronial title and aristocratic prefix. However, von Ludwig, as he now was, did perform remarkable services to botany. In 1830 he bought a piece of land in Kloof Street, Tamboerskloof, and laid it out as a garden, usually known as "Ludwig's-burg," to which visitors were admitted by ticket. In it the Baron cultivated an extraordinary variety of plants, many of them exotics which he acclimatized in Africa. He is credited with the introduction of more than sixteen hundred different plants. He was a leader in all the cultural activities of the town and in many of its business ventures; he was a highly respected citizen. His name is perpetuated both in ornithology and botany, where species are named after him. Apart from this he is almost forgotten, and his garden, deplorably, almost completely built over. It is also of importance that, during the period of the diaries, he was owner of another plot of great astronomical significance, in Concordia Gardens, a little nearer the center of the town (see note 151, September 3, 1834). There is an excellent biographical study by Frank Bradlow, *Baron von Ludwig and the Ludwig's-Burg Garden.*

A Letter from Sir John, Cape Town, to Caroline Lucretia Herschel, Hanover

| | Cape Town | N⁰ 2 |

		N⁰ 1 was sent by a
	Cape of Good Hope	ship met at the Equator going
	Jany. 21 1834.	to Bahia

My dear Aunt—Here we are safely landed and comfortably housed at the far end of Africa, and having secured the landing and final stowage of all the telescopes and other matters, as far as I can see without the slightest injury, I lose no time in reporting to you our good success so FAR. Margaret and the Children are, thank God, quite well—though for fear you should think her too good a sailor, I ought to add that the[y] continued seasick, at intervals, during the whole passage.—We were nine weeks and two days at Sea, during which period we experienced only one day of contrary wind, and not the most trifling storm, or unpleasant nautical adventure of any kind. We had a brisk breeze "right aft" (that is to say driving us forward strait before it) all the way from the Bay of Biscay (which we never entered) to the "calm latitudes") that is to say to the space about 5 or 6 degrees broad, near the equator, where the trade winds cease, and where it is no unusual thing for a ship to lie becalmed for a month or six weeks, frying under a vertical sun. Such however was not our fate. We were detained only 3 days by the calms usual in that zone, but never *quite* still, or driven out of our course, and immediately on crossing "the line" got a good breeze (the South East Trade Wind) which carried us round Trinidad [this name bracketed in pencil and "? Fernando Nᵃ" written above][19]—then exchanged it

[19] Fernando Noronha, an island off the Brazilian coast. This interpolation, evidently by a later hand, seems to have been motivated by the consideration that the ship could not have gone anywhere near (British) Trinidad. On December 8 the ship was just north of the equator, in west longitude 22°, that is, about mid-

for a North West wind, which with the exception of one day's squall, from the S. E. carried us steadily into Table Bay. On the night of the 14th we were told to prepare to see the Table Mountain next morning (NB we had not seen land before since leaving England) accordingly at Morning Dawn the welcome word "Land" was heard & there stood this most magnificent hill with all its attendant mountain Range down to the farthest point of South Africa—full in view, with a clear blue ghost-like outline, and that night we cast anchor within the Bay. Next morning early we landed, under the Escort of Dr Stewart, Margaret's Brother whom we found quite well and you may imagine the meeting. We immediately took up our quarters at a most comfortable Boarding House (Miss Rabe's) and I proceeded without loss of time to unship the Instruments. This was no trifling operation as they filled (with the rest of our luggage) 15 large Boats, and owing to the difficulty of getting them up from the "hold" of the Ship required several days to complete the landing. During the whole time (and indeed up to this moment not a single South East gale (the perpetual summer torment of this Harbour) has occurred nor indeed any wind at all violent. This is a thing almost unheard of here, and has indeed been most fortunate, since otherwise it is not at all unlikely that some of the boats, laden to the water's edge as they were, might have been lost and the whole business crippled.—

For the last 2 or 3 days we have been looking at houses, and have all but agreed for one a most beautiful place within 4 or 5 miles out of town called "the Grove"—In point of situation it is a perfect paradise, in the most rich and magnificent mountain scenery, and sheltered from all winds *even* the fierce South Easter, by Thick surrounding woods.—I must reserve for my next all description of the gorgeous display of superb flowers which adorn this splendid country as well as of the astonishing brilliancy of the Southern Constellations which the calm clear night shew off to great advantage, and wishing we had

---

way between Monrovia and Pernambuco. Thereafter, the travelers picked up the southeast trades, probably sailing a southwesterly or south-southwesterly course for some time, which carried them well over to the western side of the South Atlantic. With the change of wind to the northwest, they would then cross back over almost the whole width of the ocean to Cape Town. This course would bring them round the (Brazilian) Trinidad Islands (near 20°S, 30°W). The original entry may be presumed correct, and the interpolation erroneous.

[Written perpendicularly on the front page]
we had you here to see them must conclude with best loves from
Maggie and the children

your affecte nephew
J.F.W. Herschel

The Diary: January 24 to March 27

Friday, January 24, 1834

Rose perfectly well.—After breakfast called on . . . [his omission]
and attended the New Governor's first Levee, a rather ill arranged
affair, there being no one to announce the names. Perhaps however
this omission was intentional & politic, as the Governor could not
speak to every body, and ignorance of the person might excuse omis-
sions which might otherwise hint egotism.—Then called with D[r]
Stuart[20] on D[r] Smith[21] (about to proceed on an Exped[n] into the in-

[20] In spite of the variant spelling this can hardly be other than Dr. Duncan
Stewart, Lady Herschel's brother.

[21] Although Sir John never gives a Christian name, this must be (Sir) Andrew
Smith, the zoologist and explorer (1798–1872). Before 1825 he was in the Eastern Districts in medical charge of troops. He then came to Cape Town, where he
started a museum, subsequently handed over to the government. He was also responsible, with J. Adamson, for the establishment of the South African Institution (1829) and the *South African Quarterly Journal* (1830–1835). He was assistant surgeon of the 9th Foot in 1825, staff assistant surgeon in 1826, and staff
surgeon in 1837. At Sir Benjamin D'Urban's request he traveled among the
warring Bantu tribes to the east, and negotiated treaties with the Matabele and
Basuto. His account of the zoology of the "Expedition for Exploring Central
Africa" (1834) was published by Smith, Elder & Co. (see note 1, 1833). After
his return to England in 1837, he rose steadily in his profession to become
Director-General of the Army and Ordnance Medical Departments (1853). The
conduct of his department was the subject of criticism and official investigation
during the Crimean War. He resigned as Director-General in 1858, when he was
made K.C.B. (see P. R. Kirby, *Sir Andrew Smith, M.D., K.C.B.*). The expedition
on which he was about to start (his third in six years) was inspired by a communication made in 1833 to the South African Literary and Scientific Institution
reporting discoveries by a trading party in the interior. The Cape of Good Hope
Association for Exploring Central Africa raised funds by issuing shares, and

terior)—on D[r] Phillips[22] a Missionary (a goodly portly man who has
bought himself a thoroughly good house & seems to have thriven in
the world)—& on many more about whom I can't write (as I can't
remember). At D[r] Smiths saw a Hottentot woman's skin [words
erased] stuffed—with all the extraordinary peculiarities[23] attributed
to these nymphs by travellers. D[r] S. was in daily communication with
Capt[n] Owen here & never heard him speak of the adventure of the
Leven's people[24] seeing the Barracouta's 200 miles off so as to know

---

secured government support. Herschel drafted instructions for making meteoro-
logical and other observations. The party left Cape Town after a breakfast at the
Royal Observatory on July 3, 1834, and journeyed overland to Graaff Reinet,
whither their heavy stores had been shipped by sea through Algoa bay (Port
Elizabeth). The expedition traveled extensively in the districts now known as the
Eastern Cape, the Orange Free State, Kuruman, and Transvaal, returning to
Graaff Reinet on January 4, 1836.

[22] In spite of variant spellings in the diary, this seems to have been Dr. John
Philip (1775–1851), missionary and outstanding figure in the history of South
Africa. Born in Scotland, he was trained in London, and in 1818 accompanied
Campbell on a journey into the interior to inspect the London Missionary Society's stations in South Africa. He was appointed superintendent of all the
Society's missions in South Africa in 1822. His anger was aroused by cases of
mistreatment of natives by certain of the white settlers, and he returned to England, 1826–1828, to air his views. His influence resulted in the 1828 ordinance
giving the same privileges to all free colored men as were enjoyed by the settlers.
He first supported D'Urban, whose policy was in line with Philip's ideals of
limiting the area of white settlement and creating a belt of independent native
states. After the liberation of the slaves in 1834, and following a frontier war,
D'Urban annexed all the land up to the Kei River. Philip's hostility was aroused
and he campaigned against D'Urban in England from 1836. The annexation was
abandoned and D'Urban recalled. Philip then returned to South Africa where he
became unofficial adviser to the new governor, Sir George Napier (see note 1,
January 31, 1838), on all questions relating to the treatment of the natives. He
died at Hankey in the Cape Province.

[23] Immense development of breasts and buttocks, probably serving a purpose
similar to the hump of the camel for these arid region dwellers.

[24] We are indebted to Dr. A. M. Lewin Robinson for the following text: "It is
reported in the *Narrative of Voyages to Explore the Shores of Africa, Arabia and
Madagascar; performed in H. M. Ships Leven and Barracouta under Capt.
W.F.W. Owen* (ed. by H. B. Robinson, Vol. 1, pp. 241–243), that, on the
evening of the 6th April 1823, off Danger Pt., while sailing from Algoa Bay to the
Cape, the *Leven* sighted the *Barracouta* about 2 miles to the leeward. This was
surprising as she was not thought to be anywhere near. Every detail could be
seen, however, including well-known faces on deck, but she made no effort to join

the ship & faces, as recently published in his book (What impudent lies are spread *on undoubted authority* by *eyewitnesses* &c & believed!). Maclear accompanied us, & he, & Major Hamilton dined with us. After dinner Margaret & I drove out to look at Capt[n] Latouche's[25] house as a temporary residence till we get into the Grove.

Saturday, February 1, 1834

Rose at 7. Occupied in unpacking, distributing and arranging, (with Actinometer Obs[ns]) till 11. Then Took a camera sketch[26] of the superb Mountains in front of our house, got out a table and under the Shade of the fir trees before the house, with a gentle breeze and splendid blue sky, passed an hour & a half or 2 hours in drawing with great enjoyment.—Got up the two self registering thermometers, in a capital, shaded exposure.

M[r] Skirrow[27] called and told such miraculous instances of the primeval stupidity or rather indolent apathy of the Dutch colonists (the actual race now in possession of large tracts of country as made one's hair stand on end.

Leven. At night no lights could be seen. Next morning *Leven* anchored in Simon's Bay, where she anxiously waited a week for *Barracouta.* When supposedly seen she "must have been above 300 miles from us and no other vessel of the same class was ever seen about the Cape." The only explanation given was that the phenomenon was "doubtless attributable to natural and probably simple causes." (Footnote: 'Such effects may be produced by refraction.')"

[25] Captain Latouche was an Indian visitor.

[26] A drawing made with the camera lucida, a device permitting the operator to outline on a sheet of paper an image of a scene. In its simplest form, a plane sheet of glass, inclined at 45° to a sheet of paper placed beneath it. The operator would see, simultaneously, the scene reflected in the glass, and, through the glass, his pencil point on the paper below. More sophisticated versions, such as that so energetically employed by Sir John, incorporated two successive reflections to obviate the inversion of the image in the simple form.

[27] John Skirrow, a land surveyor and architect, who had just retired from the post of government architect and assistant civil engineer. Typical of the difficulties we have met with variant spellings of names, David Gill in his *History and Description of the Royal Observatory, Cape of Good Hope,* says at page xii, with reference to the construction of the Observatory buildings in 1825, "A most efficient clerk of works—Mr Skerrow—was then appointed, and the work proceeded." It seems certain that this was the same man. The senior editor has seen a scrap of paper recovered from beneath the demolished pier of a dismantled instrument at the Cape Observatory, signed by Lowry Cole, then governor, and by Skerrow, spelled thus.

Dispatched Stone to Town for Packages. He returned with 2 Waggon loads and this day I got the foundation Ring & fixed radii of the 20-feet & some other packages on the ground destined for them at the Grove

Sunday, February 2, 1834

Attended Divine Service at the Principal Dutch Church. Mr Hough officiating—the Town insupportably hot.

Monday, February 3, 1834

The Cow not to be heard of, after sending to the Wynberg *Pound* and another

Got up the Tube & Ladders and the heavy cases 6, 7, 9, 24, 36, as also 27 & 48 to Schonbergs[28] & lodged them securely on his premises.

Col. & Miss [? Mrs.] Prendergast[29] called.

In the Evening rode with D[uncan].S[tewart]. and Margaret to *Protea*[30] a sweetly situated house but (like all others here tumbling down). As we returned I diverged from the route & rode up a path which proved to my surprise to be a practicable passage over the Mountain range between the Devil Hill & ? the Table Mountain. Found the ruins of houses in a most secluded & romantic spot called (as I afterwards learn) Paradise.[31]

The night calm clear & intolerably hot.

Tuesday, February 4, 1834

A day of purgatory. After a cloudless calm night, very warm, so as even to induce me to spread a matrass in the open air & get a troubled sleep—about Sunrise a NW wind rose and blew pretty strong. The effect on temperature was extraordinary—The Therm. which at 4 AM had fallen to 72 had started up to 89 at half past 7 and continued rising till it reached 101½ about 2 PM. By rigorously keeping the

---

[28] The owner of "The Grove," or "Feldhausen," as Sir John preferred to call it, who first leased, then sold, the property to him. In the diary he appears as Schonberg, Schönberg, and Schomberg. Dr. Robinson says that the spelling should be Schönnberg. This last is the spelling used by Maclear (*La Caille's Arc of Meridian*, 1866, I, 21), who there says that the property was sold in 1838 to Rice J. Jones.

[29] Colonel and Mrs. Prendergast were Indian visitors.

[30] "Protea," later "Bishopscourt," was then occupied by H.F.W. Maynier.

[31] "Paradise" was the country cottage of the famous Lady Anne Barnard, wife of an early colonial secretary.

Doors and windows closed I managed to keep the large Rooms & Passage down to 81 but the moment one put one's nose out of doors the sensation was that of a furnace and a very few minutes exposure in the thickest shade of the fir trees sufficed to bring on a quick throbbing pulse—and a sense of overpowering faintness. About 5 or 6 PM the therm. had lowered to 95 & thence fell rapidly till a little after sunset The outer temperature equalled the inner when we set all doors & windows open. NB. The NW. wind had now ceased—

Captn & Mrs. Mitchell[32] called   also Mr Borland & Mr Malcolm.— Drove over to "Half way house" near Constantia

Tuesday, February 11, 1834
WYNBERG

Wednesday, February 12, 1834

Wynberg

Unpacked the Iron work of 20 ft &c and rearranged it to be ready for putting together.

Got together (with J. Stone) the foundation ring and center work of the 20 feet.

Wrote to Beaufort and dispatched, enclosed to him through Barrow[33]—letters from myself to Babbage[34] & from Marg^t to her Mother & to Peter.

Slept—Dined—Music—

Rode with Marg^t & Dr Stewart to the "Halfway House opposite the Gorge Gorge [*sic*] of Hout's Bay, and round home by Constantia Farm.

---

[32] Charles Cornwallis Michell (1793–1851), Surveyor General at the Cape. Herschel persistently spells his name wrong. He had fought with distinction in the Peninsular War, and served as Quartermaster-General in the Kaffir War (1834–1835). He resigned his Cape post in 1848. Michell's Pass in the Western Cape is named after him.

[33] Sir John Barrow Bt. (1764–1848), second secretary to the Admiralty (1804–1845), private secretary to Lord Macartney at the Cape (1796–1802). He was the founder of the Royal Geographical Society.

[34] Charles Babbage (1792–1871), English mathematician, and close friend of Sir John. While fellow students at Cambridge, they joined with Peacock to found the "Analytical Society," devoted to the reform of mathematics in England. He was a founder member of the Astronomical Society and traveled with Herschel in Europe as a young man. He was chiefly celebrated for his projects, much in advance of their time, for the construction of mechanical computing engines.

Adapted the Prism Reflector of Fraunhofer[35] to the 5 feet Newtonian,[36] which produced a very great increase of light & is every way
a capital improvement.

Thursday, February 13, 1834

Wynberg

Got the Horizontal moveable frame of the 20-feet laid down

Drew flowers & Landscapes

Sir B. & Lady D'Urban & a certain Mrs. Smith[37] called—also D^r
Philip & wife and Mrs Fairburn[38] & son

Feeling ill took an emetic

D. S. returned from Hout's Bay bringing a Bunch of most magnificent flowers of Bulbs & Heaths also Brown brought in a very
singular bulb with a red stem and flower & no leaves thus [a sketch]
which he called a "Grenadier Plant"[39]

---

[35] Joseph von Fraunhofer (1787–1826), German optician and physicist. He
worked as an optician in an institute near Munich, becoming its sole manager in
1818. He was highly skilled in the design and manufacture of achromatic telescopes and other optical instruments. While working on refractive indices of
glasses he made his celebrated discovery of the dark lines in the solar spectrum
now called after him. He mapped 576 lines, designating the principal ones by the
letters *A* to *G*, and demonstrated that they were constant in position in the light
of the Sun, reflected light from the Moon and planets, and in stars. He became
conservator of the Physical Cabinet in Munich, where he died in 1826.

[36] Classical form of reflecting telescope devised by Sir Isaac Newton (1642–
1727). At the lower end of the tube is a concave mirror which reflects incoming
light back up the tube as a converging cone of rays. These are reflected to a focus
at the side of the tube by a mirror inclined at 45° to the axis of the instrument.
In small Newtonian telescopes a totally reflecting prism can be used instead. This
is probably what Sir John was doing with his own instrument, the internal total
reflection of the prism being much more efficient than surface reflection from
the speculum metal mirrors, which were all that were available at the time.

[37] Herschel had his fair share of personal prejudice. The pejorative tone, and
the fact that the lady was in company with the Governor and his wife suggest
that this was Juanita the Spanish wife (married in the Peninsula War as a very
young girl) of Colonel Harry Smith (see note 108, June 24, 1834.)

[38] A misspelling of Fairbairn, wife of John Fairbairn (see note 93, May 19,
1834) and daughter of Dr. John Philip.

[39] For the text of the botanical notes we are indebted to Dr. Anthony V. Hall.
Of this he says, "Probably a *Haemanthus* (Amaryllidaceae), the Cape species
having a 'brush' of tightly packed red flowers that recalls the ornament on a
grenadier's helmet. Present names are *Blood Flower, April Fool,* and *Veldskoen-*

Saturday, February 15, 1834

Kept my bed all day today under the orders and discipline of D[r] Murray who had my precious person swathed up in an ointment of "Pate du Diable"! as a "stimulant" or "Rubefacient" for I know not what "Gastroenteritis" which has been bothering me for the last 10 days more or less

Read Venetian history.

Monday, February 17, 1834

Got 4 Coolies from Cape Town (viz:—Abdul (a fine Sultan-looking fellow the living image of Wests figure of the Black Noble-man in his window of S[t] George's Chapel Windsor "the wise men's offering". A noble intelligent good-natured black countenance)— January and Jacob brown men Malays—and Thom a low-looking thicklipped ugly Negro—& with their & Stone's aid got up the ladders of the 20-feet, & Side Stays & Top beam.

Today it has Rained with violence & we have the full enjoyment of what is here called a "leaky" house.—i. e. our whole roof admits the water freely and deluges walls & floors.—All vessels & utensils in requisition to catch the rain

Tuesday, February 18, 1834

With the help of my 4 Coolies Abdul, January, Thom and Jacob and Stone, got the tube of the 20-feet in its place and the Bar Motion fixed & now it begins to look telescope-like.

Made a round of Calls with Marg[t] at Wynberg, Newlands, Protea &c—& found everybody at home i.e

M[rs] E. Bird—M[rs] Wilberforce Bird—M[rs] Murray[40]

M[rs] Carey—M[rs] Steuart[41]   {M[r] & M[rs] Murchison (a brother of R. I. Murchison)[42]

---

blaar (roughly, 'shoe leather leaf'), the last referring to the two wide flat leathery leaves that grow along the ground in winter."

[40] Mrs. William Wilberforce Bird: wife of the controller of customs and author of State of the Cape of Good Hope in 1822–3. Mrs. Murray: wife of Dr. John Murray.

[41] Wife of the high sheriff, John Steuart.

[42] Sir Roderick Impey Murchison (1792–1871), British geologist. After service in the Peninsula War he forsook his passion for fox hunting, and turned to geology. His early papers were in collaboration with W. H. Fitton (see note 193, Observations, 1834) and Adam Sedgwick. He explored the geology of the moun-

Mrs Hare—Mrs Ross

also went with M. to inspect divers articles of furniture at the Grove.

In the Evening McClear called & drank Tea.

A Delightful cool day and all the beautiful rides round Wynberg Protea &c looking doubly beautiful as refreshed by yesterdays copious rain.

Protea is all but falling Down, and "leaky" so as to be hardly habitable

Wednesday, February 19, 1834

Wynberg   Got out the Tube and Gallery Ropes

Opened out Boxes

Saturday, February 22, 1834

Wynberg   Sent in Stone to bring out the Mirrors, Clock, and remaining Cases, which he did and thus we have at length got all our packages &c from England safe & sound

Opened one of the Mirrors (N° 32) all safe.—Got it in & Turned the telescope on α Crucis[43] which it shews quadruple having a small $*$ 13m nearer than the small $*$ commonly seen also on η Roboris[44]—a most wonderful object and some of the Milky Way Clusters. The Moon was too strong for the Magellanic Clouds.—The Stars very ill defined & blotty.

------

tainous regions of Europe with Sir Charles Lyell (see note 7, January 5, 1834), established the geological successions known as the Silurian and Devonian, and made investigations in Russia. Knighted in 1846, he was Director-General of the Geological Survey and Director of the Royal School of Mines, 1855. He was made a baronet 1866 and received the highest scientific honors. He founded the chair of geology and mineralogy at Edinburgh and the Murchison Medal of the Geological Society.

[43] See note 41 on Crux, Diary, December 26, 1833.

[44] *Robur Carolinum*, Charles' Oak, in which the King hid from his pursuers, an obsolete constellation name introduced by Edmond Halley, as a tribute to Charles II; dismissed as ridiculous by the French astronomer Lacaille (see note 188, December 28, 1835), who delineated the new southern constellations, 1751–1753. The stars of Robur were put back by him into the constellation Argo (the Celestial Ship), which has now been divided into several parts—Puppis, Vela, Carina, and Malus. The "star" η Roboris, which is in a magnificent nebulous region of the sky, is thus identical with η Argus, and with the modern η Carinae. Sir John was lucky enough to observe an outburst of this extremely interesting object (see diary for December 19, 1837), at one time thought to be a star of the nova family.

Sunday, February 23, 1834

Wynberg

Attended div. Service by the Rev^d—Judge[45] [his omission] at Wynberg School house the Church being repairing.—A primitive worship

NB. Every body who wishes to be sure of a seat brings his Chair.—The roof thatched & the rafters ribs & Thatch seen from below there being no cieling &c. &c.—but a tolerable congregation of Wynbergians—& school children Servants—blacks &c all very orderly & attentive.—All the black women have their jetty hair neatly tied up & dressed with high tortoise shell combs—and all are very neatly dressed.

Monday, February 24, 1834

In the afternoon Rode out with M. and D. S. towards Hout's Bay and from the Hill which flanks the pass leading into it got a most beautiful view.

Here also is a feast for the botanist—a vast variety of beautiful Plants, & flowers being to be found.—Back by Moonlight.

NB. The Moon at full[46]—as it rises it presents a round, dull blotchy human face, with broad nose sulky mouth and standing perpendicularly has just the effect of some preternatural being—Demon—or god of some barbarous nation looking down on his African territory & sniffing with sullen pleasure the scent of some bloody rite or looking down on the whole region as a scene of carnage agreeable to his nature & will.

The *European* face, is quite lost, by the reversal of its position.

On our Return, M^t walking into the Nursery Stepped on a Snake—a "Baum schlange" or Tree Snake, 18ⁱⁿ long. yellowish brown with yellow stripe down the back & silvery white belly.

Tuesday, February 25, 1834

Began making Southern working lists

Another Snake killed in the Passage just by the Nursery door in the Evening. He also was a "Baumschlange."[47] declared by the Servants

---

[45] The Reverend Edward Judge, professor of English and classical literature, acting chaplain at Wynberg. He was later the first rector of Rondebosch (March 1834–1840). Services at Wynberg were held in Glebe cottage on Waterloo Green while St. John's Church was being built.

[46] These high-flown reflections proceed from the fact that the Moon is seen inverted as compared with the northern hemisphere view.

[47] The boomslang or tree snake, *Dispholidus typus*, mottled light- to dark-olive

to be very bad poison snake—one of the most deadly &c—Had him up and examined his mouth with all possible care but could find no poison fangs unless 3 small sharp teeth on either side of his upper jaw, not jointed, and not more than 1/20 inch long, could be construed into poison-fangs. Have little doubt the snake is harmless.— But the whole place is full of snakes. This afternoon I kicked one up among the grass & was going to shoot him but the gun missed fire & he got off

Wednesday, February 26, 1834
Wynberg
Attended anniversary meeting of the Infant School, a very interesting exhibition, Made a few calls—on Baron Ludwig—Major Mitchell Col. Bird[48]—&c and returned.—

Read Mackintosh's[49] History of Engl^d
In evening got the Telescope turned on the Great Nebula of Orion— also on the Great Magellan Cloud which is a collection of detached or loosely connected Clusters and Nebulae—the Chief Cluster being a very extraordinary object.—Shewed it to Marg^t & D. S. also to Major & Mrs Ross of "the Grove" (our predecessors there who are about to leave it Mar. 1.)

Viewed also ω Centauri[50] a most superb object—entirely resolved into stars of 13 . . . 14m. All very nearly of a size & most beautifully graduating in respect of central condensation. A very large object—

---

green, 4 to 4½ feet long, back-fanged, that is, fangs situated behind the eyes. Sir John's snakes may not have been of this species. The poison of the boomslang, which includes digestive enzymes, is, in fact, very toxic, possibly more so than that of a cobra. The fangs are short and very far back, and very little poison is injected in an ordinary bite. Professor Day thinks Sir John did not look far enough back to see the poison fangs, and saw only the smaller teeth nearer the front.

[48] Lieutenant Colonel Christopher Chapman Bird (1769–1861), colonial secretary, 1814–1824. He lived in retirement in Cape Town after dismissal until 1843.

[49] The Right Honorable Sir James Mackintosh wrote this three-volume work (1830) in Lardner's *Cabinet Cyclopaedia*, to which Sir John contributed the first volume, "A Preliminary Discourse on the Study of Natural Philosophy."

[50] The most striking of the globular clusters, containing of the order of a million stars. *10'* means ten minutes of arc, or one sixth of a degree angular diameter. *13 . . . 14m* means that Herschel estimated the brightest individual stars in the cluster to be of thirteenth or fourteenth apparent magnitude.

at least 10′ diam.—Visible to naked eye as a ✳ 5.6 m. or 5m rather blotty and not sharply stellar.

Thursday, February 27, 1834

Read Mackintosh's history.

Went out to get flowers and got bit by one of the ugly savage dogs which abound hereabouts. Our next neighbour keeps 5 or 6 curs of low degree to guard his premises, consisting of a mud hut not a degree superior to an Irish Cabin, and 2 or 3 acres of wretchedly kept ground, with a few stalks of Indian Corn & Bananas.—The beast gave me a confounded snap in the leg which swelled immediately & I was obliged to have it lanced & cupped and poulticed, which kept me lying on sofas all day

In the Evening got some stargazing ["strong awning?"[51] in another hand] and commenced drawing η Roboris & the wonderful nebula which surrounds it—

Blew up a strong South Easter & Clouds.

Left off at 11. P.M.

Saturday, March 8, 1834

Drew on Drummond[52] per Hamilton & Ross in favor of D[r] Stewart, for £465:10:0 payable June 8

Thursday, March 20, 1834

Dined at Gov[t] House

(Entry from recoll[n] long after)

Tuesday, March 25, 1834

Observatory. Dined at M's [Maclear's]

Night fixing Merid[53]

Wednesday, March 26, 1834

Into Cape Town

Excessive heat Lightning

[51] These documents have been copied before (see note 190, December 30, 1835), presumably to make the transcripts lodged at the Royal Society in London but, with the exception of some of the letters, are not known to have been published. Several interpolations such as this one, often completely irrelevant, have been added.

[52] Sir John's financial agents in London.

[53] When a telescope is installed, a determination of the true north and south line, the meridian, has to be made.

Fayrer[54] called abt Lever of Contact[55]

Workmen at Pillar of Eq[1] [Equatorial]

Thursday, March 27, 1834
Dined at Sir J. Wylde's[56] in Cape Town (Hopeville Lodge) where met a very large Party.

A Letter from Sir John, Wynberg, to

Caroline Lucretia Herschel, Hanover

N⁰ 3   Rec[d] June 10. 1834. [In a different hand]
                                         Wynberg, Cape of Good Hope
                                                       March 28. 1834,
My dear Aunt
      Hoping you may have long ere this arrives, received my N[os] 1 & 2. I will yet, so far recapitulate, as not to leave a blank in our history, in case of their non arrival. It is soon said.—*Imprimis* we had a capital voyage having left Portsmouth Nov[r] 13 & landed here Jan. 15 without the slightest storm or more than a day or two of calm or unfavorable wind. It is well for us that we sailed when we did—for only think! from the 15th Jan[y] when we Landed up to *this day* no ship has come in from England!—so that—up to this morning ours has continued the latest arrival.—Today I understand the spell is broken &

[54] James Fayrer, for a time employed at the Royal Observatory, initially as first assistant (1820), demoted to "Labourer" and then dismissed. Presumably he was still picking up some kind of living round Cape Town.
[55] The mural circle of the Royal Observatory, a specialized, and now obsolete, instrument for the determination of accurate star positions, had shown curious errors ever since its installation. Sir John had proposed a scheme for measuring the wandering of the central bearing by means of a series of levers of contact (Fühlhebeln) which would bear on the rim of the circle, and, by the amplified movements of their other ends, reveal departures from true circular motion.
[56] Sir John Wylde (1781–1859), First chief justice of the Cape Supreme Court, 1828–1855. Gay and dashing, there were rumors of incest with his daughter, but his character was cleared.

two ships are come in by which we hope tomorrow to get some news of our friends, if only later by a week than what we know.—However we are told that this excessive delay is quite unexampled.

Well—we all landed safe & sound, one of our servants (Margaret's maid) excepted, who was very ill on the voyage & is still so & not expected to live but thank God Margaret & the Children were and have ever since continued in excellent health and spirits.—We were very agreeably disappointed in Cape Town, which is a far larger and better Town than we had any idea of finding it. We staid there about a fortnight, while I got the Instruments and Baggage Landed and looked out a residence. All the things were in perfect safety, and the telescopes, mirrors and all have reached their final destination without the slightest injury and in perfect order. During that fortnight we scoured the country and looked at several Houses, one of which I have taken for two years, and it is now undergoing a thorough repair (which it greatly needed—as do *all* the country houses hereabouts without exception) and we expect to get into it in about a fortnight. It is called "Feldhausen"[57] by the Dutch and "The Grove" by the English & is just six miles from Cape Town on the road to Constantia & Wynberg. You can imagine nothing more magnificent than its situation which is nearly under the towering precipices of the Table Mountain, and deeply sheltered in a forest of Oak and Fir which so effectually secure it from the South East Gales, that while it has been blowing a perfect hurricane in Cape Town I have been able to sweep with the 20-feet without inconvenience, & even to light one lamp from another in the open air

The 20-feet *is erected* in an orchard, which forms a small part only of the grounds belonging to the place, which are distributed into great squares by long shady avenues of Rich Oak or tall and solemn Pines, and either overgrown with trees or laid out in Gardens.—In the same orchard also stands (or will stand in the course of next week) the Equatorial, for which the house is now building, & will be finished by that time. The sky has been for at least 3 nights out of 5 that we have been here, nearly or quite cloudless & rich in stars, nebulae, and clusters ($\oplus$) beyond anything you can imagine & of which I trust to give a pretty good account. We are now inhabiting temporarily, a

[57] Sir John adopts a Germanic form of the correct Dutch name "Veldhuizen."

house within half a quarter of a mile of "the Grove."—Not a very good one, but having the advantage of being near enough to allow of my superintending all the repairs &c there, and to go over at night with my man Stone and work with the 20-feet at my sweeps, and return when tired, to sleep here across the rich region of flowers & shrubs which *here* is called a Heath—but in England would be a flower garden or Nursery.—

Margarets Brother Dr Stewart met us at our arrival & has been with us ever since. I am sorry to say the great enjoyment of her brother's society will not long be allowed her, as he is obliged to return this month or at least *by the first ship*, to India under pain of forfeiting his appointment. Meanwhile we ride about the country exploring in all directions, and certainly it is one of the most romantic & lovely you can imagine.—It has been pretty hot, though now the heat begins somewhat to abate. One day we had the Thermometer for several hours at 101½ Fahr! in a shady grove where no sunbeam ever penetrates. I thought we should have expired. Happily this was only one day—With that exception, and one other day when it was 89, the heat though considerable has not been more than we could bear without much inconvenience.

We have been very attentively received by almost all the resident English & Dutch families of the best note, so that if we wish it we may form a very large & good acquaintorium but we shall not enter into very much company.

[Written vertically]
My paper warns me to conclude. I trust this will find you well and comfortably situated. *Pray write often* and tell us all your news   address your letters to P. Stewart 65 Cornhill London. Remember us to D^r & Mrs Groskopff & Mrs Herschel and believe me

your affecte Nephew
JFW Herschel

[Written vertically on first page]
Monday is your niece Caroline's fourth Birthday   she can spell words of 3 letters and hem a handkerchief very neatly

[Written inverted on first page]
(P.S. I find I can make the Dutch people understand talking German to them)

A Letter from Sir John, Wynberg,
to James Calder Stewart, London

Wynberg April 13
1834

Dear Jamie—
This comes hopping—all the way from the Cape to ask you
how you do and say very well thank you. At last we have news of
you. From the 15th Jany to the 1st April no arrival from England.
We thought an earthquake had swallowed you all up—but behold!
you are all alive & well to do and as happy and merry as Crickets
(I hope not quite so noisy—a Villainous one within 2 yards of my
ear, in an impregnable position behind a skirting board is shrieking
in a style perfectly deafening).
Well now.—This is really a most enchanting place. Once fairly out
of Town & round the corner of the Devilberg, and you pass at once
from purgatory to Paradise—Climate—Scenery—Manners all change
& all for the better. The formidable South Easter is here no more than
a refreshing & moderate breeze—The Romantic Abrupt Cliffs of the
Table Range of Mountains impend over a wide spreading slope
covered with the richest woods & commanding over the flat valley
between Simons & Table Berg's—a Splendid prospect of the Alpine
Region known by the odd name of Hottentots Holland, a frowning
& Craggy scene. As I never was led to regard this spot as anything
beyond a merely tolerable retreat from the heat dust & wind of Cape
Town—Only judge of the agreeable astonishment we experienced,
both Maggie & self enjoying as we both do with such a keen relish
the beauties of Mountain Scenery.—Then again it is a region of glorious flowers and rich aromatic scents—Every week brings on a fresh
succession of wild flowers, and each more lovely than the last. Maggie
has taken to flower painting and that with such success that her collection (if she will only persevere, & that she declares she will for as
her pencil gains ease & power by practice, she relishes the employment more & more) her collection promises to be most beautiful.—

Yet people tell us that there are *no* flowers at this season! "Wait till Spring"

The nights are as glorious as the days. In the hot months and when the South Easter blows, though Clear & pure, the air is ill adapted for Astronomy—the Stars tremble, swell out and waver more formidably so that I began to fear that, though our choice of a situation had procured us an exemption from all the *mechanical* inconveniences of that hurricane-like wind, and enabled me to "keep the deck" during its utmost rage with little or no personal annoyance from it, yet I should not have been much a gainer but now that the heats are abated the nights are quite satisfactory and all doubts & fears as to the complete & effectual accomplishment of the objects for which we came, if only health is granted me, are at an end.

The 20-feet has been erected and in action ever since the 26th of February and I have at last got the Equatorial house built & ready to receive its revolving roof. The repairs of our house are not quite yet finished but I think another fortnight *must* put us in possession. Till then I can do no regular work—It is nearly half a mile from where we are and poor Maggie being now all alone, I cannot find in my heart to stay late at the Telescopes.

I say now all alone—for Uncle Duncan has at length sailed, as you so strongly urged—per Claudine I saw the last of her this morning rounding Robben Island with a fair wind & lovely weather. He was very unwilling to go by her, and I question whether he would, had not your letter decided him. Independent of the satisfaction Maggie experienced in finding a Brother in this remote Corner we owe much to his attentive kindness—in smoothing all the difficulties of our arrival & location giving us our Carte du pays & teaching us how to steer among the intricacies of our new Society clear of the shoals & quicksands into which our ignorance of local customs & prejudices might otherwise have plunged us.

As to Society here we have not yet seen much of it. All the world has called on us & we on all the world, but except dining at the Governor's and a certain Sir J. Wylde's, there our intercourse with the world has hitherto terminated. When we have a home to live in of course it will be otherwise—but in the "winter" we must expect to find it rather lonely at Feldhausen (Pray don't address to "the Grove" —I hate such Fadeurs) when most of the neighboring families retreat into Town. N'importe—neither Maggie nor myself fear solitude, especially when we can face that bugbear of the idle & empty, *together*—

Be sure & write often—and (calculating on loss & delay) do not scruple *repeating* either news or sentiments. Thank Peter for the Phil. Mag. for January (that for Dec[r] has not reached) for Poisson's[58] Paper on the Motion of the Moon and for the N[os] of the "Astronomische Nachrichten" Sent with it—also for the Diary for 1834

The French Academy have voted me a Medal,[59] and I have written to Arago[60] (Royal Observatory, Paris) to desire him to receive it and retain it till Peter can find some perfectly secure mode of getting it transmitted to London. I should hardly advise trusting to the circuitous mode of Arago to Trenttel Paris—Trenttel Paris to Trenttel & Wurtz London—T & W to Smith & Elder.—Better find a sure private hand—or call on the Secretary of the French Embassy in London.— In any case, Peter will judge what is best [?], & when decided, write

[58]Siméon Denis Poisson (1781–1840), French mathematician chiefly celebrated for his work on definite integrals, electromagnetic theory, and the theory of probability. He spent most of his life at the École Polytechnique in Paris, becoming full professor in 1806 and professor of pure mechanics in 1809. In 1808 he became astronomer to the Bureau des Longitudes. He was the author of a large number of very important treatises.

[59] Presumably a Lalande Medal for 1833. For subsequent years the awards are given in the *Comptes Rendus,* which began publication in its present style in 1834. Joseph Jérôme Lafrançais de Lalande (1732–1807), noted French astronomer, was professor at the Collège de France for forty-six years. A prolific writer on astronomical subjects, he instituted a prize for the best astronomical work of the year in 1802. The "Medal of the R.S." referred to later by Sir John was the Royal Medal of the Royal Society awarded him at the end of 1833, after his departure, for his paper "On the Investigation of the Orbits of Revolving Double Stars."

[60] Dominique François Jean Arago (1786–1853), celebrated French physicist, geodesist, geometer, and astronomer. He succeeded Lalande as professor at the École Polytechnique at the age of twenty-three. Entering politics in 1830, he advocated government aid for science. He was director of the Paris Observatory, permanent secretary of the Academy of Sciences, and minister of war and marine in the government formed after the 1848 revolution.

His researches were mainly in the field of electromagnetism and the wave theory of light. He demonstrated the magnetic properties of a solenoid carrying an electric current, and with Fresnel worked on the polarization of light. He carried out a crucial experiment on the differences of velocity of light in air and in denser media. A foreign member and Copley medallist of the Royal Society, he was a personal friend and frequent correspondent of Sir John's.

to Arago (to the above address) who on receipt of his letter will deliver it as P. shall direct or to bearer—Tell P. however *not* to forward it here but to put that & the Medal of the R.S. by in some drawer or other till my return or death which shall first happen.—

It grows late and as I must write a few lines to Beaufort in the Envelope and dispatch this tomorrow I must conclude with plenty of loves to dear Mamma and Peter & Johnnie & take the best & most affectionate wishes of

Yours ever
JFW Herschel

The Diary: March 30 to June 6

Sunday, March 30, 1834

Attended at Wynberg Church. After Service walked home. Examined a vast block of Granite in Mr Hare's grounds it seems to be a piece of the Substratum of the Table Range laid bare but is the largest block of stone *without a flaw* I ever remember to have seen being upwards of 20 paces long & 15 or 16 broad.—More granite protrudes in various parts—

Mr Hare led me through his Garden & I crossed the Heath, (a beautiful walk) to our house.—Gathered the rich yellow flower & large fleshy tubers of the Anthericum? . . . which is now coming in & is very beautiful

Mr Bell, a Nephew of Lady Catherine Bell's[61] dined with us. A young man about 20 . . . 25 with much good information & desire for more—a young man with decided pursuits and industry to make progress in them. Shewed him many objects in the 20-feet—such as η Argus—ω Centauri &c.—Il vaut la peine de l'instruire—Mais il ne voulait pas de l'instruction outre que ses propres idées—assez crudes et préjugées

Bell came   Wynberg Hill [?] [very faint pencil]

---

[61] Wife of the Honorable Colonel John Bell, Colonial Secretary. The nephew was Charles Davidson Bell, artist, later Surveyor General, and designer of the famous postage stamps, the "Cape Triangulars."

Thursday, April 3, 1834
Found Plan[etary] Neb[ula]

Friday, April 4, 1834
The Claudine arrived from England bringing letters up to Jan. 1.
All well.—This is the first arrival (except the Courier on Tuesday)
since the Mount Stewart Elphinstone in which we came. There had
been immediately after our leaving an "awful hurricane" from S.W.
followed by a series of S.W. gales so that no ship could leave for 6
weeks.

Saturday, April 5, 1834
This day the Masons finished building the wall of the Equatorial
house, gave it one coat of Plaister of "Shell Lime" & sand outside
and one of "Stone Lime" & sand inside.—The Plaster outside is essential to keep the rain from *washing the whole building away*. The
Cape bricks are so incoherent as to be incapable of resisting the rain,
which furrows them into deep channels where exposed & speedily
obliterates all traces of a building. After all it is a wretched affair.

Commenced getting up a set of shelves at the Grove in my study
In evening Captn McArthur drank Tea with us. He is about to return
to Madras per Claudine on Monday. This is the best "Indian" specimen[62] we have bagged hitherto—A plain spoken sensible honest
fellow with no nonsensical airs & graces about him,—

Sunday, April 6, 1834
Attended Divine Service at Wynberg   Marg^t taken ill & obliged
to come home after taking refuge in Mrs J^n Bird's,[63] a stiff formal old
lady reading an old family bible who barely offered the accommodation of a Chair.

Monday, April 7, 1834
Wynberg
Rode over to a Sale of furniture at Protea—a most beautiful spot—
The "Sugar tree" (Protea[64] . . .) [his omission] is now coming into

---

[62] There seems to have been some feeling that military and civilian personnel
of the East India Company stopping over at the Cape gave themselves unwar-
ranted airs.

[63] John Bird was an Indian visitor.

[64] There are 260 species of Proteaceae in the southwestern Cape, including the
weird silver tree, *Leucadendron argenteum*. Many have huge showy blossoms
containing sweet nectar sought after by various species of birds.

Blossom and as it is here in immense abundance, promises to be the richest ornament of these wilds. Nothing can be more elegant & beautiful than its flowers

Much wearied, having been in action the whole day & obliged to forego sweeping through a lovely night—This is very bad work

Tuesday, April 8, 1834

D[r] Stewart[65] announced his intention of sailing per Claudine "to sail this evening".—

Walked with Marg[t] over to the Grove Signs of activity at last—though very feeble ones—The Painters are at length got to work & a couple of Masons are grubbing at a door way!

Drove into Town.—Drew on Messrs. Drummond per Thomson & Watson[66] for £400.0.0 at 30 days sight

Called for D.S.—found the Claudine does not go till Saturday. Drove over to the Observatory. Saw M[rs] Maclear (& her Chameleons!) and the Meadows's[67]—Maclear gone to M[r] Jackson's[68] wedding with Miss Kitty Raby. (NB A hundred ridiculous stories

---

[65] Dr. Duncan Stewart was obliged (see letter to Caroline of March 28, 1834) to take the first ship to India on pain of loss of his appointment. Dr. Stewart (1804–1875), L.R.C.S (Edin), M.D. (Aberdeen), was in the East India Company's service from 1825 to 1855, being for a time Superintendent-General of vaccination. He attended Herschel at his death in 1871 and died at Tunbridge Wells (*Quarterly Bulletin of the South African Library*, 12, No. 2 [1957], 73).

[66] A Cape Town firm of shipping agents.

[67] Lieutenant William Meadows, R.N., assistant at the Royal Observatory under Maclear's predecessor, Thomas Henderson (see note 195, Observations, 1834). He worked well for him, but became soured by the time Maclear arrived, and is said to have greeted him with the words, "So, Sir, you have determined to accept this wretched appointment!" Also said to have maintained a *ménage à trois*. He left the Cape, to be briefly and inadequately replaced by Thomas Bowler (see note 110, January 28, 1835), and more adequately and permanently by Charles Piazzi Smyth (see note 151, October 11, 1835). In London he became secretary to a gas company, and remained on good terms with Henderson, in spite of his eccentricities. The diary gives evidence that Herschel was willing to trust him with financial business. The list of comings and goings of the Meadows pair to the Herschel residence which appears later, may have been the result of domestic troubles.

[68] Joseph Henry Jackson, of the Honourable East India Company, married Catherine Rabe, second daughter of John Rabe, at 25 Heerengracht by special license by the Reverend G. Hough.

in circulation about the said J.—as how the wedding was *twice* interrupted by previous claimants on his "hand & heart" &c)

Returned to Wynberg. Walked out on Wynberg heath & shot 4 Widow birds[69] (A brown Bird with long beak and immensely long tail, about size of a lark. In one specimen the tail was nearly twice as long as Bird & beak together & the beak = 1/3 of the Body neck & head.

Too much fatigued to Sweep though the night proved most lovely. NB.   D.S. reports that Maclear has detected *motion* in my 2ᵈ Planetary Nebula[70] Requires verification by further observation

Thursday, April 10, 1834

Went into town with M. to take leave of D.S. who is to sail tod[ay] [practically erased; presumably entered on wrong day]

Saturday, April 12, 1834

Went into town with Margᵗ to take leave of D.S. who is to sail this afternoon by the Claudine for Madras.

Wrote to Grahame[71]

Posted a duplicate to Arago via Nantes or Bordeaux. Had a conversation with Dʳ Smith & advised him to throw overboard all instruments & surveyings unless they journeyed so as to be able to execute the work with them very accurately

Called at the Observatory hearing that Maclear had yesterday a paralytic siezure in the face. Saw Mrs. M.—

---

[69] In modern eyes, Sir John's penchant for shooting birds is, like his occasional large-scale botanical depredations, not very attractive. Mrs. M. K. Rowan, of the Percy Fitzpatrick Ornithological Institute at the University of Cape Town, who has reviewed and corrected all the ornithological notes, defends him, by reminding us that there were no binoculars at that time, and that the only means of learning about birds was by shooting them. The name "widow bird" is usually applied to species in which a velvety black predominates in the male's breeding plumage. The combination of characters described by Sir John makes it certain that this was not what would now be called a widow bird, but was the Cape Sugarbird, *Promerops capensis*.

[70] Not a possibility; the distances are all too large for proper motion to be perceptible, even in intervals of the order of a century.

[71] James Grahame, historian and college friend of Sir John who had accompanied him on one of his early journeys in Europe. He was the author of the comic verses reproduced elsewhere (see letter to James Calder Stewart, July 17, 1834).

Sunday, April 13, 1834

Rode to the observatory over the flats & saw Maclear. His face is drawn aside & right eyelid & brow & cheek incapable of motion though not insensible.

Extract from Gardener's Chronicle Nov. 4/ 43 "Death of Drejer— Drejer. The German papers state that this botanist died last year in Copenhagen from taking strong snuff mixed with lead.—JFWH subsequent entry

---

M. himself attributes his siezure to this and he gave me the Cannister to analyse it. The lead was *visibly corroded*—and the snuff infused with acid and treated with sulphuretted hydrogen gave lead.

JFWH subsequent entry ["subsequent entry" repeated in another hand]

Monday, April 14, 1834

Rode with M. up to the Paradise & Round by Protea home—a nice long ride.

The Meadows's came.

Tuesday, April 15, 1834

Rode with M [crossed out; presumably wrong day]

The Meadows's went

Rode with M. towards Town returned her horse being rather unruly.

Wednesday, April 16, 1834

Heavy Rain.

Thursday, April 17, 1834

Set out with Marg^t on horseback to go to the Observatory. Soon after Starting she observed her horse to be getting unruly and proposed returning   accordingly we tacked homewards having gone about half a mile, the horse however kept getting more & more fretful & at last set off full speed with mine after him, but unable to overtake or keep up with him. M. kept her seat well till she reached our own gate, into which she turned at full speed, but got unhorsed in the act as soon as I came up & turned in, the first object I saw was poor Maggie stretched at length by the road side! But (God be praised) without a hurt beyond a slight scratch in the cheek, and one or two trifling bruises, as she immediately proved by rising (with slight assistance) and walking into the house—and behold, as if to give point & poignancy to the escape, while I was in the act of leading her in, who should ride up to the door but Dr Murray[72] (our Medical Attendant & an excellent Surgeon) This is the second *fearful* fall I have seen dearest M. get, in circumstances when even an alarm is held to place life in jeopardy! Again & again God be praised who protects his own where no other protection can avail. In the innumerable falls from horses I have myself experienced I remember only one in which I was able to make the slightest effort for self-preservation— it was at Chudleigh returning from Bovey Tracey with my cousin S.B.,[73] when I believe a violent effort forced the horse aside who would otherwise have trodden on me as I lay.

Got M. to bed & passed rest of the day at home. The Cape horses require constant exercise or they get very skittish

Sunday, April 20, 1834

Walked over to Wynberg to Church, but found no Service there. Returning, examined the great granite block in Mr Hare's grounds. It is a single stone without a flaw 16 paces long & 10 broad of an irregular oval form and rounded surface projecting about 6 feet above the soil, & nearly smooth. The granite is chiefly White Felspar in large crystals with some black mica and quartz in small quantity. Very hard & little appearance of weathering. It is accompanied by much other granite in loosely scattered masses similarly projecting & is doubtless only a projection of a great outcropping Granite Rock. Taking a course out of the Road towards the Table Hill I observed the outcrop of Granite masses (of similar nature) in several other parts so that doubtless the substratum of Wynberg Hill is *Granite*

Brought home a rich crop of Flowers!

Two Messrs. Hawkins[74] called.—also Col & Mrs Wade & Dr Murray. Accompanied Margaret to the Observ^y & saw Maclear whose paralysis is unrelieved. The right side of the face has lost all power

[72] Dr. John Murray was principal medical officer to the Forces.

[73] Probably Sophia Baldwin.

[74] William Hawkins, agent for the East India Company, who owned "Rouwkoop," Rondebosch, and other properties, and his brother.

of motion & what with Blistering & shaving produces a strange & ghastly effect

Monday, April 21, 1834

Got together the square of the Equatorial roof. Cut out Planks— set my room at the Grove to rights. Finished reading Lytton Bulwer's[75] England, a work of more merit than I had given him credit for producing.

Wrote to Capt[n] Bance[76] & sent him Whewell's[77] paper about Cotidal lines and a spare Nautical [entry terminates]

Wednesday, April 23, 1834

This morning we shifted our quarters, quitting M[r] Borcher's[78] house on the edge of the flats[79] & entering M[r] Schomberg's house called Feldhausen alias "The Grove".—From this day our occupation

---

[75] Lytton Bulwer or Bulwer Lytton: Edward George Earle Lytton Bulwer (1803–1873), taking surname Bulwer-Lytton on succeeding to baronetcy, later first Baron Lytton. A prolific novelist and politician, he embarked on a literary career after estrangement, and loss of allowance, from his mother, following a marriage of which she disapproved. He published *England and the English,* two volumes, in 1833, and *The Last Days of Pompeii* (discussed at length by Sir John in later entries) in 1834, as well as many other novels. He was MP for Hertfordshire 1852–1866 and Secretary for the Colonies 1858–1859.

[76] Captain James Bance, port captain of Cape Town.

[77] William Whewell (1794–1866), British natural and moral philosopher. He was a contemporary of Sir John's at Cambridge, where they formed a lifelong intimacy. Whewell was second wrangler in 1816, and became, successively, fellow, tutor, and, from 1841 to 1866, master of Trinity College, Cambridge. He was professor of mineralogy (1828–1832) and, during the first part of his career, conducted extensive researches in mechanics and dynamics, and especially in the theory of tides. Correspondence with Whewell on this subject is a recurrent theme in the Herschel diaries. Later, Whewell became interested in philosophical questions, such as those related to scientific induction. He was professor of moral philosophy at Cambridge from 1838 to 1855, and became Vice-Chancellor in 1842. Among his works are the treatises "The History of the Inductive Sciences," 1837, and "The Philosophy of the Inductive Sciences," 1840.

[78] Presumably P. B. Borcherds, civil commissioner and magistrate at Wynberg, who lived at "Weltevreden," Rondebosch, of which Sir John did a camera lucida sketch.

[79] The Cape Flats, the low-lying area containing originally many sandhills, some thirty miles broad, connecting the mountainous Cape Peninsula with the true mainland of Africa.

of the house dates, to continue 2 years certain with an option on our part to continue a third year at the same Rent—i.e. for house & garden, & outhouses &c £225.—

Sunday, May 4, 1834

Walked over to Wynberg.

Returned by Major Rogers's over the high side of Wynberg hill and examined two great knolls of granite which project from its summit. It is White Granite abounding in Felspar & black mica— little Quartz. From the occurrence of a great many similar masses no doubt the whole of Wynberg hill is granite which underlies the slate of the Table Hill. NB   from this mass of Granite got a fine view of Simon's bay—and following the road to Mr Eckstein's[80] & so round by Protea, the finest Points of View in the Cape Peninsula occur.— Caught some huge Locusts.

Returned in time to get a short ride with Margt—the first she has taken since her fall—To Wynberg & home by the high road—*very leisurely.*

Monday, May 5, 1834

Accompanied Margt to the Observatory & with Maclear who is now well enough to go about his occupations again, went through a rehearsal of the operation for Examining the Circle by the Contact-levers.

Tuesday, May 6, 1834

Feldhausen

Sent the Carriage for the Maclear's children who came with their governess, Miss Geards.[81]—to pass a few days at the Grove.

Rode with Margt round by Mr Eckstein's over Wynberg hill, down by Mr Stewarts[82] and Major Rogers's and so home—one of the most beautiful rides about here.

Wednesday, May 7, 1834

Mr Malcolm dined here

---

[80] Probably D. G. Eksteen, owner of "Kirstenbosch," now the South African National Botanic Garden.

[81] Mary Geard, governess to the Maclear children.

[82] John Steuart, the High Sheriff.

Thursday, May 8, 1834

Rose late   Rode over to the Observatory and with Maclear and Meadows went through our revolutions of the Circle with the levers of Contact

Returning rode far out over "the flats" which (like the Campagna di Roma) are anything but a dull uninteresting desert.—They are covered with the richest vegetation (heaths'—Proteas—Reeds—Grapes—Bulbs—&c) and lovely and sweet scented flowers—one small white flower is the richest vegetable perfume I ever smelt—like the Magnolia only more delicate.

Nothing can be finer than the towering Mass & graceful slope of the Table Mountain & Devil-berg.

A huge brown Falcon or Vulture came & settled within 20 yards & stood staring at me—would hardly rise.

There are two great annoyances to a horseman on the flats.— 1st. Mole-hills formed by a mole as large as a rabbit into which your horse sinks suddenly knee-deep.—2dly Ant hills, black as ink & hard as stones, in the form of hemispheres, over which he stumbles. These anthills are as it were rooted in the ground by a blunt conical point and when broken (a task of some difficulty) disclose a labyrinth of cells & passages in one of which—sits the Queen, a huge white bag with an ants head & legs [small sketches of section of an ant hill and of queen ant].

Saturday, May 10, 1834

Feldhausen

Lost the whole Morning in awkward (and fruitless) attempts to get on (without assistance) the skirt-boarding of the Equatorial roof. The little Maclears' and Miss Geards returned to the Observatory. Walked with M. into one of the devious tracks among the wild heath on the outskirts of the Grove & culled Ixias & Gladiolas! Such splendid creatures!

Sunday, May 11, 1834

Feldhausen

Maclear called, at breakfast and passed the morning here, Discussed the Circle[83] & other matters—Rambled about—Saw a Falcon & tried (in vain) to get a shot at him.

[83] This is the mural circle at the Observatory (see note 55, March 26, 1834).

When M^{er} was gone, read the Psalms Lessons &c with Maggie—
Walked out with M^{t} on Wynberg heath and gathered flowers—Dug
up some Ixia bulbs—which when cut are Red, and full of cells stuffed
with transparent Jelly or mucilage.—NB. They are tubers, not bulbs.

After Dinner wrote to Captn Smyth.[84] and Drafted off some of
Dunlop's[85] Double Stars into the General history papers.

[84] William Henry Smyth (1788–1865), sailor, amateur astronomer, and antiquarian. Son of an American loyalist, he first served in the British merchant
service, and then in the Royal Navy during the Napoleonic Wars. Holding command in the Anglo-Sicilian Fleet at Messina, he undertook, on his own initiative,
hydrographic surveys of the Italian, Sicilian, Greek, and African coasts. In 1815
he married the daughter of an English merchant in Naples, and met the Italian
astronomer, Piazzi, who gave his name to their remarkable, and eccentric, second son, Charles Piazzi Smyth (see note 151, October 11, 1835). The father
produced a large illustrated antiquarian work on ancient medals. Returning from
the wars, he took up astronomy as a hobby, settling at Bedford. He became
active in the Astronomical Society and met most of the dramatis personae of the
diaries, including Dr. Lee, who later bought his instruments and set them up at
Aylesbury. He produced quite a good two-volume description of astronomical
objects in the northern sky, under the titles *The Bedford Catalogue* and *A Cycle
of Celestial Objects*. He was president of the Royal Astronomical Society 1845–
1846, and vice-president of the Society of Antiquaries, 1851, was promoted rear
admiral in 1853 and admiral ten years later, and designed Cardiff docks.

[85] James Dunlop (1795–1848), British astronomer. He accompanied Sir
Thomas Makdougall Brisbane to New South Wales in 1821, was keeper of the
Brisbane observatory at Paramatta, 1823–1827, and finished the "Brisbane Cata-
logue" of 7,385 southern stars in 1826. Sir John presented Dunlop's work "A
Catalogue of Nebulae and Clusters of Stars in the Southern Hemisphere ob-
served in New South Wales" to the Royal Society in 1827 and for this Dunlop
earned the Gold Medal of the Royal Astronomical Society in 1828, and Sir
John's warm commendation. Observations of the same stars by Sir John and
Maclear produced an increasing degree of exasperation with the errors of Dunlop's work. In a note to Maclear, Herschel remarked:

"On the whole I am sorry to observe that this shewing is even more unfavourable
to the Brisbane Catal. than the results of any of my previous comparisons—with
other authorities—The errors follow no traceable law. Henceforth I feel disposed
to dismiss the epithet the *Brisbane* Catal as connected with this remarkable
astronomical record, & rather to designate it by the names of those concerned
in its composition—it is worthy of the age of Ulugh Beg or Tycho Brahe . . ."

When Dunlop was awarded the Lalande Medal for astronomy by the French
Academy, 1835, Sir John remarked sourly, "I wish the awarders would come
here and look for some of his nebulae and double stars." Sir John and Maclear
became so outspoken as to provoke a brush with Sir George Gipps, the new
Governor of New South Wales, who passed through Cape Town in 1837, and
they had considerable difficulty in extricating themselves with grace. Gipps

NB.   α Crucis[86] is Binary.

17 Hydrae[87] is in Motion having changed quadrants since Mayer's[88] time.

Monday, May 12, 1834

An announcement at breakfast that Geoffrey the black Cook had taken French leave Proved a false alarm as he returned about noon. Got up partly the corner-pieces of the Equatorial roof to carry the skirting board.

A most delicious day, unclouded Sun yet not too warm—Marg^t was occupied all the morning drawing flowers in the open air on the Stoop.

Rode with M^t on the flats, Called on Mr Rogerson[89] (formerly Slave Protector) and Col^l Cheap.[90]

Returned to Dinner & Stars.

Sunday, May 18, 1834

Rose late. Rode out on the Camp Ground to call on Mr Hawkins Senr—Saw his "Belier Hydraulique[91]

NB.   Extraordinary bearing of his Son a boy of 9 or 10 years

Called on Mr Hawkins Junr. found him & his brother & Major Smith[92] just sitting down to Dinner   Joined them

Home at Sunset

Monday, May 19, 1834

A tremendously hot wind from N. or NE accompanied with hot Sun.—The thermom, rose to 83 and the air most oppressive.

(1791–1847), was Governor of New South Wales from 1838 to 1846 and did much to promote the exploration of that colony. His financial policies made him very unpopular there and ultimately led to his resignation.

[86] See note 41, December 26, 1833.

[87] There has been no significant motion in more than a century.

[88] Johann Tobias Mayer (1723–1762), German astronomer, professor at the University of Göttingen, 1751; superintendent of the University observatory, 1754, until his death. His work on the motion of the Moon, and the lunar tables which he computed were developed for use as a practical method for the determination of longitude at sea by the Astronomer Royal, Maskelyne, being only superseded by Harrison's invention of the chronometer (see note 22, November 30, 1833).

[89] Ralph Rogerson, collector of taxes, lived at "Vredenburg," Rosebank.

[90] Colonel Cheape was an Indian visitor.

[91] Hydraulic ram.

[92] This Major Smith was an Indian visitor.

Worked at the Equatorial Roof & got on part of the Canvas.

Dr Phillip, Mr Fairbairn,[93] Mr Innes[94] and Dr Adamson[95] called Mr Malcolm dined with us

Monday, June 2, 1834

Rode over to the observatory in the dark to dinner and got beset by Van Renen's Dogs,[96] a hungry savage pack of 5 or 6 large hounds & curs of low degree, who had all but eaten me & my horse too—only Diis aliter Visum[97]—These confounded Dogs are the pest of this country—After dark they are really furious and mean mischief by their attack

Met at McLears.—Mr Brink[98] (Auditor Genl) and his *wife who* gave us a touch of Colonial politics, the burden of her story being the partiality of þ^e English Gov^n to the English & exclusion of the Dutch from office (NB her husb^d is a Dutchman).

Tuesday, June 3, 1834

Walked forth with Marg^t in the Afternoon on Wynberg Hill &

[93] John Fairbairn (1794–1864). (We are indebted to Dr. A. M. Lewin Robinson for the text of this, and the following two notes, referring to three of the most distinguished of Cape Town inhabitants at this time.) A South African editor and advocate of press freedom, Fairbairn was born in Scotland. He was co-editor with Thomas Pringle of the *South African Commercial Advertiser* (founded by George Greig) in 1824 and again from August, 1825, until 1859 by himself. He married the second daughter of Dr. John Philip. His newspaper editorials on political and social questions had great influence. He was nominated a "popular" member of the Cape Legislative Council in 1850 but resigned almost at once. On the grant of representative government in 1854, however, he entered the first Cape Parliament, where he sat till 1863.

[94] James Rose Innes, professor of mathematics at the South African College, was appointed first Superintendent General of Education at the Cape Colony, 1839–1859, under a scheme in which Sir John Herschel played a leading part. He was the grandfather of Sir James Rose Innes, Chief Justice of the Union of South Africa.

[95] The Reverend Dr. James Adamson (1797–1875) was one of the most brilliant men ever to come to the Cape of Good Hope. He was largely responsible for the establishment of the South African College, where he lectured in physics and other subjects. He came to Cape Town in 1827 as minister of St. Andrew's Church, and was secretary of the South African Institution, 1829–1850.

[96] Daniel van Renen was the owner of the brewery (De Brouwery, Papenboom, Newlands), a notably handsome building, illustrated by Sir John's sketch.

[97] "It seemed otherwise to the Gods."

[98] The Auditor-General was the Honorable P. G. Brink.

Heath—There is a spell on that Hill & Heath—one never tires of it, it never twice comes upon one in the same aspect.—This Time, got into thick Jungle of Protea & shot Widow Birds, and talked over Miss E's[99] Helen.

Wednesday, June 4, 1834

After dispatching the Carpenter &c—Took Gun & Baskets & Vol. 2 of "Helen"—and started for adventures.—

Trotted to M[r] Ecksteins & then tried a Clamber to see if the Table Hill could be scaled by the Cloof[100] (Gorge) behind his house[101]— Deuce a bit—with infinite toil forced a passage through the thick tangled vegetation, up the right bank of the streamlet—w[h] runs to Ecksteins, but when the Talus was surmounted and the 3 first stages of the rock, found all further progress barred by mural precipices.— so returned after reading 3 or 4 Chapters, in a wolfs' nest.—Good Heavens! What a view from this point over Simons Bay & the flats with those noble hills beyond!

Thursday, June 5, 1834

Finished reading Miss Edgeworth's new novel. It is a real pleasure to see an author at the last *not* breaking down, but going on with a brilliant blaze instead of fuming & smouldering in the Socket. Such an expiring blaze is "Helen"

Friday, June 6, 1834

Rode into Town & called on Sir E & Lady Ryan[102] just arrived per Zenobia In way called on M[rs] Menzies[103] to give her Miss Edgeworth's new Novel. Called also on Mr Ross & M[r] Templeton.—

---

[99] Maria Edgeworth (1767–1849), novelist, supporter of education for women. She was active in the relief of sufferers in the Irish famine. *Helen, a Tale* (3 vols.), which she wrote, was published in 1834.

[100] The South African word *Kloof* means a ravine or glen.

[101] If this is Skeleton Gorge, the present ascent lies on the left side, technically the right bank of the stream.

[102] Sir Edward Ryan (1793–1875), Chief Justice of Bengal, 1833–1843, later Civil Commissioner. He was at the Cape on sick leave.

[103] Mrs. Menzies, wife of William Menzies (1795–1850), judge of the Cape Supreme Court, 1828–1850. He was the "first great interpreter and expounder of Roman-Dutch jurisprudence as it has been handed down to us." On circuit in 1842, he crossed the Orange River and hoisted the British flag. (A. A. Roberts, *South African Legal Bibliography*, pp. 371–372.)

A Letter from Sir John, Feldhausen, to
Caroline Lucretia Herschel, Hanover

Feldhausen near Wynberg
Cape of Good Hope
June 6. 1834
[In a different hand] Rec^d August 19   N⁰ 4

My dear Aunt

I hope you have got the letters I wrote you announcing our arrival & comfortable settlement here, or that at all events you have heard of us & our proceedings from Cousin Mary or Brother James. We are at length completely settled, in an excellent house, & in a most beautiful situation, and have all so *snug* about us that we are only reminded by the warmth of the season (now the depth of winter) the splendour of the sunshine and the profusion and beauty of the flowers that we are not in England, enjoying a *fine summer* in some rich romantic part of the country.—We are all quite well, and indeed the Children seem to flourish & thrive upon the pure air and sunshine, so well that I cannot but hope their little constitutions will be permanently cherished & strengthened by their visit to Africa. We have, around us several agreeable resident English families, and might, if we pleased be always in Society, but independent of my own particular views in settling here, neither Margaret nor myself have much enjoyment in general & indiscriminate company, though we by no means intend to live the lives of mere recluses. For example the other evening we were at a very gay & indeed splendid fancy ball given by the Governor's Lady where all the company were in some costume or other. I wish you could have seen dear Maggie in this glittering throng—(though I say it who as her husband perhaps should not say it)—decidedly the lovliest person among a whole assembly of beauties—(for the Cape Ladies are remarkably handsome).—Well, I am not going into a description of Balls & routes.—I have quite enough to do to describe the Stars & keep my engagements with them on fine nights—that is to say almost every night.—The 20 feet has been in activity ever since the end of February and as I have now got the polishing apparatus erected and three mirrors, one of which

I mean to keep constantly polishing, the sweeping gets on rapidly. I had hardly begun regular sweeping when I discovered two beautiful Planetary Nebulae, exactly like planets & one of a fine Blue colour. I have not been unmindful of your hint about *Scorpio*.[104] I am now *rummaging* the recesses of that constellation & find it full of beautiful Globular Clusters ⊕, ⊗ &c. A few evenings ago I lighted on a strange Nebula of which here is a figure! and since I am about it, I shall add a figure of one of the resolvable nebulae in the Greater Magellanic cloud, as below [Figure 2]

The Equatorial is at last erected and the revolving roof (upon a plan of my own) works perfectly well but I am sorry to say that the nights in which it can be used to advantage are rare, even rarer than in England, as in spite of the clearness of the sky the stars are ill defined and excessively tremulous.—But a truce to Astronomical details, though from time to time I shall continue to plague you with them.—

Pray write to us through P. Stewart 65 Cornhill London.—We long to see your handwriting   I hope soon to send you a word or two in little Caroline's. She is getting on with her reading and can hem and sew very neatly. Your little nephew William begins to say Papa, and runs about and makes friends of every body. James & Stone (my Astronomical blacksmith, a most excellent and useful man) are his great allies and the black Servants are delighted with him.—Farewell. Margaret desires to add her kindest regards to those of your affectᵉ Nephew.   J.F.W. Herschel

[On front]

Pray remember me to Dʳ and Mʳˢ Groskopff[105] and Mʳˢ H.—also to Mr. & Miss Beckedorff Dr Mühig[?] and Mr. Hausmann &c.

---

[104] In a postscript to a letter of August 1, 1833 (*Memoir and Correspondence of Caroline Herschel*), Caroline remarks, "I wish you would see if there was not something remarkable in the lower part of the Scorpion to be found, for I remember your father returned several nights and years to the same spot, but could not satisfy himself about the uncommon appearance of that part of the heavens. It was something more than a total absence of stars (I believe)." This is the region of the center of the galaxy with the more distant stars highly obscured by interstellar dust and a great concentration of globular clusters.

[105] William Herschel had a brother, Dietrich, who had three daughters, of whom the youngest, Caroline (i.e., thus a cousin of Sir John's), married a Dr. Groskopff.

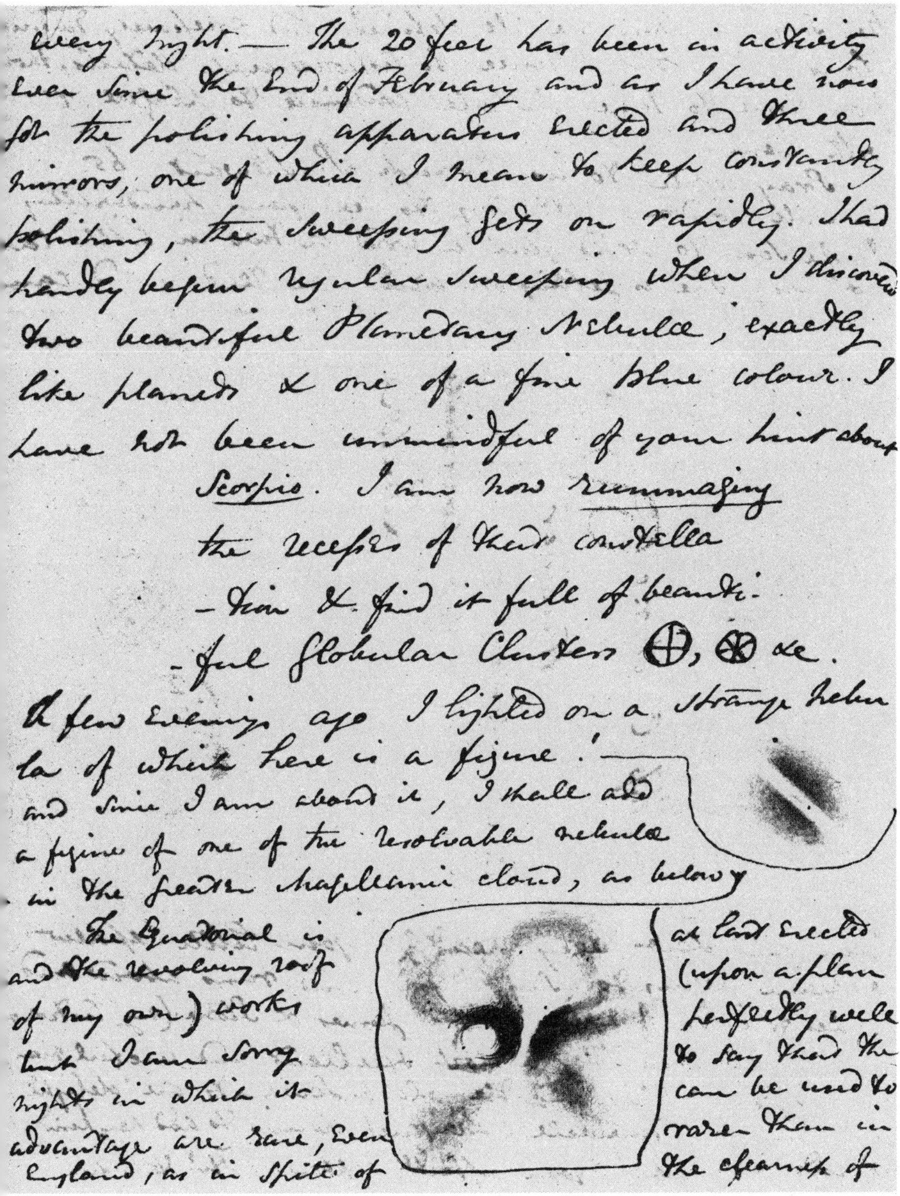

Figure 2. Sketches of the extragalactic nebula NGC 5128 (above) and of the nebula 30 Doradus in the Large Magellanic Cloud from Sir John's letter to Caroline Herschel, June 6, 1834.

The Diary: June 8 to July 17

Sunday, June 8, 1834

Finished letter to Capt[n] Horsburgh and despatched it (it contains Statement of my Met[l] Obs[ns] & conclusions drawn during the voyage hither—

Walked with M[t] out on Wynberg hill.

A *Perfect* Astronomical Night.—I hereby retract all I have said in disparagement of the Cape Atmosphere.—Such tranquility & definition of stars equals anything I have had in England:

NB.   Signs to be attended to
        Perfect calm all day
        Large chequered Mackarel Sky about Sunset
        The Hott[ento][t] Holland Hills remarkably distinct
        Moderate Rain the night before.

Tuesday, June 10, 1834

Marg[t] went with M[rs] Menzies to the Play,[106] performed by the Officers of the Garrison for the benefit of the Charity schools—

Tuesday, June 17, 1834

Dined at Sir J. Wylde's—Present Sir Benj[n] & Lady D'Urban, Sir E & Lady Ryan & 2 Misses R.—M[r] M[rs] & Miss Petrie[107] &c, &c &c

Thursday, June 19, 1834

Busied all the Morning with rectifying the 3 self-registering thermom[rs] of the African Exped[n] and writing to D[r] Smith—also began the astronomical part of the "instructions"

A fine observing night—worked with Equatorial at review & measures till 3

Tuesday, June 24, 1834

A great fire on Wynberg Heath, close to the Outskirts of the Grove. Piped all Hands to Extinguish it and succeeded, by making lanes

---

[106] Possibly a private performance of a bill given publicly by the officers on July 11, consisting of "Katherine and Petruchio" (adapted from Shakespeare by David Garrick), followed by "The Haunted Inn" and "Bombastes Furioso."

[107] W. Petrie was Deputy Commissary General at the Cape.

among the Brushwood to windward and heading it towards the wind, beating out the flame which runs along the dry grass & throwing on Earth.—To leward it was stopped by a road, and to aid the road a strip to *windward* of the road was purposely set on fire as AB. This burns, but in a less furious & more manageable way and stops at the road so that when the fire X reaches AB it ceases for want of fuel [References are to a small lettered sketch showing how the counter-fire was lit]

Where it got among Large Proteas and Copse wood the flame roared & rolled in huge volumes forming a superb conflagration but it was easily got under though a vast tract was found to be burned. After this it was amusing to see the white necked crows coming in crowds to catch up the half roasted tortoises carrying them aloft & dropping them to crack the shells. (This I saw them do)

Dined at Col¹ Wade's—Present Sir B & Lady D'Urban, Col & Mrs Smith,[108] Col & Mrs Munro,[109] Dr Smith & Lieut Edie, Capt Dyason or Tyson.

---

[108] Probably Sir Henry George Wakelyn Smith (Bart) (1787–1860), cele-brated British soldier, usually known as Harry Smith. He saw service in South America and in the Peninsular War. There in 1812 he married, when she was 14, Juana María de los Dolores de León, who campaigned with him under all con-ditions and was adored by the troops. (See note 37, February 13, 1834, which suggests that Sir John may have found the lady unusual.) Smith served in the United States and was horrified at the sight of the burning of the Capitol at Washington. He fought at Waterloo as a Brigade Major. Coming to the Cape in 1828, he commanded a division in the Kaffir war of 1834–1836, and at this time made his famous ride of six hundred miles from Cape Town to Grahamstown on the frontier in less than six days. With the territorial annexation of D'Urban he became Governor of the new province of Queen Adelaide and acquired immense influence over the native tribes. When the annexation was abandoned, Smith was removed from command. Later, campaigning in India he defeated the Sikhs at Aliwal and received the thanks of Parliament and a baronetcy. He became Governor of the Cape in 1847, and after further trouble with the Boers and war with the Kaffirs was recalled in 1852. He actively supported the grant of re-sponsible government to the Cape. Toward the end of his life he held a military appointment in London. He and his victory in India are commemorated by the towns Harrismith, and Aliwal North in South Africa, and his wife by the towns Ladysmith in Natal and Ladismith in the Cape.

[109] Lieutenant Colonel Alexander Munro commanded the artillery. Lieu-tenant, later Captain, William Edie, of the 98th Regiment, went on Dr. Andrew Smith's expedition in July. "Captain Dyason" was probably Charles Tyssen of the 75th Regiment.

Sunday, June 29, 1834
Walked with M. on the Wynberg Hill

Monday, June 30, 1834
Dined at the Mess of the 98th at a Dinner given by Major Gregory
& the Officers of the Reg[t] to Capt[n] Edie & D[r] Smith—met there M[r]
Oliphant,[110] M[r] Menzies,—C. Bell &c

Thursday, July 3, 1834
At Daybreak rode over to the Obs[y] to take leave of the Heads of the
African Expedition (D[r] Smith, Lieut Eady 98th—Chas Bell Esq,—
Young M[r] Burrow[111] (and M[r] Grant & M[r] Grahame who accompany
them as far as Latakoo[112])
    After a hospitable breakfast at the Obs[y] given by Maclear (at which
I found Margaret who had passed the night there) Rode out some
way on the flats with the Waggons with Baron Ludwig—who fired
a salute, his gun missing fire 4 or 5 times to every discharge
    Accomp[d] M[t] into Town to a Fancy Bazaar got up by Lady D'Urban
Called on Fell[113] & *rowed* him about his exorbitant bill (NB he
charged £6:15:6. for hanging 6 bells.
    Returned to the Grove—prepared Guns & then rode back to
Observatory where I slept, making a few Obs[ns] with the Circle (NB.
M[c]Lear got an obs[n] of my last Planetary Neb[a] discovered 2 nights ago.

Friday, July 4, 1834
Rode with M[c]Lear & Meadows out to the hills between Cape Town

[110] Honorable Sir Andrew Oliphant, 1793–1859, Attorney General, 1828, and
Chief Justice of Ceylon, 1839.
    [111] "Eady" should be spelled "Edie." Charles Bell, the artist, has already been
mentioned (see note 61, March 30, 1834). Young Mr. Burrow was presumably
the son of the Reverend Dr. Burrow, chaplain to the forces and one of the secre-
taries of the Cape of Good Hope Association for Exploring Central Africa.
    [112] Latakoo (or Lattakoo, Lithakoa) was the chief kraal of the Batlapin tribe.
According to A. Sillery (*Bechuanaland Protectorate*), it is correctly "Dithakong."
"Old Lattakoo" was something under forty-five miles northeast of the Kuruman
River. The tribe moved to the site of the present Kuruman (in the northern Cape
Province) in 1817 and this became "Lattakoo" or "New Lattakoo." (Footnote
by Dr. Lewin Robinson.) (See note 21, January 24, 1834.)
    [113] John Fell and his son were tin-plate workers and bell hangers in Hout
Street, Cape Town.

& Stellenbosch—Gathered bulbs &c   Shot a pheasant, & nearly lost our way on returning to the Obs^y in the Dark.—Came home much tired having only had a small *roll* which we got from a passing waggon to divide between ourselves three & our 3 horses from 7 in the Mor[nin]^g till 6 at night—
Found Mrs Meadows here who remained all night—

Saturday, July 5, 1834
Mr Meadows called & took back M^rs M to the Obs^y

Monday, July 14, 1834
Baron, M^rs & Miss Ludwig   M^r & Mrs Meadows Mr Wheatley[114]
Mr Skirrow   M^r & Mrs & Miss Petrie   dined with us
After Dinner got 20-ft on the Moon but it clouded & got low so that nothing could be made of it. NB. Moon's age 8th day—The 9th & 10th (for *after* dinner gazing) would be preferable.
M^r & Mrs Meadows remained all night.

Wednesday, July 16, 1834
Watched the carcase of a dead horse on the heath until 3 AM in expectation of a wolf (the wolf tracks were large and distinct about him this morning)—stood sentry with loaded rifle in a bush for 4 hours but not a wolf appeared. Left Brown & James on the look out —They watched till daybreak but not a wolf did they see. NB   a dull but strange & solemn watch, had it been held in expectation of *human* adventures.

Thursday, July 17, 1834
A rainy day. Kept house & read Dr Phillips's[115] memorials & correspondence about the civilization & Christianizing of the Frontier tribes & interior
Wrote to J.C.S.[116]—long letter

[114] Mr. Wheatley was an Indian visitor.
[115] Dr. John Philip (see note 22, January 24, 1834). The book was his *Researches in South Africa,* two volumes, 1828.
[116] James Calder Stewart. The tone of Sir John's letters to him suggest a warm brotherly relationship between the two.

A Letter from Sir John, Feldhausen, to
James Calder Stewart, care of P. Stewart, Cornhill,
London, per Euphrates

Feldhausen near Wynberg
Cape of Good Hope
July 17. 1834

Dear Jamie—

Here we have been these 6 months and Maggie has had all our Correspondence to herself while I have been a mere bysitter and well wisher and Enquirer after news. And what is worse little or no news is come! We are in fact ignorant (except from the Newspapers) that the Elphinstone's arrival at the Cape is known in England. That London was not (on the 26th April) swallowed up by an Earthquake, or roasted & Eaten by the radicals the same Papers also inform us—but not a word of any Mortal's handwriting in reply to the letters we dispatched the day after landing or since!—Well—it is no use fretting, as we are sure they are on the road, and will be all the more welcome for the long abstinence—but truly this want of intelligence of all that we care to hear about at home is a grievance worse than we bargained for in coming hither. In all other respects (*almost* all) we find our residence here more comfortable than from any previous information we should have supposed possible, and this being a time of the year when we were threatened, at our first settling here—with many evils attendant or to be attendant on our local situation—it is satisfactory to be able to declare them all bugbears.

Duncan left by the Claudine on the 13th April, and we are now beginning to keep a sharp look out for letters announcing his arrival at Madras. You may imagine better than I can tell you how gratifying this meeting was to me. And yet now that it is over and gone like a dream I sometimes feel as if I wished it had either lasted longer, or was yet to come, for a reason independent of the mere pleasure of the thing. Much of my time during his stay here was occupied with the erection of the instruments & the other details of our *location* and these occupations not only deprived me of much of his society but, by the exhaustion they produced, in that burning season unfitted me for the enjoyment of much that remained. Thus I experienced that uncomfortable feeling that a golden opportunity was passing, of which I could but very imperfectly avail myself; and since his departure have often felt disposed to blame myself for not having made more & steadier efforts to break through that veil of reserve which conceals mind from mind, and make 3 months do the work of as many years in improving acquaintance into ripened friendship. Nothing could surpass his useful kindness, and active attention to our comforts, the good effects of which we feel & shall continue to feel as long as we continue resident here. We owe in fact to the activity of his enquiries and knowledge of the country ways of going on, the very roof over our heads (in the literal sense of the word—for it was at his suggestion that I was led to insist on having the house new roofed without which, I now clearly perceive it would not have been possible to live in it)!

From one of your letters, in which you mention having been with Jones, and having made [Sir James] South the subject of your conversation, & from some of your remarks thereon, I gather that the perverse and naughty spirit which is I grieve to say taking stronger & stronger hold on him, and will end in not suffering him to retain a Single friend—has somehow or other (I cannot divine how) [been] manifesting itself in relation to myself.—Dear James—Whatever he may say or do—let me intreat you to take no notice of it. Be assured he cannot harm me. I have not a point that is tangible, either in reality, or in the opinion of any mortal whose opinion is worth a straw, by any attack, open or covert—or by any insinuation as to motive or conduct, of Sir J. S. Therefore let him alone —and above all let no feeling of indignation for my sake lead you into any personal communication with him. Believe me he is not a subject worthy of *your* anger.—Leave his chastisement (should he merit it) to less refined & inferior hands, and to the recoil of opinion which anything he may do will be sure to provoke.

I dare say some good jokes have been let off about this our quixotic trip.—If so do in pity to our dulness impart them.—Here I set you the example in the shape of some epigrams from Madeira [name, "James Grahame," inserted by a later hand in pencil]

1.

Stargazing Knight Errant, beware of the day
When the Hottentots catch thee observing away!
Be sure they will pluck thy eyes out of their sockets
To prevent thee from stuffing the stars in thy pockets

2.

If Herschel should find a new star at the Cape,
His perils no longer would pain us
He will salt the star's tail to prevent its escape
And call it "The Hottentot Venus"

3.

Herschel! tis amazing————that you should go stargazing
Among the Hottentots————Like a guinea among groats
I fear some Hottentot————will bisect your learned throat
To put you out of tune—for bewitching the Moon!

— &c &c.

I have not many adventures to relate—As to bisecting of throats
there is not much business in that line stirring in these parts. I was
ass enough a few nights ago to stand sentry, rifle in hand, in a bush
stirring neither hand or foot & hardly fetching breath—in view of
the carcase of a dead horse—on the heath outside of our premises, in
hopes of shooting a wolf! whose tracks were visible in the sand
beside it. But alas! no wolf came and after standing 4 mortal hours
in the cold the winding up of the adventure was a vow to let the
wolves devour the horses live or dead at their will & pleasure in
future. The Cape "Wolf"[117] as they call him by the way is not a wolf,
but some kind of hyaena—an arrant coward as ever robbed a pigsty.
Maggie has been very busy of late painting flowers, and you will
I hope ere long receive some specimens of her really beautiful per-
formances in that line, by the hands of M[rs] Warren[118] (the Admiral's

[117] Professor Day remarks, "There are two possibilities: (i) the aardwolf
(Proteles), a smallish, somewhat degenerate carnivore, feeding largely on insects, particularly termites, as well as carrion. (ii) the hyaena, which is not such
a coward as popularly supposed. In Central Africa today, the hyaena hunt down
living game, and at times, lions, as well as jackals, feed on the hyaena's kill."
[118] Wife of Rear Admiral Frederick Warren.

lady) who sails with her Husband for England (being recalled) in a
week or 10 days. She has written (by this opportunity—the Euphrates) a long letter to Miss Baldwin, containing true and faithful accounts of self & children to which letter, as I doubt not Mary will
forward it to Grandmamma, I shall refer for family news.

The Stars flourish, and in spite of all my attempts to thin them and
as Graham calls it, stuff them in my pockets, continue to afford a rich
harvest.

The Instruments are all up & in full activity. The revolving roof
works well, and answers in every respect, which is fortunate, as I do
not think I *could* have got it constructed here. Workmanship here is
bad, dreadfully dilatory—and *abominably* expensive. There is no
competition in most of the ordinary useful trades. The tradesmen feel
this and are not only recklessly exorbitant in their charges but insuf-
ferably unaccommodating and provoking. As Emigration seems so
much the rage at present I am astonished that small tradesmen who
can barely live on a moderate Capital in London, should not come out
and *settle* here, either in Cape Town or at Wynberg (which though
a considerable neighbourhood, and rising by degrees into a place of
some consequence, by the Erection of new & good houses) is entirely
destitute of shops (except Butchers & Bakers). Plumbers, Glaziers,
Painters, Bell Hangers & Furnishing Ironmongers, &c &c would I
should think be sure to succeed as the creation of houses for the increasing population, of a kind adapted to their increasing wealth is
a matter of first necessity, and as the place grows more English in its
tastes, & habits (as it does, by no slow degrees) the need of English
comforts & accommodations becomes more & more felt.

Do you ever look at the Cape Papers. That Edited by Fairbairn
"The South African commercial advertiser" is excellently well conducted. Fairbairn's *leading Articles,* are in fact for the most part,
admirable essays on practical legislation, conceived in the best spirit
of temperate, philosophical reasoning, and with a generality of view
quite remarkable in the limited circle of a Colony.

We don't go out much, Though not confined by stress of Weather
Though now the depth of Winter, I have known few English sum-
mers comparable to it either for warmth, or brilliancy of Sky & Sunshine.—Since the beginning of this month we have had the Thermom^r
one day at 82 and generally nearer 70 than 60—A few nights ago at 2½

AM it marked 72 on my observing desk.—Our garden in front of the house is all glowing with Roses and Aloes in rich flower, & the plains are enamelled with blue Ixias and lovely pink Oxalises!

Well—loves to you all from all, which will not cool in crossing the Atlantic.   Adieu!

Your aff^{ly}
JFW Herschel

The Diary: July 18 to November 15

Friday, July 18, 1834

Note to Fell further disputing his bill—(Copy) Finished Drawings N°1. N°2, N° 3, N° 4 of our House & grounds to send to the good people in England

Made out by Pacing a ground plan (in the rough) of Schonberg's Estate so far as it concerns the right understanding of the *country* scenery of it   Read Wilhelm Meister[119]   A thick rainy day, such as we often have in England but very seldom here.—

Dispatched letters per Euphrates

Sunday, July 20, 1834

Maclear spent morning   Rode round in dirⁿ of Hout Bay and Botanised on the hill side   Equatorial till 3 AM

Monday, July 21, 1834

rode out with Stone across the pass above Constantia in the direction of Hout's Bay—Saw some beautiful new Proteas—took two Camera Sketches, & admired the magnificent Scenery—*there is no* Constantia Berg so high as Capᵗ Owen[120] lays down in his chart—nor any hill so very conical *where* he places that—Deep in the gorge of Houts Bay there is a very fine cone visible from the Sea   Equatorial till 3 AM

---

[119] A novel by the celebrated poet, philosopher, novelist, and scientist, Johann Wolfgang von Goethe (1749–1832) which had "an immediate and lasting effect on German literature."

[120] See note 24, January 24, 1834.

Tuesday, July 22, 1834

At home all day

A Mr Goree[121] or Gorny called to ask me to write articles for him in a Cape-made periodical "Literary and Scientific". declined on plea of want of time

Wednesday, July 23, 1834

Began to look a little into the Laboratory and the 1st process being to lay in a supply of alcohol for vegetable analysis, I commenced by distilling Cape Brandy which is tolerably strong & cheap (2s 6d. per Gallon). This required of course 1 rough distillation in a Copper Still one gentler one from Salt of Tartar or lime in a glass retort & Water bath—and for absol. Alcohol a final process from Min. Lime.

[NB.—To mix Min Lime with Sand ? by fusion, to get over the annoyance of powdering it.] [his square brackets and query]

Rode out to call on Sir J. Truter.[122] Not at home.—Met Ryan & his Daur and asked him to pass Monday at Feldhn (He comes).—

A N. Wester coming on. Noticed the extraordinary contest between that & the S.E. on opposite sides of the hill   [A small sketch of the Table Mountain and Devil's Peak skyline with arrows indicating directions of air flow and formation of lee waves]

Dined & Slept

NB. For a long while past the propensity to sleep after dinner has been *irresistible,* not a mere nap, but a long deep sleep of 1½ or 2 hours—shaken off with difficulty, returning unless thoroughly got rid of, and only to be conquered by decided action, and going into the open air.—

Worked at the Equatorial adjustments[123] & got the Alt. right.— Stars dreadful owing to the conflict of the winds   the NW gaining the upper hand by very slow degrees

Thursday, July 24, 1834

Meadows called   Ludwig called   A Note from McLear enclosing

---

[121] This was William Gorrie, schoolmaster, and editor of *The Cape Cyclopaedia,* 1835.

[122] Sir John Truter (1736–1845), former Chief Justice, 1812–1827. He was the first South African to be knighted.

[123] In an equatorially mounted telescope not only must the polar axis be aligned truly north and south (see note 53, March 25, 1834), but it must also be inclined at the correct angle (altitude) so as to point to the true pole of the heavens.

a letter of Captn Smyth in which he says that Miller of St Johns[124] has found *also* de son coté—that the Mean height of Barom depends on the Latitude. Worked in Laboratory. Finished outline of N° 6. Sketch.[125] Rain coming on at night the house began to leak and Stone & myself had a good 2 hours work "providing for the exigencies" by all sorts of temporary expedients—

Friday, July 25, 1834

Chemically employed. preparing a stock of Alcohol.—NB. 65000 gms. Cape Brandy yielded 20,000 good alcohol [this line erased] from Cape Brandy. A dreadful lack of means for applying heat neatly.

Saturday, July 26, 1834

Lady Ryan called.   See Sweeps   Copying Sw[eeps] into Cape Bk Dined at D[r] Murray's (ordered carriage [erased] carriage at Dixons)

Sunday, July 27, 1834

Passed the Day at home.   read Prayers with M.   Gathered some bulbs & planted   Swept till ☽ rose

Monday, July 28, 1834

M[r] E. Gordon[126] called   Drove out with M. to Wynberg & half way house & gathered flowers   Sir E & Lady Ryan & 2 Miss Ryans dined with us

Tuesday, July 29, 1834

Hitchcock came & tuned Piano   Lady D'Urban Miss D'Urban & Miss Stedman[127] called. A superb Day—Gathered Flowers and bulbs for the Bulbgarden. Set sweep[128]   Read Quarterly Review No Events Measured γ Virg[129] before dinner   Swept from dark to Moon rising & 1 hour beyond up to 23[h] ST[130]

---

[124] Of St. John's College, Cambridge.

[125] With the camera lucida (see note 26, February 1, 1834).

[126] Mr. Evelyn M. Gordon was an Indian visitor.

[127] Miss Stedman traveled with the D'Urbans, presumably as a companion to Miss D'Urban.

[128] That is, adjusted the direction of pointing of the twenty-feet preparatory to observing.

[129] γ Virginis: famous binary star with period of 171 years which passed through periastron (closest physical approach of components) in 1836.

[130] ST means sidereal time, that is, time as indicated by the position of the stars relative to the observer. Not the same as solar time, which defines the po-

Wednesday, July 30, 1834

Rode over to Observatory Botanising & shooting by the way. Shot a large Gull—The Splendor of the flowers in the flats about the Camp Ground is amazing   Swept till 3 AM

Thursday, July 31, 1834

Rode with Maclear   1ˢᵗ to Wynberg hill to the great granite stone near the top,[131] from which took a set of Bearings—Thence down through Wynberg—called on Mʳ & Mʳˢ Stewart [?Steuart] and thence round by Farsfields [Versvelds] and young Dryern's [?Dreyer's] to the "Look out" on Houts pass & into the gorge beyond.

Got a sketch and took a set of bearings. Returned dined and Swept (a broken & irregular Sweep).

Friday, August 1, 1834

Passed the day at home   No Events.   Mʳˢ Maclear remained here with Margᵗ   Swept

Saturday, August 2, 1834

Mʳˢ Maclear Left Feldhausen   Rode into Town to attend Meeting of the S A Lit & Phil Institⁿ[132] at which I was elected an ordinary member, & their chosen President for the ensuing year.—After the Meeting examined the birds & some live Chedar Tigers[133] & jackalls   NB. a Strong South Easter & the hill has its Table Cloth[134] on it   Came home very much tired partly with riding in the winds eye wʰ I find

---

sition of the Sun, because of the progressive motion with the seasons, of the Sun against the star background. At 23h ST a certain stellar constellation would be on the meridian. Star positions east and west (known as Right Ascension) are defined by the sidereal time at which they appear on the observer's meridian.

[131] The Wynberg Stone of Sir John is now known as the Hen and Chickens.

[132] The South African Literary and Philosophical Institution, of which Adamson was secretary. It was the precursor of the Royal Society of South Africa.

[133] Because it was not suggested that these animals were of local origin, it was conjectured that the "Chedar Tigers" were cheetahs, of which the natural distribution does not extend so far south. Although of the dog and not the cat family, they do look tigerish to the untutored eye. Professor Day disagrees, and thinks the animals were leopards, which are indigenous, and were usually referred to as "tigers," particularly by the Dutch. There are, of course, no true tigers on the continent.

[134] Local name for the orographic cloud on Table Mountain produced when the southeaster blows, often when all the rest of the sky is cloudless.

very fatiguing, partly from sweeping late the night before and not taking it out in the morning.—So went to bed at 9

M. kept her bed all day with the Rheumatism

Sunday, August 3, 1834

South Easter moderate—Marg^t was persuaded by the beauty of the day to rise & take a drive out on the flats where the flowers the Sun the blue Sky and the *copious* balmy breath of the S. E. formed altogether a combination approaching ones ideas of Paradise

Monday, August 4, 1834

Drove over with Marg^t to the flats and gathered flowers Blue Gladiolus &c &c   Swept

Tuesday, August 5, 1834

Drove out with M. to the flats below Wynberg and then round by the half way house gathering flowers Above the half way house is a rich habitat of that singular looking flower the Disa longicornis[135] which ought to be called Disa Shiptonia so exactly is it like the grotesque figure of "Old Mother Shipton"[136] [a small sketch of the flower; see plate 16] Returned.—M^rs & Miss Hare called   Mrs Blair[137] & Mrs Hamilton

No Sweeping but kept awake half the night by the orgies of the detestable Dogs with which this country swarms.—There are 6 or 8 about the premises barking like furies

Wednesday, August 6, 1834

Drove over with Marg^t to the flats to gather flowers. Commenced experimenting with the roots &c of the Cape   began with the yellow-rooted flag so common on the flats [erased] heath here and the yellow Anthericum roots (See Chemical book)

Copied sweeps (one page) Sketch N° 7. of the Telescopes Swept till 20^h½ ST when it clouded

Thursday, August 7, 1834

Worked at Sketch N° 7.

---

[135] *Disa longicornis* is a misuse of Thunberg's name for the "Moederkappie" (Mother's sunbonnet), *Disperis capensis,* a Cape ground orchid with peculiar horned flowers.

[136] A half-legendary early Tudor witch and prophetess.

[137] Mrs. Blair was the wife, either of William Thomas Blair of the East India Company, who retired to the Cape about 1830 and was a member of the South African Literary and Philosophical Institute, or of Colonel Blair.

Friday, August 8, 1834

Chemical Expts. Contrived a Charcoal lamp in w^h the supply of air shall be cut off by water [sketch of a lamp chimney with air entry ports round its lower end partly lowered into a water bath]

Copied Sweeps. Swept till 20^h ± when it clouded

Saturday, August 9, 1834

Chemical Exp^{ts}—Walked out with M^t and gathered wild flowers— The Pink ixias with green backs now begin to peep up in the paths The Blue Ixias are almost over their new bulbs are nearly complete The White Ixias are in full luxuriance and are Magnificent and most profusely spread. The Pink Oxalis (Purpurea) is gradually going off & becoming less frequent   The [entry terminates]

Copied Sweeps   No sweeping—Mackarel drift from N W in Afternoon followed (as once before) with heavy rain

Sunday, August 10, 1834

Rained hard all day

Monday, August 11, 1834

Rained hard great part of the day and very cloudy & wet the rest   Put off Party for tomorrow till Saturday

Tuesday, August 12, 1834

Put off the D'Urban's Menzies & Wades till Saturday, the morning being still doubtful—

Maclear called at Breakfast and mentioned some facts relative to astronomical matters in general.

Turned out a superb night! Such definition of Stars.—Got *at last* a dependable series of Meas.[ures] of γ Virginis w^h is fortunate this being if I remember the day of perihelion[138]

After a few other measures with Equat^1—went to bed till the Moon set—then rose & swept over the small Magell[ani]^c Cloud—then turned the Telescope on Jupiter & Mars which are in near conjunction

Wednesday, August 13, 1834

Before Sunrise Directed telescopes to Jupiter & Mars in near conjunction   Saw them both in the same field of the ring Micrometer[139]

---

[138] Probably a mistake for periastron, moment of closest physical approach of the components.

[139] A device for measuring the separation and position angle of double stars,

of the Equatorial   Shewed them to Mr Mrs Miss Schonberg—Marg^t also got up to see "the sight" which was really very beautiful Shewed it to James & Stone [sketch of Jupiter and Mars] Worked at Drawing, N° 7 of the Table Mountain & the telescopes the Hill being most beautifully distinct at Sunrise!

Took some Azimuths with Compass & went to bed till 11½ A short day, as after tea I slept 2^h and then sate up for Stars but the Sky was full of Cirri and a halo round the Moon, and in the Equatorial the Stars were thus [sketch of fuzzy spot to illustrate star image spoiled by bad seeing] and finding that ζ Aquarii could not be seen double, gave up.

Thursday, August 14, 1834

A lowering morning with [unfinished]

Dined at Gov^t House. Met M^r & Mrs Murchison,[140] Sir E. Lady & Misses Ryan Major Gregory &c &c

Friday, August 15, 1834

Unwell and lay in bed till 1 PM! & then idled about till dinner time doing little or no good.

M^r & Mrs Murchison

M^r & Mrs Evelyn Gordon

Captn & M^{rs} & M^r & Miss Hare

dined with us.—

Saturday, August 16, 1834

Sir Benjn & Lady D'Urban, Miss D'Urban Miss Steedman [Stedman], Major Dutton Col & Mrs Wade M^r & Mrs Menzies

Dined with us—the ostensible object was to see Stars & Moon but a fierce N. Wester prevailed & neither M. nor Star would peep out.

Sunday, August 17, 1834

The N Wester still strong with Rain & squalls—Read Lessons & Psalms with M and kept in doors all day

Monday, August 18, 1834

Mr Stewart [?Steuart] called   Marg^t went to the Observatory to pass a few days with Mrs Maclear

---

necessarily with a small field, thus emphasizing how very close together in the sky the two planets appeared.

[140] Mr. and Mrs. Murchison were Indian visitors (see note 42, February 18, 1834).

Tuesday, August 19, 1834

Passed the Day at home

Wednesday, August 20, 1834

Drove over with Caroline & Bella to the Observatory to see their Mamma—Took a set of angles from Observatory roof with McLears theodolite

Thursday, August 21, 1834

At the Observatory. A wet dull & rather cold morning.—Wandered about and gathered such! Gladiolus'es and Such! Homerias[141] and shot some yellow Birds called here Canaries[142] (and not unlike in size shape & colour)

After a Confab. with McLear & examining the "Skeleton forms of reduction" left by Fallows[143] w^h M. has condemned rode Home in the wet.

NB. Took a set of bearings from the Observatory Roof

Pursued some sort of Falcon, of a grey colour about size of a large Pigeon but could not get a good shot at him

NB. The Oak trees now all [erased] rapidly throwing out their leaves   Some trees are in considerable foliage while others close by & of same growth shew no signs

Friday, August 22, 1834

Maggie returned from Observatory   Prepared a new Chart Index   At Night worked at the Equatorial

[141] *Gladiolus* is well represented at the Cape. Many of the garden gladioli known throughout the world have been bred directly from the Cape species, which, in the wild, are smaller, but often quite spectacular in patterning and brilliance of color. *Homeria* is a less-well-known Cape genus in the same family (Iridaceae).

[142] Probably the Cape Canary (*Serinus canicollis*), or *Serinus sulphuratus*, the Bully Seed-eater of today. *Serinus (Crithagra) flavientris* seems less likely, though not impossible.

[143] The Reverend Fearon Fallows (1789–1831), first holder of the post of H. M. Astronomer at the Cape. Established the Observatory, 1827, on a desolate site overlooking Table Bay, convenient for making signals to ships, but still infested with leopards, snakes, and hippopotami. He died following an attack of scarlet fever and is buried on the Observatory grounds. Although Gill, in his *History and Description of the Royal Observatory, Cape of Good Hope* (p. viii), specifically states that Fallows was born on July 4, 1789, and was, therefore, only just over forty-two when he died, it does not seem to have been noticed that his tombstone says "Aged 43 Years." Perhaps the not-unusual phrase "In the forty-third year of his age" is implied.

Saturday, August 23, 1834

Rode with Mr Stewart [Steuart] (High Sheriff) to Hout's Pass, and climbed Constantia Berg, whence took a semi-Panorama—

Returning viewed Wynberg Church of which the newly erected gable and all the buttresses on one side are tumbled down, the effect of the last 2 or 3 days (not hard) rain—The other gable will go as the Arch (too flat) is cracked

Mr Ross, Miss Ross, Miss Morton, Major Cloete[144] Mr Hawkins Junr & his Brother dined with us and in Evening shewed them the cluster near the Nubecula Minor,[145] and the Moon—After they were gone got some obs^ns with the Equatorial

Sunday, August 24, 1834

Drove out with Maggie to Wynberg to view the ruins of the Church & on the flats

Monday, August 25, 1834

Rode out to observatory with Camera & Drawing board to accomp^y Maclear to Block-house[146] to take angles & a Panorama, but found him suffering with an attack something like the *"little* Cholera" & had passed a night of great pain & was still too much indisposed to go—

Called on the Meadow's & returned home.—A N. Wester coming on

D^r & Mrs Adamson   M^r & Mrs Fairbairn   Prof^r Innes Dined with us

F. entertained us with anecdotes of Frontier Murders tyrannies & oppressions—half which if true are enough to justify any degree of outcry for a revisal of the system under which they exist From his account the Frontier is still in the same state of violence *as in the* most barbarous ages A Black & Stormy night

Tuesday, August 26, 1834

A Violent Rainy Day & NW

Kept in doors & prepared Colcothar[147] &c for Polishing

---

[144] Probably to be identified with Colonel Abraham Josias Cloete, who led the force which relieved Port Natal in 1842. Major Cloete was then town major of Cape Town.

[145] Another name for the Small Magellanic Cloud.

[146] Probably the King's Blockhouse at about nine hundred feet on the north-east of Devil's Peak where the mountain becomes precipitous.

[147] A polishing compound: red oxide of iron, jeweller's rouge.

Wednesday, August 27, 1834

Wind Rain Hail & Storm

Kept home and worked in the Laboratory

Thursday, August 28, 1834

A Bad day—in the Ev[en]ing it blew a Storm—A Heavy rain and Night of the most pitchy darkness.—

Marg[t] not feeling well sent in for D[r] Liesching[148] who came out in the Evening & returned there being no necessity for his stay

Strapped down the Equatorial roof and made all tight for the night expecting damage.

Friday, August 29, 1834

Examined premises as to the effect of storm last night

No leakage in the roof of house   Part of coach house unroofed   A fine Guava tree blown down full of fruit   (NB. the fruit is desperately bad—*positively nasty*) Got in a lot of Roots of Anthericum & other tubers & put to dry

Saturday, August 30, 1834

This morning put the long focussed Mirror on the polisher and with very sharp colcothar and slow motion gave it about 1½ or 2 hours polishing. As the Pol[ishe][r] at first starting had no figure and did not touch in the Middle, I expected to find the figure destroyed. The polish was most brilliant & not in the least scabrous—and to my great surprise & satisfaction I found on trying it at night in a Sweep that I have thus luckily blundered out one of the finest figured mirrors I ever beheld.

Sunday, August 31, 1834

Attended divine Service at Wynberg School house.

The other *gable* of the New Church fell during the bad weather of the week but this was owing to a crack in the *too flat Arch* (which it was evident before would soon give way—See Aug 23) The wretched materials—Bricks hardly warmed through and cement of incoherent mud miscalled clay, render it the height of absurdity to attempt *arches* or indeed anything but cones and pyramids.—Yet no farther than Constantia is stiff clay of the best quality and as for Building Stone Wynberg is based on granite & covered with granite blocks & Sandstone & slate are close at hand.

[148] Probably Dr. Louis (i.e., Carl Ludwig) Liesching, 1786–1843, son of Dr. Friedrich Ludwig Liesching, then an old man.

Tuesday, September 2, 1834

A Violent Windy & Rainy day

Kept home, and worked at Bance's (or McLeod's) meteorological register

Procured from Schönberg files of Cape Gazettes containing Meteorological registers from various parts of the Colony, kept at the Drostdy's[149] by order of the Govr at the Instance of Col Bird. Though irregular & some of them very carelessly done they may be of use

Wednesday, September 3, 1834

Completed Curve of the Barom^r from Bance's Register.—Attended 1st ordinary meeting of the Phil Society as Presid^t—The Meeting was thinly attended & nobody (after the routine matters of domestic arrangement were discussed) had any paper, or communication to make—So I stated (verbally) the results of my examination of Macleod's (or Bance's) Barometric obs^{ns} with a view to get organised a system of Meteorological communications and to get up a talk.—

Read Life of Alcibiades (Plutarch)

Read part of Campbell's Travels in S. A^a.[150] he seems to have

---

[149] A landdrost was a magistrate whose official premises were called a drostdy.

[150] The reference is to the Reverend John Campbell, minister of Kingsland Chapel, London. He wrote a considerable number of works on travels in southern Africa, of which the two most important are *Travels in South Africa, undertaken at the Request of the Missionary Society* and *Travels in South Africa, undertaken at the request of the Missionary Society, being a narrative of a second journey in the interior of that country.* The first describes a journey of inspection of the missions of the London Missionary Society made between October, 1812, and November, 1813, through Bethelsdorp, Graham's Town, Kaffraria, Lattakoo, Griquatown, and Namaqualand. Theal commented on the author's "simplicity and credulity," with respect to this work. The second journey, starting from Cape Town early in 1819, was the one on which Campbell was accompanied by Dr. John Philip (see note 22, January 24, 1834), and by Messrs. Evans and Moffat. The missions of the Cape Colony and Kaffraria were visited. Campbell also made a further journey, starting early in 1820 and going to Lattakoo, Griquatown, and the modern territory of the Transvaal and possibly South-West Africa, but the topographical descriptions are obscure. Campbell met a number of chieftains, including Cornelius Kok and his son, Adam Kok, and "Africaner." He returned to Cape Town after ten months' travel. He regarded the folklore of the native inhabitants with much contempt. Campbell was born in Edinburgh in 1766, and was ordained in 1804, then becoming pastor of Kingsland Chapel, a post which he held until his death in 1840.

have gone on an hour but Maclear came in & I took it off to shew him

Thursday, September 11, 1834

M. had a quiet good night no fever & going on well. Dr Liesching attends her & Mrs Maclear proves a most kind & considerate nurse

Friday, September 12, 1834

M. still going on as well as possible

Saturday, September 13, 1834

M. still going on uninterruptedly well. D^o Baby who is a quiet well behaved little creature—sleeps half the day & all night and *very* seldom cries. Heaven be praised *therefor*!

Saturday, September 20, 1834

Walked out & called on Mrs Menzies, where met Lady C. Bell who gave a letter from Lady A.M. Donkin.[155]—Delivered to Miss Dick to hand over to Miss Ross who sails with her per Ld Hungerford a note & Parcel from M^t to D.S.[156] at Calcutta.

Called on Mr Huddleston found him suffering under loss of 6 teeth. (having had 7 out not long before)

Sunday, September 21, 1834

Read Prayers with Maggie at home   Mrs Huddlestone[157] called Read "Geschichte einer Schönen Seele" in Göthe's Wilhelm Meister— the history of a gradual conversion—or rather of an increasing intensity of devotional feeling from youth to age

Monday, September 22, 1834

Prepared for further Sweeping.—

Chemical proceedings continued—Commenced examination of the Orobanche (? Hyobanche) Sanguinea[158] a red flower spike which

---

[155] Lady Anna Maria Donkin, daughter of the first Earl of Minto, second wife (m. 1832) of Sir Rufane Donkin, Acting Governor of the Cape in 1820 at the time of the arrival of the British settlers at Algoa Bay. He laid out the city of Port Elizabeth, named for his deceased first wife.

[156] Duncan Stewart.

[157] Mr. Huddleston was an Indian visitor.

[158] *Hyobanche sanguinea*, a fleshy leafless parasite which grows attached to the roots of other plants. Included either in the family Orobanchaceae, or in the antirrhinum family, Scrophulariaceae.

first projects above ground with a long thick acrid fleshy brittle root—
Sir B & Lady D'Urban called. Marg[t] got into the Drawing Room
today for the 1st time since her confinement.

Friday, September 26, 1834

Drove out with M. onto the flats & gathered flowers—Such Watsonias! Rode over to Observatory to see Maclear whom I found ill in
bed still suffering from internal obstruction, suffering occasionally
severe pain & lamenting over the still stand of his Astronom[l] work
(so he considers a momentary relaxation)

Monday, October 13, 1834

M[r] White called. Baby quite well.
Sketched out a plan for the comparison of the mean results of the
R.S. Meteor[l] Journal with the Cape.—At 1½ Accompanied Marg[t] to
dine at the Observatory.—Met there Mrs. Thomson wife of Col.
Thomson[159] & Dau[r] of the late M[r] Stoll—A nice little Dutch woman,
evidently much harassed by the late afflictions of the family

M[c]Lear quite recov[d] Attributes his illness to taking Snuff impreg-
nated with Lead! from a lead Canister. Gave me specimens for analysis—the inside of the box evidently much corroded

[In another hand] See April 13th

Singular appearance of Clouds flowing down bet[n] Lions Head and
Table M[n] [rough sketch of skyline and cloud structure]

Monday, November 10, 1834

Set out with Margaret, M[r] Withers, Hannah[160] and Baby Margaret-
Louisa, Somai[161] and Braun[162] for an excursion to Stellenbosch, &c.
For a more particular acc[t] of which see travelling diary.

Tuesday, November 11, 1834

Stellenbosch to Paarl. See Travelling diary.—

Wednesday, November 12, 1834

Paarl. Climbed the Paarl Mountain. Wagonmaker's valley[163] (M[r]
Retief's)—returned to Paarl.

[159] Lieutenant Colonel R. Thomson was an engineer officer; his father-in-law
was the Honorable J. W. Stoll, Treasurer and Accountant General.

[160] Nursemaid.

[161] Somai (Zomai), a coachman.

[162] Brown, a coachman or groom; presumably "Braun" is a misspelling.

[163] Wagenmakers Valley is the town of Wellington.

Thursday, November 13, 1834
Paarl to Frenchhook. Crossed the Pass & returned to Field Cornet Hugo's
Friday, November 14, 1834
Returned to Paarl [erased] by the way of Banghook[164] & Drachenstein [165] to Stellenbosch

Saturday, November 15, 1834
Left Stellenbosch & returned to Feldhausen.

A Letter from Lady Herschel, Cape of Good Hope,

to Caroline Lucretia Herschel, Hanover

Rec[d] Dec[r] 14 [Pencil note]
Cape of Good Hope
Sept[r] 29th 1834

N° 5

My dearest Aunt,
    I claim the privilege of writing to you this time, as my news more particularly belongs to the lady's department, which you will easily believe when I tell you that we have been made so happy during the last fortnight by the arrival of letters from England & *Hanover*, & a *third little daughter* from—I don't know where exactly, but Papa has welcomed the little stranger very kindly & took such affectionate care of her Mama, that at a fortnight's end she is almost as well as ever, & very happy in being able to catch an excellent opportunity of writing to her dear Aunt Herschel to tell her of the pleasure Herschel had in receiving your letter dated [omitted] in answer to his N° 3—I am sorry you did not receive N°s 1 & 2, & this uncertainty of communication is the worst of water carriage, but while an equal share of good news as hitherto has to pass between us, we must be thankful for the cheering intelligence even now & then—We actually screamed with joy when your letter tumbled from out of a parcel of

[164] Banhoek.
[165] Drakenstein (Dragon Stone).

others, & were not long before we *devoured* its contents—We admire
much your Christian strength of mind & composure in looking forward to that event which must happen to all, but I have a presenti-
ment that you will yet live to see *some* of us, at least if their earnest
prayers are granted, & whatever happens, you may have this con-
solation that the *moral example* of your character has left an impression, not to be effaced & is calculated to be of as lasting utility, as
those talents which all the world admires—And yet I do hope to see
you "in the Body", & talk with you in *reality*—You wish to hear about
Herschel, & in truth I like to write about him—at this moment he is
observing, & the bell of his telescope tells me he is very busy—he
came in a minute ago to say that it is the most brilliant night he ever
saw, & he has had many such of late—His sweeps seem hitherto more
productive of Nebulae than Double Stars, & the Mirrors are kept in
the highest state of polish, by about ¼ of an hour's rubbing on the very
finest powder, so that the figure is not destroyed, & the process can be
repeated *very often*—The winter has been most favourable to him &
he feels already, that his voyage hitherto has been for good—The
Literary & Scientific Institution of the Cape have made him their
President (very wisely,) & he is setting them all to work to observe
the Tides, & collect Meteorological Obs[ns] &c, & I think he ought to
consider the members, as *learned Operatives* under his superintendence—Nothing can be better than his health during the whole winter—indeed he looks ten years younger, & I doubt if ever he enjoyed
existence so much as now for there are not the numerous distractions
which *tore him* to pieces in England, & here he has time to saunter
about with his gun on his shoulder & basket & trowel in his hand—I
sometimes think we are *all* too happy, & life goes too smoothly with
us, but we don't forget from whom all our blessings come, & know
that we must be prepared for the cloudy day when *it* comes—Caroline & Isabella are growing great girls, & continue very good, I must
confess, for they don't shew any naughty dispositions or sulky tem-
pers, though Bella is very warm & hasty, yet a smile is very soon seen
on her little round face—Caroline reads very nicely, but promises to
excel in singing, for already she has a remarkably sweet & correct
voice, & learns an air from once or twice hearing it—She is all day
at the piano forte, & Papa loves dearly to hear her rude attempts at
harmony—Bella is nice little pet with no originality in her composition, but is content to imitate Caroline in everything & thus she has

learnt her letters, without my knowing how—But little William is my pride & he is indeed a noble boy with a more animated face than a pretty one—he is bold, high spirited & active, yet tractable & obedient as a lamb—quick in observing & understanding *everything*, but does not speak one word yet—he is tall, slender & very fair, with light curly hair, & large deep sparkling eyes placed *just* between a short round face, & a high round forehead—his health is quite robust   Papa says that the little stranger who was born on the 10th Sept is the prettiest of all his Babies, & he pleads hard that she sh^d be called Margaret—Mrs Maclear, the Royal Astronomer's lady, who is a favourite friend of mine will be one God Mama, & perhaps Lady Ryan, an old mutual friend of Herschel's & my Uncle's in Calcutta who is visiting the Cape for her husband's health, will be the other   Uncle Duncan, of whose safe arrival in India we have just heard must be God Papa—All friends in England were well by our last acc^{ts}   As my hands don't feel very strong yet, I am almost afraid you can't read my scribble, but my *wishes* are, that I could *save your* eyes   Herschel desires his most affec^t love to you & I beg to add that of your sincerely attached & respectful Niece

Marg^t B. Herschel

The Travel Diary: An Ascent of Table Mountain and an Excursion to Paarl and Franschhoek

[Sir John kept a pocket notebook in which he jotted down his observations, thoughts, and experiences while traveling. These jottings were made on horseback, in coaches, and under all kinds of unfavorable circumstances, more often than not with a rather blunt pencil. He interpolated his notes with sketches of all kinds and made extensive use of abbreviations and symbols. Some portions of these notes are thus almost impossible to reproduce. This is the case with his notes of an ascent of Table Mountain made on November 8, 1834. He left Feldhausen at 6:20 A.M. with Maclear and someone called Agar. They repaired to Mr. Farsfield's (Versveld's) near what is now called Constantia Nek, and picked up Mr. Steuart, the Sheriff. On horseback they proceeded in a general northerly direction, ascending Table Mountain from the rear or southern side. The notes are full of sketches of rocks seen on the mountain. These are often bizarre, being outcrops of sandstone deeply eroded by wind into all kinds of strange shapes. He noted the presence of *Disa grandiflora* (now called *Disa uniflora*) near a "fountain or puddle." The riders evidently kept to the east side of the massif, making height steadily and picking their way between hummocks. At one point they saw Devil's Peak to the right of the Table wall above "Menzies Gorge" (to be identified with the modern Newlands Ravine?) and turning left came to the "great flat top of the hill" at 10:46 A.M. (probably the point now known as "Maclear's Beacon"), from which they went to the edge and had a fine view of Cape Town, Table Bay, Robben Island, and Green Point. "Thence to the Kloof where the ascent from Cape Town enters—a vast and deep Cleft of w^h took sketch." This is the top of Platteklip Gorge dividing the northern face of the mountain. "Thence to the West side of the hill, to look down on Camp's Bay with Mr Oliphant's[166] house, *such an atom!* On this side is a vast perpendicular precipice & a view of Hout's Bay and the range of Mountains to the Slangkop & Cape of Good Hope Returned, Lunched in view of the Town & got water from a small lake on the very top level of the Table Hill. Deposited in good order a self registering thermom on a ledge, lying on its back & covered with stones piled on it." He sketches the situation. At 1:44 P.M. they began the descent, gradually working over to the east side of the mountains, going through swampy land ("2.14 Dismal swamp") and being enveloped in clouds. At 3:06 P.M., entering on "the Gorge," he saw Pink Disas; at 3:45 P.M. they were at "one tree Hummock where the descent to Constantia begins," and at 5:00 P.M. at Mr. "Versfields," and thence home.

After a day's rest, he took the whole family, including the two-month old baby on a much more extended and adventurous journey into the mountainous regions of the Western Cape. For the most part, the record of this trip is coherent enough to be reproduced as he wrote it.]

The Travel Diary

Monday, November 10, 1834

10 55 A.M.   Started from Feldhausen in a Waggon & 8 with Mag-

---

[166] The Attorney General of the Cape.

gie & Little Louisa (in Charge of Hannah) our Coachman Zomai & Brown the groom riding & leading a
couple of saddle horses for occasional mounting.—Mr
Withers just arrived from Calcutta per Aurora joins
the party taking a seat in waggon or riding as best
agrees with his strength.

12	40	Reached the 1st Sand hills
2	30	Outspan[167] in a waste of sand hillocks overgrown with brushwd and interspersed (sparingly) with blue moreas & magnificent rose coloured Disas!!
3	35	En route Direct for the Eldaberg whose tall & mossy down contrasts strangely with sharp rugged pinnacles [A sketch of the Helderberg Mountains] In immediate front are the farm houses on the Eerste River Sandvliet (Cloete, Ford & Brink)
4	15	Started 3 Korhahns[168] in succession—a very Large Bird with white wings & Dark head & body of a slow heavy flight (NB. It is the best *game* of the Cape)
6	20	Reached Konigsburghs at Stellenbosch, after a most desperate jolting which *unscrewed* both screws of the cisterns of my Barom & let out all the mercury without breaking any glass.

[No date, but space indicates that this now Tuesday, November 11,
1834]

3	5	En route from Stellenbosch
3	20	A most superb view of the Amphitheatre formed by Simonds berg French Hook Donkersberg and Elder-berg embracing Stellenbosch[169]
4	20	Klapmus—over a rough & hilly road, latterly bordered

[167] To release the draft oxen from their harness and make camp.

[168] Korhaan is the name applied to any of several species of bustard (Family
Otidae). The biggest, which are of enormous size, are usually called "Pou." Sir
John's birds were probably the Black Korhaan (*Afrotis afra afra*), the common
korhaan of the western Cape.

[169] Simondsberg = Simonsberg (after Simon van der Stel, early Dutch governor); Elderberg, Eldaberg = Helderberg; Klapmus = Klapmuts; Paarl (literally "pearl") is a small town of the Western Cape vine lands dominated by
Paarl Rock, a domed granite hill.

[?] partially with cornfields clean kept & well looking. The hills are mostly covered very deep with Alluvium of sand & clay. About Klapmus farm the Hedges of the Vineyards are white & red Rose. It is close under the Simondsberg or a low prolongation of its ridge

6   35   Paarl

Wednesday, November 12, 1834

7   0   En Route on foot with Mr Withers for top of Paarl hill
7   20   In Cleft bet^n Paarl & Diamond
8   0   On Paarl. A most remarkable Dike cuts across it like the top of a brick wall. General Direction from ESE to WNW ($\theta = 287°$, $\theta' = 108°$) breadth 12^{in}–14^{in} in broadest part but thins off & subdivides & reunites thus —See next page
            [A sketch of the dike structure some two hundred paces long, dividing into two, with one branch passing through two small ponds measuring six paces and five paces, respectively, separated by two paces]
8   45   Summit of Paarl a Desperate scramble on bare round granite a noble view—
            angles taken as follows per arc compass
            [Angle measures and sketches of various landmarks taken during next half hour omitted]
9   15   En route Down Paarl
10 5   At Hotel. after a Desperate broil and scramble
10 55   En route from Paarl to Wagenmaker's Valy
11 17   Cross the Berg River
1   15   Wagenmäkers Valley. Fine orange trees tolerable viney^{ds} au Reste a regular Humbug
3   0   Left Wagenmakers Valley a place where along the course of the stream there are trees, vineyards & houses —27 lovely houses seen at once! neat white Dutch built Crossed Berg River here M^t got on horseback and we rode together home to Paarl, diverging a little up the P. hill to get an overlook of the P. valley w^h is extremely beautiful & of the Town or Village which is a pretty consid^{ble} straggling place with a Church.

Houses all good, purely white, thatched, scroll gables
& Fronts & embowered in oak, fir & vineyards.

6   ¼   Found waggon arrived before us.—
           Dined & took moonlight walk Splendid meteor with
           a long brilliant train

Thursday, November 13, 1834

A M

9   10   En Route for French Hook after walking before break-
           fast a little way up the Paarl hill & getting a sketch of
           the Village & Valley.
           Our Dutch landlord a stupid old paralytic blear eyed
           old fellow meaning to be civil in his way (NB. knows
           how to charge) his wife a younger woman evidently
           does *not* mean to be civil but *tolerates* her guests—
           her slave Mary a hardy intelligent good humoured
           girl is worth them all put together.

9   35   Cross the Berg River, a dreadful jumble over rough
           stones

11   23   Cross River (Berg) in the French Hook, a fine flat &
           might be rich extensive vale embowered in Romantic
           Mountains all *green* (except the Precipices) with
           shrub & heath. All sorts of sugar bush here & there a
           house in recesses of the Drachenstein hills. The great
           Buttress Hill at the Corner & the Simonds berg beyond
           it shut up the Pass—Drachenstein Sandstone Strata
           Dipping about E French Hook—Outspan at M^r
           D'Ewet's [name erased] (Hugo's)

3   —   Lunched & sketched. & looked for bulbs. There is a
           Church & a [?] school here. All Dutch, no French
           spoken or understood.

3   45
           The Sandstone here in the M^n to right of the Pass dips
En route
           at ∠ 70 or 80 thus making a saddle while further down
           the valley the beds are horizontal
           [A sketch of the folded strata]

3   57   Begin ascent of the French Hook pass the 1st Rampe

leading to the main ascent which slopes up in a very engineer-looking way at a gentle angle across the face of a high mountain.

3   25   On the great Rampe. Fine view of the Valley of French Hook. A great bed of Pisolitic Iron *very blue* & like a conglomerate of peas & Beans with their pods with dirt between. It looks so blue that ? Phosphate. Took specimens The general mass of the rock is a hard siliceous stone. I found one block of regular siliceous Pudding

4   0   Rock apparently Felspar & quartz no mica—Strata dip to SSE at $\angle = 50°$ $\pm$

4   10   Bad news. Brought by Brown. No accommodation at Holmes's

4   15   On 3ᵈ Rampe Culminating. Superb view down valley. *Outspanned.* Sketched the Ascent we came up & Descended, Mᵗ & Mʳ Withers Rode on but found the other side much less Picturesque than that we came up Descended—& reached, after a most dreadful jumble

6   0   Mʳ Hugo's where we found good accommodation   H. is field-Cornet.[170] There has been here a great mortality among children, 9 having died of Hooping cough. The house resounds with coughing—this we did not know till too late to start again, on account of little Margt—but had we been ever so inclined, the horses were too tired to go on.

8   0   A great hymn-singing breaks forth from the back part of the house. Men women & children at the top of their voices

9   0   Got dinner or supper & went to bed. A Rainy night. Forgot to mention that we were caught in the 1ˢᵗ shower, while exploring a path between 2 hedges of peach trees, full of fruit—towards Mʳ Hauman's[171] pretty farm house on the hill ½ mile off

[170] Fieldcornet, a rural administrative and military official under the government of the Dutch East India Company. When the British took over, they kept much of the Dutch organization and many of their personnel.

[171] Mr. E. Hauman was resident elder of the Dutch Reformed Church at Franschhoek, a place which takes its name from the settlement of French

Friday, November 14, 1834

7	0	Rose. A Rainy cold dreary morning. A theft has been committed in the night on the baggage of another Traveller who stopped here. All his things but clothes on his person stolen
9	45	Started, the waggonners declaring that if we delay the Berg River will become impassable. All looks heavy, the roads are soppy & floundering & the Mountains have their cloudy veils on. However, the River was passed easily.
11	40	A fine waterfall behind the great buttress shoulder of the Drachenstein (so after all I suppose we must call the huge abrupt guardian mass which forms the corner (SW) of the French Hook valley)
12	0	Let the waggon go on & took horse, that I might gather bulbs at leisure. Got plenty at a rushy, stony meadow at foot of Drachenstein just after crossing a deep ford near some farm houses. The Flowers here are truly magnificent & I soon filled basket with bulbs. The view of Rock scenery is also very fine & this part of the road is altogether the best worth seeing of any we have come.—A Farm house called Bang-Hook is especially pretty. Curious interview with 2 young Dutch farmer Dandies. "What are you going to do with all those flowers?"—To plant them in my garden—"They wont grow"—Perhaps not this year but I expect them to thrive the next. "Where are you going?" To Stellenbosch "Who are you?" That is a question I don't answer till I know who asks it—pray who are you?— "I'm ..." (Arend, I think he said) I'm Herschel—good morning. They sheered off rather disconcerted but after a while returned as I was arranging my bulbs & after standing by awhile looking on of which I took no notice, began in a much more civilized manner to ask whether I thought the country pretty—Finding their note changed, I changed mine & we had a good long

---

Huguenot refugees there. All knowledge of the French language has now disappeared.

chat & parted very good friends. Got wet & rode to
Stellenbosch & put up at Konigsbergs where the rest
of our party had arrived before me   Descending the
slopes after passing between Simondsberg and Drachenstein got a fine view of the range of Mountains
over Simon's Town & to the Cape of Good hope The
Country seems fertile but wants trees & inhabitants.
Noticed another great bed of pisolitic Iron-oxide on
one of the great water worn slopes, curiously situated
among deep clay in a water-cut Ravine

Saturday, November 15, 1834

8	30	Walked out to the River. It has cut a deep channel in the slopes of the hills & discloses the materials of which that slope is formed—viz: a huge confused mass of water rolled stones large & small, from half a ton to sand—all smoothly rounded, as if by a debacle or long succession of Torrents
10	30	Walked out of Stellenbosch to avoid the horrible jolt-ing just at the skirts of the Town
1	20	After slaying 3 snakes in road reached our former out-span place but now we go on
1	48	Outspan in a grassy recess among white sandhills over-grown with bush
2	45±	En route Mr Withers & I riding
6	20.	Reached Feldhausen found all well, children laugh-ing welcome &c

The Diary: November 16 to December 25

Sunday, November 16, 1834

Mr Withers remained at Feldhausen.

Walked over to Ryans to ascertain what he had done in my absence
relative to Schonberg's application for Money.—He has not *acted* but
has taken information & advises taking security of a Bond & Mortgage

Tuesday, November 18, 1834

Offered Schonberg £3000 for Feldhausen estate out and out. Seems ready enough to jump at the offer but requires till "tomorrow" to "consult his old Mother."—

[Here follow four lines scored through so as to be illegible]

After a while he came & asked me to walk round the near Boundary which we did.—During this walk he appeared to be still more ready to agree to the Sale.—In the afternoon he came again & asked me to look at a plot of the waste ground near Cloete's Spring which he half stipulated for, half begged for himself to build & reside on— and enquired if 3 months would be too long for himself & family to remain on the premises while the house was building[172]

In short we may consider the bargain verbally agreed on.—Further he stipulated to cut 50 fir-spars which he says he has already sold but gave his promise to cut no more timber on the Estate.

Dined at Col. Bell's. Rest. Col & Lady Cath. Bell Dr Murray— Dr Grant[173]—

Wednesday, November 19, 1834

[Ten lines indecipherable due to heavy scoring.]

Met Schönberg by appointment at Messrs. Cadogan & Reids Notaries & in Mr Reids presence both parties signed "Articles of agreement" hastily drawn up by myself by which (subject to the title) the bargain is completed. The Mortgages amounting to about 1900 £ are to be paid off out of the purchase money.—In consid^n I advanced him 124:10 being half a years rent—to be taken as rent if the purchase should not be completed.

Dined at Sir E. Ryans at Stellenberg.

R. advises particular caution lest Schönberg should prove a Crown Debtor.[174] In consequence I wrote tonight to Cadogan & Reid calling on them to take distinct steps to the effect of ascertaining that no Crown Encumbrance exists

Thursday, November 20, 1834

The Meteorological Committee met at Feldhausen when we re-

---

[172] Schönnberg did build the house and called it "Herschel."

[173] Dr. Grant was an Indian visitor.

[174] Settlement of debts to the Crown (the government) takes precedence over private indebtedness.

solved some resolutions & organised a plan of action and dined to-gether—viz Major Cloete—Dr Adamson—Mr Chase[175] Cap^n Bance—Mr Maclear

Friday, November 21, 1834

Fairbairn & his wife—with Mr Wright a Missionary at Griqua Town, and Waterboer the celebrated Chief of þe Griquas[176] dined at Feldhausen also Baron Ludwig and Mr Maclear.

Monday, November 24, 1834

Rose at 7 to prepare letters and box to go by Aurora.—Saw Schon-berg and handed him my ground plan of his allotment and of the Estate.—At 1 PM. went regularly to bed and slept till 5 to prepare for Sweeping. Dined and Swept over the Nubecula Major.—

Tuesday, November 25, 1834

A Rainy warm day. At home. Letter writing Drawing & reading "Indian Annual" M. ill of face ache.—Gladiolus Natalensis

Monday, December 8, 1834

Repolished Mirror (the short focused one)

Thursday, December 11, 1834

Worked at the Mirror   see Polishing Book.—
In Even[in]g a Dinner party at home   Col Bell—Lady C. Bell
Mr Oliphant   M^r and Mrs Murchison   M^r and Mrs Hutchinson[177]

Friday, December 12, 1834

Went into Town with Schömberg to see Watermeyer & negotiate Bills for 1100 w^h W. has agreed to take. Also to attend the S.A. Col-lege Exam^n as Innes's Assessor. For *this* find I have mistaken the day being next Friday

---

[175] John Centlivres Chase (1795–1877), a secretary of the South African Institution, and secretary of the Cape of Good Hope Association for Exploring Central Africa.

[176] A group of mixed blood, originating under Adam Kok, who gathered half-breed Bastaards and Hottentot Grigriquas round him in the western Cape in the late eighteenth century (Eric A. Walker, *A History of Southern Africa*). The Griquas had, by this time, moved across the Orange River and established, partly under missionary tutelage, several settlements under various leaders, of whom Andries Waterboer was one of the most outstanding.

[177] Alexander Hutchinson, notary, attorney, and proctor.

Saturday, December 13, 1834

Very hot. Staid at Home

Tried Equatorial & sat up till 2½ AM Watching opportunities but the ✱ s continued ill defined & telesc would not act

NB   Peacocks exceptionally troublesome at night

Sunday, December 14, 1834

A violently Hot day, Clear & Cloudless.

Kept close in doors—read prayers with Margt & the little bodies—about 5 took a drive on to the flats—returned

Mr. Murchison dined with us & looked at the ☽ & Jupiter—with the new polished mirror

All night kept awake & watching with fleas & the noisome Peacocks with whom at 1st Dawn a regular Battle & succeeded in disloding them after which got a few hours rest.

Monday, December 15, 1834

Rather Cooler. Drew on Drd[178] Draf [erased] Bill of Exchge. Letter F for 200. Schönberg via Meadows

Friday, December 19, 1834

9–10   Went into Cape Town

10–2½   Examination of Mathematical Class at the S.A. College under Profr Innes to whom I acted as "Assessor"— i e sitting by to judge of merits & to put occasional questions

Saturday, December 20, 1834

9 AM–10. On Road to Cape Town

10–3—Examination of Mathem¹ Class at S.A. College

Monday, December 22, 1834

9–10 On Road to Cape Town

10–3   Examn and distribution of Prizes at S.A. College after which a Speechification—

Tuesday, December 23, 1834

Dined [illegible words] Midnight at wh hour I rose, with the Horary Meteorological obsns[179]

---

[178] Messrs. Drummond, 48, Charing Cross, London, the financial firm with whom Sir John dealt.

[179] Sir John proposed a scheme for taking hourly meteorological observations for at least twenty-four hours at the equinoxes and solstices.

Went to bed for 3 hours

At night a fire spreading in the grounds of the Vineyard[180] alarmed us. All hands to help put it out. The wind was gentle & the vegetation not yet thoroughly Dried by the Summer heat, or it might have been a serious conflagration—It is attributed to "bad boys" (rascally slaves) who burn it to get up afterwards the Dry half burnt sticks to exchange at Canteens for Brandy.

Swept till 6ʰ½ ST.

Wednesday, December 24, 1834

Wrote to Babbage

The fire gaining ahead again—Got the men to work and cut it off by making a lane across its *intended* path to windward. N B It may be stopped with the greatest ease to windward especially if you can get to the wind side of a ditch.—Clear all small stuff so as to *see the ground* bare if it is only a hands breadth & press back the high bushes on the wind side & cut or break those to leward & fling the broken branches *in*wards, a good way.—Then keep watch & flog out the flame.

Thursday, December 25, 1834

Mʳ Oliphant called.—Just as we were going to Church a nearer alarm of fire accompanied with dense smoke & a copious fall of ashes obliged me to hasten out call all hands and set to work. It was running down a *conductor* a line of Proteas by the Protea road burning with great fury. We got it cut off here by making a broad lane in it's way & tearing up all the small brush wood & then we headed it up towards Protea and got it effectively under.

Planted 27 seeds of the Acacia[181] . . . [his omission] from N[ew] S[outh] W[ales]

---

[180] The Vineyard, Newlands, adjoined "Feldhausen" on the northwest.

[181] Dr. Hall remarks: "This is certainly one of the earliest references known on the planting of Australian Acacias at the Cape, an interesting point for Cape botanists, as some have invaded, and, in many places, entirely suppressed, the local rich flora. None of the chief offenders comes from New South Wales, however."

The Diary: Occasional Memoranda and Observations

[Most of the diaries have printed sections, such as "Monthly Summaries" at the front, "Occasional Memoranda," "Observations," at the back. These have been filled up with miscellaneous notes. These are not particularly connected in theme, but are of value as showing the breadth of Sir John's activities and correspondence, the kinds of things purchased from England and the prices paid, and the increasing interest which he took in botanical matters.]

Occasional Memoranda, January

Correspondence
Herschel. Miss C. Hanover—Nº 2. In post at Cape Town Jan. 25
Mʳˢ Stewart. Nº 2 In post from Cape Town. Jan 17. per Iris.
Mʳˢ Jones. Nº 1. Jan 17. per Iris from Ceylon
Mʳˢ Stone. Nº 1—Jan 25.—per.
F. Baily[182] Esqe. Nº 1. ⎫
                          ⎬ Per Thames . Jan 30
Revᵈ W. Whewell. Nº. 1 ⎭

Mʳˢ Goodall— ⎫
Miss Tunno.— ⎬ Per . . . Feb 26 1834 Under cover
Mʳˢ A. Baillie ⎭ to Beaufort per Barrow. Posted by myself

C.C. Stewart ⎫
Mʳˢ Meer   ⎬ p. Marquis of Huntly
Miss [erased] ⎭

[Page crossed through]

182 Francis Baily (1774–1844), British astronomer, and, with Sir John, one of the founders of the Astronomical Society in 1820, of which he was four times president. He revised the star catalogues of Flamsteed, Lalande, and Lacaille, and produced simplified tables for the reduction of aberration and nutation. Vice-president of the Geographical Society, trustee of the British Association, vice-president and treasurer of the Royal Society, he was a gold medallist of the Royal Astronomical Society, 1827.

Occasional Memoranda, February

P.Stewart—
Mrs Stewart—
Mrs Meer—
Babbage

Miss Smithson
Miss Rackham
Stultz.—
Troughton.[183]—
Painter.—

Enclosed to Miss S.
& P.S.

Enclosed to Captn
Beaufort through
Barrow, per
"Duke of Buccleugh"

Captn Beaufort.—

Mrs Goodall

Miss Tunno

Mrs A. Baillie

Captn Beaufort.

Enclosed Feb. 26 1834 to Captn
Beaufort and under cover to Barrow.

[Entries crossed through.]

Ordered from England for Miss Smithson's box.

Painter—Strand—24 p^{s} ladies' gloves—6 p^{rs} gentlemen's
               estimated price—   £ 6. 0. 0.
Mrs Rackham—2 Corsettes—   6. 6. 0
Stultz—2 Coats—2 Waistcoats—   14. 0. 0.
Troughton—Self registering Therm—   15. 0
Miss Smithson—Dress. & Millinery
    Bordeaux, 6 doz Claret—to be paid here

[This entry in Lady Herschel's hand]

[183] Edward Troughton (?1753–1835), the instrument maker and inventor, in 1809, of a method for the accurate graduation of circles. He engaged in a long drawn out dispute with Sir James South (see note 7, April 10, 1833).

Occasional Memoranda, March

Correspondence—

Miss Gwatkin ⎫
⎬ p. post
Miss H. Smith ⎭

Mrs Hall— ⎫
Mr Mackintosh ⎪
Miss Smithson ⎬ under cover to Beaufort & Barrow
Mrs Allan ⎭

Lady A.M. Donkin ⎫ under cover to Hon. G. Elliot
Mrs Stewart ⎭

Miss Baldwin ⎫
⎬ under cover to Adm^{y} [This entry scored through]
M^{r} & Mrs Airy[184] ⎭

Ordered from England by P. Stewart's box.—

		£		
Miss Smithson—Riding Hat—Estim Price . . .		£ 1.	10.	0
Miss Baldwin	Flannel 20 y^{ds}—	2.	10.	0
	18 p^{rs} shoes for child^{n}—	3.	0.	0.
	pelisse for Boy	1.	0.	0.
	pomatum, Soap, spunge	1.	10.	0
P. Stewart—6 Camel hair brushes—		1.	5.	0.
D. Stewart *from India*—4 room mats—		20.	0.	0.

[184] Sir George Biddell Airy (1801–1892), celebrated British astronomer. He was educated at Cambridge, where he was senior wrangler and first Smith's prizeman in 1823. He was Lucasian professor of mathematics at Cambridge, 1826; Plumian professor of astronomy and director of the Observatory, 1828; and Astronomer Royal, 1835–1881. He re-equipped the Royal Observatory at Greenwich with instruments of his own design, and established a magnetic and meteorological department in 1838. His autocratic direction of the Observatory has become legendary. He was awarded the Gold Medal of the Royal Astronomical Society in 1846 for his reduction of all the lunar and planetary observations made at Greenwich. President of the Royal Society, 1872–1873, and five times president of the Royal Astronomical Society, he first met Sir John during the winter of 1823–1824, which he spent in London with Sir James South.

Tippets & muffs—	5.	0. 0.
rugs—humps [?] &c &c—	5.	0. 0.
(work table to be sent in due time)		

[The list in Lady Herschel's hand]

Occasional Memoranda, April

Arago—under cover to P.S.
Arago—direct to France

Beaufort
P.S.   per Beaufort
J.S.   per Barrow   per Mary Ann
J.C.S.

Grahame Nantes—Direct for France

[This entry scored through]

Occasional Memoranda, May

J.C.S.
          (MBH)   per Eliza Jane
P.S.

Captn Smyth (per Beaufort) by Eliza Jane. &c

Mamma
          May 10.—(Margts.) per Mary Ann [erased] Eliza Jane
James S.

Mrs Meer (MBH) announcing Nanson's death per Mary Ann
Duncan S. A parcel by City of Edinburgh
May 30th Mama—(MBH)—p. Marianne—
          —Stratford—(J.F.H.) p. -do-

[This page scored through]

Occasional Memoranda, June

June 8th   Mrs. Stone (J. Stone)
           Horsburgh—(J.F.W.H.)—
           Miss Herschel –do–   Adm^y cover p. Zenobia
           Mrs. L. Wilson (M.B.H.)—

Mʳˢ Fitton (MBH)—
Mʳˢ Delavane (–do–)—
Miss Small   (–do–)
Miss Baldwin (–do–)
P. Stewart   (–do–)

(in post)
no cover
p. Zenobia.

(ordering Music forgot & 4 pʳˢ
  India rubber shoes)—

[These entries, in Lady Herschel's hand, are scored through]

Occasional Memoranda, July

July 12. Dispatched to Post to go by Salus
  1. —M. to Mʳˢ S.
  2. —H to Struve[185]—(Nᵒ1)
  3. —M & H to P.S. enclosing N° 2 of this list
  4. —H to Beaufort See Duplic. Book
  5. —Chase's receipt for Chronometers
The whole under Cover to Beaufort
          under Cover to Barrow.

July 18 1834.—In post for "Euphrates"
  1.   MH to M. Baldwin
  2.   MH to Mrs. Moorsom
  3.   J.F.W.H. to J.C. Stewart.
  4.) ⟨ J.H. Nelson. 26 Savage Gardens
  5. ⟨) Enclosure. Letter to Copeland.

For list of Drawings sent home per Mʳˢ Warren   See "Observations"
—at end of the volume

[185] Friedrich George Wilhelm von Struve (1793–1864), German astronomer, who surveyed all the available sky for visual double stars and published the results in his classic *Mensurae Micrometricae*. He attended the University of Dorpat (=Tartu, Estonia), where he became professor of astronomy and mathematics, 1813. He was summoned by the Emperor Nicolas I of Russia to superintend the building of the famous observatory of Pulkovo near St. Petersburg, of which he became director in 1839. He was the first of an astronomical dynasty which includes the late Otto Struve, first director of the McDonald Observatory, Fort Davis, Texas, to whose memory the eighty-two-inch reflecting telescope was dedicated, 1966.

July 20th In Post for 'Euphrates'
   1. M.H. to P. Stewart—

ordering a supply of letter & note writing paper, Talc, & Wax— & fine drawing board—also six p[rs] of Holland boots, 3 p[rs] bl[k] prunel slippers & one p[r] bl[k] satin slippers from Grantham, Windsor—reminding or reordering 6 p[rs] shoes from Moore for H. ordering a dozen col[d] smock frocks for little William, & giving hints about the usefulness of sending child[ns] shoes, stockings, gloves & flannel—

[All but the last paragraph scored through; last paragraph in Lady Herschel's hand]

Occasional Memoranda, August

Aug. 11. Received. 3 Packets under Admiralty cover and a letter.
          from J.C.S.
   1st Packet.—f[m] Airy containing a letter from A. dated Apl 24
                    a letter from M[rs] A to M
                    his book on Gravitation
   2[d] Greenwich OB[sn] N[os] 2 3 4 5—1833, &
                    Notices of R & A.S.
   3[d] Foster's Pend[m] Exper[ts]—

Sep. 13. In post per Sir E Paget.
   1. to D[r] D. Stewart. Hourah Calcutta.

Sep. 13. In post per Thomas Snook
   1. To M[rs] Stewart Care of P.S. 65 Cornhill
   Notice of Birth of Baby

Sep. 13. Per M[rs] E. Gordon who goes by the Paget. for D.S. at
          Calcutta.
   1. Letter Do D.S. announcing birth of Baby
   2. Letter from Miss Prendergast to Marg[t]
   3. Gladiolus (Hispidus?) M's drawing
   4. View of the Telescope & Table Hill indian ink Sketch JFWH.

Sep. 13. Rec[d] per Hungerford a letter from P.S. with enclosures.
(date July 1. 1834). and one from Miss Smith date June 30.

[Second paragraph and part of the fourth paragraph scored through]

Occasional Memoranda, September

(for Mem. of Sept 13th See last page)
Sept. 29th, Enclosed to Captn Beaufort—per Claudine—to sail
    Sep 30.—
    1. To P S. from Maggie—
    2. To Miss Herschel (No 5) 376 Braunschweiger Strasse Hanover—
       Enclosed in P. S's note
    3. To Mrs Stewart.—Enclosed in P. S's—from Margt
    4. Whewell. Trin Coll.—(apart for Beaufort to forward)

Sep. 30 In charge of Major Hamilton to deliver to Mrs S.—to sail
    per Claudine this day—A parcel containing
    1. Drawings for which See "Observations" over 2 leaves.—
    2. A Note to Mrs Stewart from Mt [entry crossed out]
    3. A collection of Notes, to Mt for Mrs S's reading
[First paragraph scored through]

Occasional Memoranda, October

Oct 21. Enclosed to Beaufort per Mount Stewart Elphinstone,
    1. To Jones (Kings Coll. Lond.) from JFWH
    2. To Robt Grahame Whitehill to be forwarded to Jas. Grahame—
       (JFWH)
    3. To Mrs Stewart—from MBH.
[This entry scored through]

Occasional Memoranda, November

Dispatched by the Revd G. Withers passenger per Aurora a box Tin
lined (Receipt acknowld by Mrs W) containing as follows—
Addressed to P.S. 65 Cornhill
    1. Wheatstone's[186] Symphonion for repair—
    2. A Gnadenthal[187] box of Olive-root containing

[186] Sir Charles Wheatstone (1802–1875), British scientist, working mainly
in the field of electricity. He improved the electric telegraph, and worked on the
development of submarine cables and magneto-electric machines, and on de-
vices for measuring electrical quantities. He was a professor at King's College,
London, 1834, FRS, 1836, and knighted, 1868. An instrument called the sym-
phonion, invented by him in 1839, was a combination of piano and harmonium.
    [187] Gnadenthal = Genadendal, formerly Baviaan's Kloof (Baboon Ravine),
the Moravian Mission, still in existence, founded by Georg Schmidt, 1737,

1. Letters for Mrs S. and for Revd C.S. [entry crossed out]

2. Bushwoman's hair

3.   —D°— Bracelets.

for   4. Sea-wool from Oliphant's River[188]

Mrs S.   5. Silver chain Madagascar manufactory

3. 16 Specimens stuffed birds for M B^n and a note for M Baldwin.

4. 6 Coloured Drawings for Mrs S. See list over 2 leaves.

5. A bottle of Insects in Spirits for H. Griesbach.

6. A Bundle of Everlasting flowers.

Also per Aurora Letter to Drummond

Sent to Mr Withers Monday Morning Nov. 23. 1834.—

Sent also to Mr Withers a Certificate of Mrs Nanson's Burial to be by him forwarded in England to Mrs Monkhouse.

NB. Mrs Monkhouse to be written to about it.

Wrote to Mrs Allan &c &c to Revd C. Stewart

[This entry scored through]

Sent to Captn Clarke, at Mrs Gambiers boarding house & receipt by him acknowledged—2 Boxes containing Mrs Nanson's Clothes & Effects—every atom, for delivery in England to Mrs Monkhouse

---

Letters sent by Captn Ireland of the Royal William to post in England Sailed Nov 24 or 25

1. Drummond & Co. (Duplic)

2. Mrs Stewart (News of þ^e Mercury being heard of)

3. H. Beaufoy Esq. (Duplic)

[List of letters scored through]

Ordered—2 doz Calico day shirts for H. ⎤
  —Cotton Stockings   ⎥
  —Black silk neck handk^s   ⎥ Estimated price
1 doz French Cambric po. [?] handk^s   ⎥ £ 20.
  —White silk stockings   ⎥
    to come in Alfred's box   ⎦

---

abandoned, 1743, and recommenced, 1792. It is in the mountains about one hundred miles east of Cape Town.

[188] In Afrikaans, Olifantsrivier, or Elephant River.

Occasional Memoranda, December

Wrote to Capt. Macarthur franked by Sir E. Ryan

Wrote to Duncan by Sir E. Ryan, who also took charge of a small parcel for D. containing a black silk whalebone Cap—3 p^{rs} gloves—& some bulbs gathered in our excursions for Mrs Holmes—They also took £20—of which £5 was due to D. for freight of Mats &c—the rest to be expended in presents & orders—wrote also to Mrs Grahame thanking her for the Children's toys—

Herschel wrote to James & I added a Parag^h—went by the Platina, Simon's Bay—enclosed to Peter.

Decr 18th—Wrote to Miss Augusta Mackenzie ⎫ enclosed
                                                                       ⎬ to
                                        Miss Prendergast ⎭ Peter

Decr 22nd—to Miss Baldwin—p. post
               to Miss Monkhouse ⎧
               to Mrs Prendergast ⎩

Sent on Duncan's letter to James— ⎧   all by the Lord
              to Mrs Beckwith[189] ⎪   Lynedoch
Decr 24   wrote to Mama   ⎬   [written vertically]
              —Mary Carter   ⎩

Herschel wrote to Mr Babbage

received Duncan's letters dated Sept 6 & Octr 4th recounting his wedding—

[Whole entry in Lady Herschel's hand]

Observations

List of Drawings &c sent home per Mrs Warren per Iris.

5 *Indian Ink Sketches*—

N° 1—View from A (in Ground plan p. 4. plan book) corner of the Stoop looking outwards towards the Devils Peak along the N W Avenue

N° 2. —View from B (plan p. 4) of the front of the house.

N° 3 —View from C, near the house end of the N W Avenue, of the North Corner of the house and along the S E Avenue.

---

[189] A cousin of Sir John's on his mother's side.

N⁰ 4 —View from D (plan page 4) about the middle of the S W
Avenue looking inwards

N⁰ 5 —View along N E Avenue from the house door on stoop.

3 Ground Plans—

N⁰ 1.   Plan of the house [Figure 3]—exact facsimile of the plan p
2 in the Plan Book.—p. 2.

N⁰ 2.   Plan of the Premises—facsimile of the plan in p. 4. Plan
Book. (Lettered alike as to the letters A. B. C. D. N. P.
Q. S. Z.)

N⁰ 3.   —Plan of the Hills from Table Bay to false Bay [Figure 4]—
a rough map. Laying down the situation of Feldhausen.

1 Paper of Explanations.

12 Coloured Drawings of Flowers by M.B.H.

N⁰ 1. Disa grandiflora—Red	
2. Disa . . . (scarlet, long nectaries) [his omission]	
3. Gladiolus Punctatus (Brown)	Large size
4. Oxalis Versicolor (White, Pink bordered)	$7^{in}00$
5. Oxalis Sulphurea (yellow)	by
6. Phlomis Leonurus (Orange Red)	5.75
7. Oxalis Purpurea (pink rich)	

N⁰ 8. Gladiolus Brevifolius (pale pink)	
9. Blairia Ericoides (pink)	Small size
10 Aponogeton Distachyon (white)	$6^{in}00$
11. Disa (? Graminifolia) blue	by
12. Lobelia Triquetra (blue) [a small sketch]	4.65

1 Tin Can of Bird-skins—Enclosing some for J. Stone's wife

List of Drawings sent home by Major Hamilton per Claudine.
Sep. 30. 1834.

2 Indian Ink Sketches.

N⁰ 6. View from E (in the ground plan p. 4) (of Plan Book) the
front of the house seen looking *towards A*, or to the near end
of the North West Avenue.

N⁰ 7. View from F in the ground plan p. 4. looking towards Pussie's

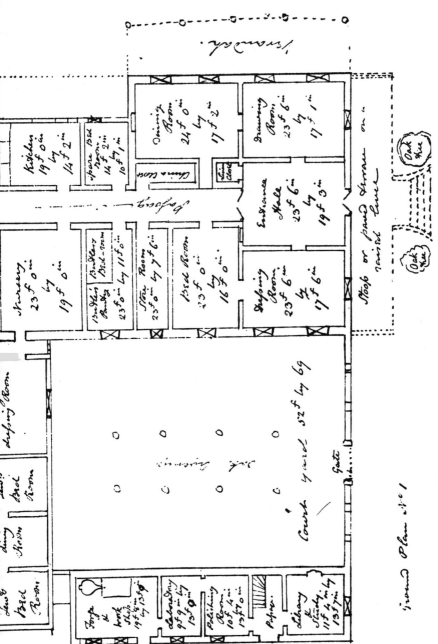

Figure 3. Sir John's plan No. 1 of "Feldhausen." (By permission of the South African Library)

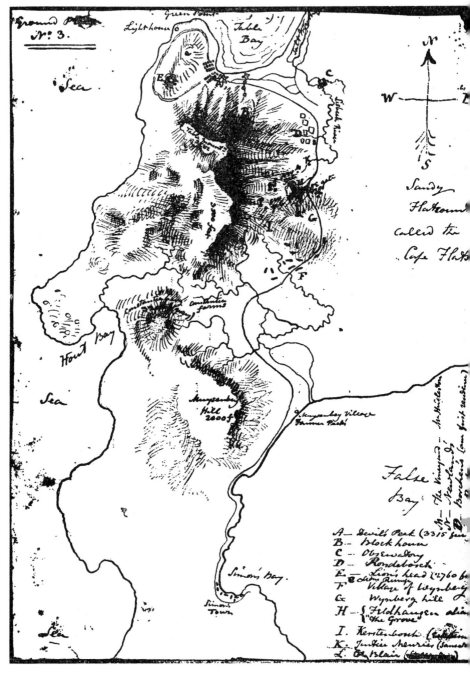

Figure 4. Sir John's plan No. 3 of the Cape Peninsula. (By permission of the South African Library)

Corner in the Table hill—so as to take in both the 20 feet and Equatorial house, with a careful portrait of the Devil-berg —Pussie's Corner and Table Hill. (NB. This intended for a Frontispiece to my Catalogus Generalis Nebularum)[190]

1 Paper of Explanation
     8 Coloured Drawings of Flowers by MBH.

Nº 13. —Bignonia Capensis (*red*, Trumpet flower) ⎫
Nº 14 —Watsonia Marginata (Long *Pink*)   ⎪
Nº 15 —Homeria (1. Collina, yell.)   ⎪   Large Size
              (2. Flexuosa, Red)   ⎬   7^{in}00
Nº 16. —Trichonema Alba.   ⎪   by
Nº 18. —Babiana? . . . (Earliest, Blue)   ⎪   5^{in}75
          [his omission]   ⎭

Nº 19. —Gladiolus Watsonius. (Scarlet)   ⎫
                                          ⎪   Small size
Nº 20. —Rafnia Triflora (yell. Papil^s) ⎬   (6.00
                                          ⎪   4.65)
Nº 21. —Disa Longicornis (Old Mother Shipton) ⎭

List of Drawings sent home by Mr Withers per Aurora.—
6 coloured drawings of flowers by MBH.

Nº 22.   Hypoxis Elegans   ⎫
Nº 23 —Watsonia . . . Brick yellow, purple anthers   ⎪
              —large flowers—alternate   ⎪
[his omission; a small sketch; uncertain to which flower ⎪
          it applies]   ⎪   All large
Nº 24 —Disa Graminifolia (White, Flats)   ⎬   size
Nº 25 —Gladiolus Angustus (3 deep pink *spades*   ⎪   7^{in}00
          or horse shoes—gath^d with Mr Petrie.   ⎪   by
Nº 26 —Oncidium . . . Yellow—Ludwig   ⎪   5.75
          [his omission]   ⎪
Nº 27 —Ornithogalum Thyrsiflora[191]   ⎭

[190] This drawing did indeed become the frontispiece to *Observations at the Cape of Good Hope.*
[191] Number 17 is missing. All these drawings were donated by the Herschel family to the South African Public Library, Cape Town. Dr. Robinson points out that the numbers of the flower sketches at the South African Public Library do

Sent home by Lieut Worster "Prince George"
Given to him Jan. 24. 1835—to sail tomorrow

1. Flower Drawings by M.B.H.—Size C
   N⁰ˢ 28, 29, 30, 31, 32, 33,
2. Indian Ink Sketches JFWH N⁰ˢ 8, 9.—
3. A few small parcels of Seeds.
4. Specimens of leaves made into gloves, Dolls,
   night caps &c from Tulbagh.
5. An open unfolded Letter from Mᵗ to Mrs. Stewart
   Jan. 28ᵗʰ sent per Exmouth

1835   Cousin Mary ⎫
         Mʳˢ James Thomson │
         Miss Turner   │   all per
         Miss Gwatkin   ⎬   Barrow.
         Miss Richards   │
         Mrs Goodall   ⎭

                                           in tin case open ⎫
         Mʳˢ Stewart   ⎱   per Lieut Worster ⎬
         J.C.S. Exmouth ⎰ Jan 26.   per Prince George ⎭

---

not agree with this list, though Numbers 2, 5, 7, 8, 9, 11, 12, 15, 20, and 21, are
also depicted.

Dr. Hall has provided the following list of modern names for the species depicted in the flower sketches:

No.1. Disa Grandiflora = *D. uniflora* (Orchidaceae), Red Disa.

No. 2. Disa . . . = *D. ferruginea?*, Cluster Disa.

No. 6. Phlomis Leonurus = *Leonotis leonurus* (Solaraceae), Wild Dagga.

No. 9. Blairia Ericoides = *Blaeria ericoides* (Ericaceae).

No. 10. Aponogeton Distachyon = *A. distachya* (Aponogetonaceae).

No. 12. Lobelia Triquetra = *L. comosa* (Campanulaceae).

No. 13. Bignonia Capensis = *Tecomaria capensis* (Bignoniaceae), Cape
Honeysuckle.

No. 15. Homeria Collina = *H. breyniana* (Iridaceae), tulip.

No. 16. Trichonema Alba: evidently a white *Romulea* (Iridaceae).

No. 18. Babiana . . . = probably *B. hiemalis* (Iridaceae).

No. 22. Hypoxis Stellata = *Spiloxene capensis* (Amaryllidaceae), sterretjie
(i.e., little star).

No. 26. Oncidium . . . : this name for an American orchid genus had not been
used for Cape species.

$\left\{ \begin{array}{l} \text{P.S.—} \\ \text{C.S. encld to PS.} \end{array} \right\}$ per Prince George Jan 26.

H $\left\{ \begin{array}{l} \text{Babbage} \\ \text{Beaufoy}[192] \end{array} \right\}$ Jan. 24. per Exmouth B.

Wallick.[192]   Jan 26. 1834.

$\left. \begin{array}{l} \text{D}^r \text{ Fitton}[193] \\ \text{Capt Hall.}[194] \end{array} \right\}$ B. Jan . . .

$\left[\begin{array}{l} \text{1 D}^r \text{ Stewart—Calcutta (M.B.H.} \\ \text{2. do. do. do.   (J.F.W.H.) S.} \\ \text{3. M}^{rs} \text{ D. Stewart (enc. in N}^o \text{ 1) (M.B.H.)} \\ \text{4 D}^r \text{ Wallick—Calcutta (J.F.H.) S.} \end{array} \right.$ $\left. \begin{array}{c} \\ \\ \\ \\ \end{array} \right]$ per
Sherburne
via
Bombay—

[Entries in Lady Herschel's hand]

Cover to Barrow Adm^y	$\left[\begin{array}{l} \text{1 M}^r \text{ Airy—Cambridge (J.F.W.H.)} \\ \text{2. M}^{rs} \text{ Airy—(encl. in do)—(M.B.H).} \\ \text{3. Miss Herschel—(encl. in N}^o \text{ 4)} \\ \qquad \text{(J.F.H.)} \\ \text{4. P. Stewart—Cornhill (J.F.H.)} \\ \text{5. M}^r \text{ Henderson}[195] \text{—Edin}^b \text{ (J.F.H.)} \end{array} \right.$	$\left] \begin{array}{l} \text{Feb}^y \text{ 21}^{st} \\ \text{by} \\ \text{Duke of} \\ \text{Buccleugh} \\ \text{Capt}^n \\ \text{Henning} \end{array} \right.$
per Post	$\left\{ \begin{array}{l} \text{1. M}^{rs} \text{ Stewart—encl in N}^o \text{ 2 (M.B.H.)} \\ \text{2. Sundry notes enclosed} \\ \qquad \text{Peter Stewart—Cornhill (M.B.H.)} \end{array} \right.$	

[192] Colonel Mark Beaufoy F.R.S., or Lieutenant George Beaufoy R.N., of Bushey Heath, near London, members of the Astronomical Society. The letter to Wallick should be dated January 26, 1835.

[193] William Henry Fitton (1780–1861), B.A., Trinity College Dublin, M.D. Cambridge, geologist, FRS 1815. He was a supporter of Sir John against the Duke of Sussex for presidency of the Royal Society. He was elected president of the Geological Society, 1828.

[194] Captain Basil Hall (1788–1843), a naval officer, friend of the Herschels, commanded an expedition to China in 1816, and attained the rank of post captain in 1817. He made extensive journeys on the continent of Europe, in North America, and in Egypt, was a prolific correspondent with wide interests, including astronomy, and for some time was a fellow of the Royal Astronomical Society. His obituary appears in *Memoirs of the Royal Astronomical Society*, Vol. XV, 1846, at page 365, immediately before that of Thomas Henderson, and just after Sir John's lengthy memoir of Francis Baily.

[195] Thomas Henderson (1798–1844), Astronomer Royal for Scotland, and pre-

NB. Capt[n] H. takes a small bottle of Phosphoric Insects
to D[r] Buckland Cam.[196]

NB. Ordered from Peter—Carbon paper for Herschel Beethoven's
Music in No[s]—French & German spelling books for children—
2 copies of Barbaulds Hymns in Prose & Miss Edgeworth's
"Early lessons" & "Parent's Assistant"—

[These entries in Lady Herschel's hand]

A Letter from Sir John and Lady Herschel, Feldhausen,

to James Calder Stewart

                                        (Feldhausen) [editor's hand]
                                             Dec[r] 6th 1834

[Sir John's hand]

Dearest Jamie—

News of you from Peter—your Portrait (a speaking likeness—
in that language that needs not words) your letter from Genoa—and
a letter about you and about nothing else than you from Plana[197]—
all reached us the day before yesterday—It will rain you next.—All
hail! This is luck. I only hope it will not be a forerunner of your own

---

decessor of Maclear as H. M. Astronomer at the Cape (1832–1833). Beginning
life as a lawyer's clerk, he turned to astronomy and during his brief tenure of
office at the Cape measured, but did not publish until too late to secure priority,
the first stellar parallax (of α Centauri). He disliked the Cape and customarily
referred to the observatory as "Dismal swamp," a phrase taken over by Herschel
in one of his travel diaries.

[196] William Buckland (1784–1856), English divine and geologist, a fellow of
Corpus Christi, Oxford, who systematically examined the geology of Britain. He
became reader in mineralogy at Oxford, 1813, and first reader in geology, was
elected FRS 1818, and twice was president of the Geological Society. He was the
author of *Reliquiae Diluvianae*, one of the apologetic Bridgewater treatises (as
was also William Whewell) designed to prove the truth of the Biblical flood.

[197] Giovanni Antonio Amedeo Plana (1781–1864), mathematician, astronomer, and specialist in geodesy and the motion of the Moon. He was director of
the Turin Observatory from 1813.

self in propria persona for *that* however we should delight in a sight of your own proper phyz—we do heartily deprecate, at least if India were to be your ultimate destination. Well—and so you found Plana just what I always held him to be—a kind frank friendly being. Just what I found him on presenting myself at his door unannounced—unintroduced—His arms of brotherhood were opened to me at once and in half the time it would have taken to go through the grimaces and thaw the first crust of ice which envelopes ordinary hearts—his had expanded like a flower under the genial influence of his sunny temperament and native good feeling—Don't (like your own self) attribute all the attentions you say you rec^d from him to extraneous causes—I judge by his letter that he has felt more than a second-hand interest in you and I love him for it.

Delightful as is the Climate of this place, and thoroughly comfortable as we have made ourselves here, I cannot help envying you your winter in Florence. As for Rome, if you see it with my eyes, avoiding Ciceroni, keeping aloof from antiquarian discussion, and rejecting all the exaggeration of talkee-talkee among dilettanti or mawkish fashionable loungers, virtuosi & pretenders—You may in that case safely risk yourself among the Aristocracy then and there assembled, secure in a panoply, and seeking a nameless and boundless enjoyment in those scenes gray with the mist of years, to which no temptation that gaiety or dissipation can hold out can offer an equivalent. It is in solitude that Rome must be enjoyed—Go but go alone the while and home returning soothly swear was never scene so sad and fair—

Probably you have met Captn & Mrs Hall. They have written us most interesting letters during their Sojourn in Italy.

Of our news—First & best—the Mercury is heard of! all safe, but weatherbeaten. The news is not *yet* confirmed, but it came in so distinct & circumstantial a form that we can hardly doubt it.

Then too Maggie and all the brats are well (face aches excepted). The New Baby is the most *silent* unresisting healthy creature you ever heard of. Carry, Bella and the Boy Bill well & hearty. The latter grows unruly & self willed.—

The Stars now flash merrily (though for the last 3 months they have perversely hid their heads to my great annoyance) Yesterday however they cleared and all is once more in Activity. I have on the whole got on *very well*.

This place is full of Bankruptcies—there is a commercial panic—

but among all the evil produced it has given rise to some noble displays of generous feeling of which more when room & more time for I write to Catch the Courier.

Sir E. Ryan Sails on Sunday per Zenobia. His health which was dreadfully shattered is now thoroughly reestablished and so seems to have been (by the voyage hither, and a much shorter stay) that of Mr Withers whom I dare say you enumerate among your Calcutta acquaintances.—

Your Portrait goes in care of Ryan as I can't guess where this may hit you—perhaps in Egypt or Tunis—I will only say—Sis felix ubicumque mavis[198] and recommend myself to your thoughts—(in Golden Genoese Sunsets—or in Boeotian fogs) with all good wishes from

Ever yours with warmest affection
JFW Herschel

[Lady Herschel's hand]

Darling Jamie—Let me add my very hurried blessings, & my very prettiest thanks for the pretty seal which now seals this as its first experiment, & which I kiss actually sometimes not having your own sweet cheek to put mine on—Mr Deas Thompson takes this to Simon's Bay—So good bye my pet—Your picture is tolerably good— The Ryans are gone—

Your own affec^t Sister
Marg^t B. Herschel

James C. Stewart Esq.

J.F.W.H.
20 Dec /34
rec^d 27 July /35
ans^d 28   ″   —

[198] Sis felix ubicumque mavis: May you be happy wherever you wish.

1835

The Diary: January 1 to February 22

[This volume is laid out differently from the others, having dated spaces for entries on one side and blank facing pages. Longer entries are often continued on to the latter.]

Feldhausen Cape of Good Hope

Thursday, January 1, 1835
Miss Geard and the little Maclear's spend the day with us

Friday, January 2, 1835
Accompanied Maggie into Town & called on Dr Philip to congratulate on his return

Saturday, January 3, 1835
Sir Bladon & Lady Capel,[1] Mr and Mrs Murchison, Mr Drummond (one of Sir B. C's Midshipmen) and Mr Malcolm dined with us and in the evening viewed Stars &c. Mr Malcolm (who is just returned from Latakoo) remained the night.

Sunday, January 4, 1835
A great fire on Devil's Hill. I saw its commencement just before breakfast. After breakfast while viewing it from 20 feet,[2] it broke out at a lower point among van Renen's protea woods. I am certain *that* was not by communication from the fire above.

Rode with Mr Malcolm to a point commanding a view of the conflagration. A truly sublime spectacle. The Mountain had precisely the appearance of the representations of Vesuvius during the course of a great lava current—Streams of lines of bright flame. Volumes of smoke from the slopes and fiery columns mounting among the steep

[1] Rear Admiral the Honorable Sir Thomas Bladen Capel, K.C.B. Commander-in-Chief, East Indies, May 30, 1834–July 26, 1837. He was later promoted to admiral and made G.C.B.
[2] Not from a distance of twenty feet, but from the site of the reflecting telescope of twenty feet focal length.

inaccessible crags & a vast Drifting mass of Smoke gliding off in the vapour-plume under the influence of the S.E. wind.

The advance of the fire to windward was most imposing—the rush, the crackle—the glowing flames seen by glimpses through the trees like the gleams of sunrise in Martin's Pictures[3] & the rolling volumes of flame when it conquered some fresh mass of Wood, surmounted by ragged flakes & sheets of flame & sparks—The wiry flames in Martin's Hell are true to nature where the fire is large & the flame has its own way.

At night Monograph of Neb in Orion.[4]

Friday, January 9, 1835
Wrote to Lord Adare

Monday, January 12, 1835
Stargazing at night till 3 A.M.

Tuesday, January 13, 1835
Marg[t] went into Town to get little Louisa vaccinated by Dr. Murray. Wrote to Hend[er]son. Dispatched letter to Dr Fitton with one of Margt[s] to Miss Smith (in an admiralty cover)

Wednesday, January 14, 1835
A Packet arrived from D.S.

Friday, January 16, 1835
Mr Ebden[5] called asking me to join in a testimony in Honour of Fairbairn[6] to support him against the attack made on him by the

---

[3] John Martin (1789–1854), English artist specializing in Biblical scenes, such as *The Fall of Nineveh* (1828) and *The Deluge* (1837).

[4] He made a special drawing of the great gaseous nebula in Orion, which has many stars involved in it.

[5] The Honorable John Bardwell Ebden, merchant, financier, and member of the legislative council, lived at "Belmont," Rondebosch, now St. Joseph's College.

[6] Dr. Robinson, in addition to many other historical notes, writes the following:—

The declaration in support of John Fairbairn appeared in the *South African Commercial Advertiser* for 28th January 1835, but without Sir John's signature. It resulted from an attack on Fairbairn in the *Grahamstown Journal* of January 2, 1835, signed by 335 settlers. This charged him with misrepresentations in his paper, which, with his visit to the Frontier in 1830, were claimed to be among the causes of the confederacy among the Kaffir chiefs which threatened the

Grahame's town people. Assented most readily, regarding as I do, Fairbairn as a model of newspaper editors or at least as a most useful and valuable writer; who never caters for base passions & always advocates the broad & high moral side of every question

Sate up very late (till 5 A.M.) and inter alia Drew up a proposed form of Resolution relative to Fairbairn & enclosed it to Ebden in a letter of both of which kept a Copy.

Tuesday, January 20, 1835
Swept till Moon Rose

Thursday, January 22, 1835
South Easter the most violent said to have been for 20 years— really tremendous   Swept till Moon Rose

Friday, January 23, 1835
Lieut Worster dined here. After dinner Swept and found certain nebulae which he saw

Sunday, January 25, 1835
Went to Church with M. at Rondebosch Sent carriage for Mrs Maclear to appear at Church. After Church, accompanied Mr & Mrs M. to Observatory where took tea.

Monday, January 26, 1835
M. Letter writing all day & night

Lieut. Worster called T.T.L[7] w[h] he did but past

[Last words erased—as if the Lieutenant had outstayed his welcome]

As we had just finished our ev[eni]ng dinner Major Macdowall[8] called & staid till 6±

Colony. (The 6th Kaffir War broke out in December 1834). The suppression of the *S.A. Commercial Advertiser* was urged.

Fairbairn commented (January 10) that few of the 335 signatories ever read the *Advertiser*, for only 25 copies went to the Eastern District, and that, if they had, they would have known that he had been urging the government to devise some new plan for the security of the frontiers. In his strongest leaders (August 27 and 30, 1834) he blamed, not the settlers, but the state of the law, the weakness of the government, the laxity of Frontier authorities, and the "bloodthirsty disposition of certain individuals," as responsible for the situation.

[7] To take leave.

[8] Major McDowel was an Indian visitor.

Mᵗ called on Mrs Blair & Mrs. Stewart.⁹

Tuesday, January 27, 1835

M. continues her perverse system of letter writing which seems endless

Admired with Margᵗ the magnificent display of the Cloud rolled by the N W wind over the Table Mⁿ (NB. The sky falling) like a vast mountain of snow falling over in wreaths or cotton bales developing, then rolling down in Avalanches & as it were rising again after rebound at a distance & drifting out over the valley across our Zenith, rapidly as they approach, but v. slowly at first and as if emerging from some vast cauldron in Pussy's Hollow. A most extraordinary & all but supernatural effect of light shade & perspective [sketch of cloud over Table Mountain].

Wednesday, January 28, 1835

Sent letters in for Exmouth M. called on Mrs. Menzies H. reduced Sweeps—Evaporated Sol of Agaric & other chemicals. Box arrived per Sherborne. Opened it while M was gone to Mrs. Menzies.

M. still letter writing!—will this go on to Eternity?!

NB. Last night M. dreamt that "Dr Brewster was dead, after being Eulogized the week before in Church"

το γαρ θ'οναρ εκ Διος εςιν¹⁰

Now then fiat Experimentum

Thursday, January 29, 1835

Accompanied Mr Schömberg to Cape Town and executed bills G. H. I. J. to the amount of 350 £ by which Mr. Reitz's mortgage on the Feldhausen Estate is discharged.

Conferred with Neethling about *his* mortgage.

M stupid & lay in Bed, evidently tired out with letter writing—complains of ill health &c

Found in No 28 of the Phil Mag.¹¹ a notice by Rumker¹² of the

⁹ Presumably Mrs. Steuart, the wife of the sheriff.

¹⁰ Homer *Iliad* i, 63: καὶ γάρ τ᾽ὄναρ ἐκ Διός ἐστιν (for a dream, too, is from Zeus).

¹¹ The *Philosophical Magazine*.

¹² Carl Ludwig Christian Rümker (1788–1862), German astronomer who served in the British Navy, 1807–1817, was director of the Hamburg School of Navigation, 1819–1820, and was astronomer at the Paramatta observatory, New

Halley[13] Comet's place for Decr Jan Feb . . . April.—Set 20 feet on it. At Twilight swept for it. and then, turning the telesc. out of Merid followed & thoroughly examined the region where it is. Could find no trace of it—Afterwards went on with the Neb. in Orion with a new polished mirror (w^h was also used for the Comet)

And finished with a Southern sweep.—NB This mirror is exquisite both in figure & polish.

Friday, January 30, 1835

M's letter-writing seems to have exhausted itself, at least pro tempore.

Repolished the Shortfocussed mirror by *mere* rubbing with "Polish-

---

South Wales, 1821–1830. He produced a preliminary catalogue of southern stars, 1832, and a catalogue of 12,000 stars, 1846–1852. He died in Lisbon.

[13] Edmond Halley (1656–1742), British astronomer and physical scientist. Halley was educated at Oxford, and, becoming friendly with Flamsteed and Hooke, assisted with the design of the Royal Observatory at Greenwich. He left Oxford without a degree and went to the island of St. Helena to make astronomical observations. It was at this time that, in an excess of Royalist zeal he defined the now-obsolete constellation of Robur Carolinum, but it led to his obtaining an MA degree from Oxford. He was friendly with Isaac Newton, and predicted the return of the comet, now called after him, in 1758. This was an event of high scientific importance since it showed that comets, long thought to be erratic portents, were amenable to physical laws. Halley's comet, last observed in 1910, has a mean period between apparitions of 76.2 years, but these can be displaced due to perturbations of the comet's motion, particularly by Jupiter. Apparitions of Halley's comet have been identified back into the earliest historical times. One was expected in 1834, and, as the diary shows, was eagerly awaited by the astronomers of the time.

Halley had very broad interests, and made important contributions in the fields of magnetism, meteorology, and oceanography. The connecting link is probably the attempt to find a means of navigating the globe, other than by the determination of longitude (see notes 22, November 30, and 55, December 26, 1833). Halley undertook the first expressedly scientific voyage (1698–1700) as commander of the pink, *The Paramour,* and made observations of the kind listed above. In 1703 he became Savilian professor of astronomy at Oxford, and in 1720, Astronomer Royal. Starting in the sixty-fourth year of his life he undertook, successfully, to observe the motion of the Moon throughout a cycle of its nodes (18 years). His other scientific contributions include the discovery and analysis of the proper motions of stars, and the acceleration of the Moon's motion. He was the originator of the proposal to use observations of the transits of Venus across the solar disc (very rare events) for the determination of the solar parallax.

ing paste" & a leather—as an experiment. It is so tarnished by depo-
sition of moisture it cannot be wiped

M. greatly refreshed by a ride round by Newlands & the Brewery—
the Vapours of innumerable letters written & received begin to Dissipate themselves. We have now hopes of her

Saturday, January 31, 1835

Finished the "Last days of Pompeii"[14] Actinometer observed at intervals from 10 till 3 the sky being fine & blue to a wonder, & the Dry-
ness of the Air extreme

Mr & Mrs Hutchinson called—H says. "Fairbairn would have
united all voices in the colony in his favor but for his paper of
Wednesday last"[15]

Drove out with Margt to the flats towards the Camp Ground[16]
whence a fine view of the T. Hill

Ludwig[17] came back, his report of his proceedings is ludicrous
enough. He says—"Plenty drink at the Cape (i.e. in Cape Town) but
no money & nothing to eat"—comrades bad—&c. in short he is glad
to get back—(L. garden boy) Swept till 12. when the wind suddenly
chopped round to S & the sky became overspread.

Sunday, February 1, 1835

Attended divine Service at Rondebosch with M. M. went over to
Observatory.

I walked back home.

Swept till 11[h] Sid T.

[Remarks continued on opposite page of diary]

---

[14] *The Last Days of Pompeii*, a novel by Lord Lytton (see note 75, April 21,
1834), was published in 1834.

[15] Dr. Robinson remarks that it is a little difficult to know what inspired this
statement. Fairbairn's leading article on Wednesday, January 28, speculated on
the consequences which might follow if the Whigs were in opposition in Britain.
He thought that they would urge improvements in the colonies, hastening of
compensation money to be paid after the liberation of the slaves in 1834, and
voting of funds for the relief of distressed settlers—but the Cape must not be a
divided people. He also published a signed letter thanking those who had defended him against the attack in the *Grahamstown Journal*, and reviewing his
past advocacy of freedom of the press, the slave question, and the vagrancy law.

[16] A common or open space which still exists, giving a fine view of T. (i.e.,
Table) Hill.

[17] The garden boy, not Baron Ludwig.

Last days of Pompeii

Heroine a trifle insipid. Primitive Xtians not very invitingly depicted *rather* fanatical, & fierce (*for* Xtians)

Witch of Vesuvius—effective except in her disappearance. Her glimpses of the lava however poetically imagined, physically impossible.

The Girl madly strong with the excitement of the Gladiators ("Ho Ho for the Merry Merry Show") however shocking, true to nature, as a form of vulgar love of the horrible & the exciting heightened by an excitability skirting on insanity.

A general want of *good* characters. In one place L. B's notions of the political Summum bonum peep out & are *chaotesque* his idea seems to be that a great nation is a great evil & that the perfection of human society is to be only attained by breaking down States into the smallest fragments that will hold together i.e.—to *disintegrate* to the very verge of *decomposition.*

Monday, February 2, 1835

Rose at 12.—

Chemistry.—Analysis of the Mucilage of the Hottentot fig fruit.[18]

Lieut Eady called, on his return from the expedition with Dr. Smith's party by reason of an accident & near loss of right hand.

Swept till 12^h Sid T.—

Tuesday, February 3, 1835

Further Examination of the Hottentot fig mucilage—Mr. Murchison called P.P.C.[19] Rode into Town to call at Gov^t house and to attend Mrs Murray's[20] funeral (A). Joined M^t at Mrs Rabys. & returned with her after setting down Mr Hutchinson at his home & taking up M

Captn Wanson of the Sherborne dined here. A cloudy night prevented the examination of his sextant which seems to have puzzled him tho' I find no fault in it.

[Opposite page; the reference is to the point (A) above]

(A). Tuesday.—At the Cape all male friends *generally* are invited to attend funerals & join the procession. They are shewn into the

---

[18] *Carpobrotus edulis* (Alizoaceae).

[19] *Pour prendre congé*, to take leave (T.T.L.).

[20] Mrs. Eliza Murray, wife of Dr. John Murray, died on February 2.

largest room in the house where they wait till the corpse is hearsed & the Chief Mourners arranged to follow—then all fall into procession (on foot or in Carriages) In this case we went all in Carriages which being confusedly arranged & *not announced* caused the fracture of several mine among the rest. Arrived at the Church yard little order is observed no processional line kept & there (the service over) they break up & depart each to his own home.

Wednesday, February 4, 1835

Went in to attend the Meeting of the Phil & Lit Socy. and a committee of the S.A. exploring Exped[n] to examine Smith's 1st remittance of specimens. Chiefly Birds & Beast Skins.—

Accomp[d] Maclear to Eady's[21] who shewed us some of his Curiosities collected on the way home Among the rest 2 beautiful tame squirrels which stand on end [tiny sketch of a squirrel] thus and leap up vertically. A natural magnet.—Silicified Bones &c

Mr Malcolm Drank Tea here PPC. *per Sherborne* A remarkably cold night for the time of year

Swept from Moon Setting to 11½ ST.

Thursday, February 5, 1835

In Even[in]g rode out with M far down on the flats. Scarlet Watsonia Made a long nights Sweep, and the night being most superb— the mirror brilliant and the zone swept (147 148 140) the richest perhaps in the heavens—attained *the sublime of Astronomy*—a sort of *ne plus ultra*. For particulars vide Books—but it is an epoch in my Astron[l] life.

Friday, February 6, 1835

Rose very late Wrote a long letter to McLear about the verification of the Circle[22] by the use of Collimators

---

[21] Captain William Edie, who had been on Dr. Smith's expedition.

[22] He is still concerned with the instrumental errors of the mural circle at the Observatory. He now proposes to study its behavior by means of collimators. These are telescopes, one to the north and one to the south, fixed horizontally and focussed to infinity, aligned with each other and with the telescope of the mural circle, when the latter is horizontal. With the mural circle telescope out of the way (which can be achieved by turning it vertically and looking through ports in the sides of its tube), the two collimators are lined up on each other. Looking through the eyepiece of either, the observer will see the focal plane of the other, and moveable cross wires in their focal planes can be adjusted so that each coincides with its image as seen in the other. When the telescope of the

Chemicals a litter—Spent day wholly at home

Saturday, February 7, 1835
Chemicals [next words erased]

Sunday, February 8, 1835
Made Church at home being a desperate pouring Rain—which lasted all day—wrote to Maclear about a fresh microscopic Exam^n of the Circle Wrote to Henderson Arranged all my correspondence

Marg^t commenced a *regular* "Baby book" on an "entire new system"

Monday, February 9, 1835
Rain a little in Morn[in]g but on the whole a tolerable day—Got out & gardened—Laid down as an Exp^t a long young shoot of the Great Bamboo to see if it will root & propagate.—The Canes do so very readily.

Rode over to Observatory
Back & Equatorial work till 3 when the stars grew tremulous

Tuesday, February 10, 1835
After an early dinner or Tiffin[23] (with Mrs Stewart [? Steuart] & her children who had come out from a Vaccination party in Cape Town) rode with Maggie up to Hout's pass and home by Moonlight
Equatorial work in evening but it soon clouded.

Wednesday, February 11, 1835
Went in to Town to "beg off" attending the meeting of SA Exped^n Subscribers to sell Dr Smith's specimens, being a mere commercial affair—

---

mural circle is turned alternately north and south, a cross wire in its focal plane can be brought into coincidence with the image of the cross wire of the collimator. If the instrumental axis is tilted, there will be a constant difference of setting with telescope north and telescope south. In the case of this particular instrument, since, as was discovered many years later, the bearing was loose, there was presumably an erratic difference between the pairs of readings. In fact, the mural circle had been damaged when first shipped to South Africa, and was never persuaded to perform satisfactorily. In February 16, 1835, begins another series of references to the circle, this time to a plan to observe the wandering of the end of the axis by means of a fixed microscope. (See also note 55, March 26, 1834.)

[23] A light meal, a word most used by those with Indian connections but in fact derived from the obsolete English *tiff*, to eat between meals.

Rectified 20-ft Collimator

Equatorial work with a superb defining night till Midnight when the Clouds became too thick to go on

Observed ϵ Argus double

Thursday, February 12, 1835

Outlined for M[t] a most beautiful Pink Amaryllis[24] which I had watched growing some days.

Rode with M after dinner to call on Lady D'Urban at Wynberg— hot Dinner & then out—Then round by Versfield's and home.

Observed ϵ Argus, T Argus & λ Argus all double & all just alike in pos[n] & dist. This startled me &, examining the matter carefully, I discovered the whole to be an illusion, originating in the Object glass w[h] gave appendages so like stars that any one might be deceived.#

[Continued on opposite page]

Feb. 12. Thursday continued

This arose from an attempt to readjust the glass, by loosening the cell & shaking it.—I succeeded in correcting or at least so palliating it by another shake as to be no longer troublesome

Friday, February 13, 1835

Occupied the Morning in planning out & in part clearing a Walk through the NW. shrubbery, which proved a fatiguing job & thoroughly tired me

M. at work with her Pink Amaryllis the most beautiful flower I have seen at the Cape

A S.Easter—in early part of Ev[en]i[n]g Clouds but these cleared & left blue sky & steady pretty strong SE wind

Equatorial work till 2 AM when the stars grew tremulous. Therm 63.

[Opposite entry for February 14]

Pelargonium Ciliatum?[25] in flower and the old leaves still green

---

[24] *Amaryllis belladonna*: Belladonna Lily (Amaryllidaceae). This widely cultivated Cape species has a magnificent umbel of pink, fruit-scented, trumpetlike flowers.

[25] Pelargonium Ciliatum?: the present name is *P. longifolium*. Herschel's plant may have been a late-flowering form. Most of the garden forms of *Pelargonium* (popularly, geranium) have been bred from South African species.

Sunday, February 15, 1835

Attended Divine service in Rondebosch Ch<u>ch</u>

Monday, February 16, 1835

Rode over to Observatory to observe with Maclear the central dis-placement of Mural Circle by a microscope on my proposed plan. At 3 Maggie joined us Dined & came home terribly late by feeble Moon-light in a hard Southeaster no joke—lost a superb sweeping night by this Expedition—NB never quit the observatory henceforth later than 4 PM

Tuesday, February 17, 1835

A day at home, occupied chiefly in Gardening.

Cleared more of the Shrubbery walks.

M. Drove out in Evening to call on Lady C. Bell, & others.

Worked a little at Monograph of 30 Dorad.[26] Observed tonight. Stars not over well defined. Looked for Comet not found, in or near Rumker's place. Exam<u>d</u> η Octantis n s d.[27]

Wednesday, February 18, 1835

Went into Town to Anniversary meeting of Infant School. Adam-son made a good speech—Returning deviated to the woods behind Rondebosch—there no ["no" on end of line] noticed a falcon stupefied with eating too much dinner, perched on a tree. Rode as fast as I could home for a gun—got it—returned—and shot him at 20 or 25 paces—but he flew clean off with the whole charge in him

Thursday, February 19, 1835

Rode over to Observatory and observed with Maclear with a Cen-tral Microscope the displacement of the Mural Circle.—Botanised in returning & dug 27 roots of the Great pink Amaryllis*—Returned and being a bad night & much tired *re*tired early—

The Centre of the C[erased]

[Continued on opposite page] *Thursday. 19.*

The upshot of our examination is that the Centre of the Circle de-

* Amaryllis Belladonna

[26] 30 Doradus, the magnificent gaseous nebula in the Large Magellanic Cloud (the Tarantula Nebula) drawn in his letter to Caroline of June 6, 1834.

[27] Not seen double.

scribes a triangular or horned Curve thus [a rough sketch] with very unequal velocity the great excursions lying wholly between 0° and 180°

Friday, February 20, 1835

At home all day. Wrote to Airy—Captn Henning[28] of the "Duke of Buccleugh" called with a parcel from Dr Stewart to Maggie

Got a couple of hours sweeping zone 159 but almost as soon as Cap^{tn} Henning took his place in the Gallery it clouded & began to rain

Saturday, February 21, 1835

A dull cloudy & *quite cold* day. Therm 66 at Midday a Black S. Easter![29] Dr Murray called & says it is blowing fiercely in Town.— He is on the point of setting off for the frontier

Sunday, February 22, 1835

Staid at home & wrote to my Aunt No 6 and to Airy No 3. Capt Henning called—Made up a little packet of Specimens of the luminous Mollusca caught Dec 10 near the line to send by him to Buckland.

Maggie went into Town to hear Dr. Philip preach.

Swept till 2 AM.—

[Opposite page]

Bulbine falcata[30] the flower stalks are now 5 or 6 inches high above ground & the flower heads swelling

Albuca Viridiflora[31]—New leaves have shown themselves & are now 3 or 4 ^{in} high

Eriospermum Lanceaefolium[32]—flowers tall & long & slender 6–8^{in} high

Haemanthus Coccineus[33] flowers some in perfection, but others withering no leaves yet shewing themselves

Pelargonium Ciliatum[34] putting forth young leaves

---

[28] Presumably identical with the captain of the *Windsor,* the ship on which the Herschels returned to England.

[29] Normally a fair-weather wind, the southeaster sometimes brings low clouds, occasionally with rain.

[30] *Bulbine falcata* R. & S., now known as *Anthericum falcatum* L. f.

[31] *Albuca Viridiflora*, now known as *Albuca spiralis* L. f.

[32] *Eriospermum lanceaefolium* Jacq.

[33] *Haemanthus coccineus* L.

[34] *Pelargonium Ciliatum*: now know as *Pelargonium longifolium* Jacq.

A Letter from Sir John and Lady Herschel, Feldhausen, to Caroline Lucretia Herschel, Hanover

N⁰ 6

Feldhausen near Wynberg
Cape of Good Hope

Feb. 22. 1835   Rec^d May 11 [In a different hand]

[In Sir John's hand]

My dear Aunt.

Your letter N⁰ 2 of Septr 11 arrived with a large budget of English News by the Zenobia and I cannot tell you what pleasure it gave us both to see your fine handwriting, as firm and distinct as ever, and to read your kind and unaffected strait-forward sentences. The account you give of the heat of last summer at Hanover is really surprising, and you must have felt it severely, though I hope not permanently. We are all here *perfectly well*, the threatened damps & rains of what they call winter here, having proved anything but injurious. Marg^t is still nursing *the new baby* who grows fat and *never cries*. Little Caroline can now read pretty fluently and begins to use her needle to some good purpose—Bella is learning of her, and little Willie runs about and shouts and climbs and bangs everything about with extraordinary energy and perseverance. He is in excellent health though very thin and is growing very like his Grandfather, especially about the mouth. For my own part I never enjoyed such good health, in England as I have done since I came here.—The first coming on of the hot season affected me a little (odd enough with colds and rheumatisms) but it soon went off.

The Stars continue to be propitious, and the nights which follow a shower, or a "black South Easter" are the most beautiful observing nights it is possible to imagine.—I have swept well over Scorpio and have many entries in my sweeping books of the kind you describe— viz: blank spaces in the heavens *without the smallest star. For Example*—RA.[35] 16^h 15^m—NPD[36] 113 56—a field without the smallest star

[35] Right Ascension, that celestial coordinate corresponding to longitude on the earth. It is measured by the sidereal time at which the given object is on the observer's meridian.

[36] North-polar distance: angular distance from the north pole of the heavens.

RA. 16ʰ 19ᵐ—NPD 116° 3'—*Antares* (α Scorpii)
      16   23 —114 25—to 114 5—field entirely void of stars
      16   26 —114 14—not a star 16 m.—Nothing!
      16   27 —114 0—D°—as far as 114 10—

and so on—then come on the Globular Clusters—then more blank
fields—then suddenly the Milky Way comes on as here described
(from my Sweep 474. July 29. 1834).

"17ʰ 28ᵐ—114° 27'—The Milky way comes on in large milky Nebulous irregular patches and banks, with few stars of visible magnitude,
after a succession of blank fields and extremely rare stars above 18ᵗʰ
Mag.—I do not remember ever seeing the milky way so decidedly
Nebulous, or indeed at all so, before . . ."

Altogether the constitution of the Milky way in its whole extent
from Scorpio to Argo Navis is extremely curious and interesting. I
have already collected a pretty large Catalogue of Southern Nebulae
for the most hitherto unobserved, but my most remarkable object is
a fine Planetary Nebula[37] of a beautiful greenish-blue colour, a full
and intense tint, (not as when one says Lyra is a *bluish* star &c) but a
positive & evident blue, between indigo-blue and Verditter Green.
It is about 12″ in diameter, exactly Round or a *very* little Elliptic,
and quite as sharply defined as a planet. Its place is 11ʰ 42ᵐ RA and
146° 14′ NPD. My review for Double Stars goes on in Moonlight
nights, and among them I may mention γ Lupi & ε Chamaeleontis, as
among the Closest and most interesting.

I have been hunting for Halley's comet by Rümkers Ephemeris in
Taurus but without success, though in the finest sky, quite dark, and
with a newly polished mirror. (By the way I should mention that I
have not had the least difficulty in my polishing work, and my mirrors
are now more perfect than at any former time since I have used
them). My last Comet hunt was Feb. 18—I shall however continue to
look out for it. Pray mention this to Schumacher, who is Rümker's
next door neighbour.

Speaking of Profʳ V. Schumacher, I shall within a day or two con-
sign *to him*, at Altona, for you, a case of Constantia Wine of the best
quality that is to be got here (and right good it is) hoping it will

---

[37] The modern designation of this object is NGC 3918.

reach you safe and prove to your liking. I should have dispatched it before but was in expectation of a ship direct for Hamburg touching here in her way from Singapore. I shall rely on his kind offices to receive and forward it to you, as I know no one else at Hamburg, and I shall give orders that he shall be put to no charges by it.—Peter Stewart has forwarded to me the Göttingische Gelehrte Anzeigen" with Gauss's[38] reviews of my essay on the study of N. Phil[39]—and my Astronomy for which I am much obliged to you. In Both there are useful hints which I shall take advantage of in future editions. I am sorry to hear of Prof[r] Harding's death.

And now before I conclude, do let me once more intreat that [you] will not again *send back* the money which you draw from Mess[rs] Cohen & Co, and which ought to be devoted entirely to your own personal comforts. I do assure you it grieves me to the soul to hear of such repeated sacrifices on your part and Maggie bids me add her intreaties to mine that you will henceforward *retain*, and *use*—those means which are at your disposal—(for both our sakes)—in providing those comforts and conveniences which your increasing years render necessaries.—And so God bless you, for us both—and believe me ever   Your affect nephew

J.F.W. Herschel

[Lady Herschel writes on page 1 upside down on the top]

My dearest Aunt—Let me say God bless you too—& add one thing to satisfy you on something you mentioned in your last letter—My whole energy will be bent on making Carry especially a very Hers= chel—viz—a wise man's daughter—distinct in head & thought as in penning letters & figures, & no wife ever had a better assistant than I have in dear Papa—Y[r] affect. Niece—

M.B. Herschel—

---

[38] Johann Carl Friedrich Gauss (1777–1855), German mathematician and scientist, founder of modern mathematics. He is usually ranked as one of the three greatest mathematicians who have ever lived.

[39] Natural philosophy.

The Diary: February 23 to May 17

Monday, February 23, 1835

Dispatched a Case of Constantia to my Aunt via London & Hamburg per Prof Schumacher   wrote about it to S.[40] and to P S.[41] to whom enclosed Bills of Lading (or at least ordered them to be incl^d by T & W[42]

Tuesday, February 24, 1835

Swept till 2. Disc^d my 4th   Plan Neb.

Wednesday, February 25, 1835

Major Macdowall [McDowel] called #
Miss Crybbace spent Evening & Night here
Swept till 2

[Entry continued on opposite page]
Wedn 25th [erased]
Macdowall [McDowel] gave us a taste of his notions of the Sequence of Events—He resolves all (shortly & simply) into an overruling & ALL-superintending-&-directing Providence—This doctrine carried out into its minute detail is *equivalent* TO (I don't say *identical* WITH) necessity.

Quoad nos it comes to the same. & his illustrations were to the point—In a Massacre for example—some are sent to kill—some to be killed—good again—who can gainsay.—But then say I in a Tyranny —some are sent to be the tyrants—some the slaves—In the great feast of Nature some to eat—some to be eaten—& so forth. Now when it can be said of a religious—or moral—or Metaphysical theory that granting its truth, or denying it, makes not the slightest point of difference in any one matter of practical conduct—the best & shortest way seems to me to be—to move the previous question.[43]—*My theory*

---

[40] Schumacher.

[41] Peter Stewart.

[42] Thomson and Watson, shipping agents, Cape Town.

[43] A parliamentary motion that the pending question be put to an immediate vote without further debate or amendment and that a defeat of the question

in such points is the only maintainable one—to resolve all into one principle & that one to be "The sequence of Events"—Show me an event that has ever happened otherwise or in contradiction of this principle!!!

Thursday, February 26, 1835
[Written vertically with reference to following entry]
#The names of Truter & Bestandig figured in this relation Arcades ambo[44]
worked
Maclear came to make a sweep (See opp page)—At Supper he regaled us with a rich account of his being challenged by a French officer[45] de Sa Majesté—Officier de Marine, for uncourteous conduct (which by his own account appears to have been the fact #

[Entry continued on opposite page]
Thursday Feb 26—
Maclear came this Ev[en]ing to make "The Great Sweep" with me. The Night turned out glorious! Clear as crystal and pure as Æther. A finer night for Definition was never seen. So we had it all our own

---

has the effect of permitting resumption of debate. Normally it is a maneuver for terminating a discussion.

[44] Arcadians both.

[45] After a night's observing, Maclear was told that there were visitors to the Observatory, and, sending a message that they be shown into a room to wait, he descended. He found them interfering with the instruments, and evidently spoke sharply. The visitors left, but later a note appeared—"written on a slip of dirty paper & sealed with a bit of wax with the impression of the thumb!"

23[d]. Feb 1835
"Sir!
As I have been at the Royal Observatory being a lover of the Science & being insulted by your appearance If it is your intention by so doing and as you are Officer of the navy and I Commander of the French man of war the Madagascar at present in Table Bay I wish to have satisfaction for your improper conduct & request for an answer
I am, Bosse,
Commander of the French Ship Madagascar"
The duel was not fought. Maclear debated the matter with himself, considering the interests of his family and the terms of his insurance—which evidently did not cover this kind of contingency. As much as anything else, he was influenced by the low social standing of the Frenchman's seconds, Bestandig and Truter, one of whom was the baker's assistant. Captain Bosse was talked out of his homicidal intentions by the intervention of a British naval officer.

way & Maggie, Maclear & I had a perfect astronomical regale. We took in my two Planetary Nebulae (the 1st & the Blue one)—the *sublimest* part of the Milky Way—the Great Nebula (η Argus) and the Superb Cluster[46] which follows it, and β Crucis with its Sanguine Companion all with a fresh Polished Mirror in perfect condition

Retired at 3½ AM

Friday, February 27, 1835

At 9½ AM. went into Cape Town with Schömberg to execute the transfer of the Feldhausen Estate, which was accordingly completed (all but the *official* formalities)—i.e. Mr Schömberg signed the transfer which I began to think would never happen.—So now we are bonâ fide and "per legem tenae" Cape Proprietors—My!!—We know what we are—but we know not what we may be. Swept till 10^h 17^m ST. when finding myself & Stone regularly worn out and that to go on a fresh sweep must be set gave in reluctantly, the night being very fine

[entry continued on opposite page]

Friday—Lobelia Triquetra[47] in profusion, in flower on Wynberg Slope—NB   its milky juice is full of elastic Gum

Sunday, March 1, 1835

Planted out bulbs of the Small Ornithogalum   Mr. Trotter of the Bengal Civil Service called with an unintelligible Sanscrit paper from Prinsep.[48]

Went to Rondebosch Church,[49] too late, in at 2nd lesson.—Returning called at Mrs. Blair's

Monday, March 2, 1835

After dinner took bearings from the Road at end of the SW (Newlands) Avenue. Returning about dusk found Mr & Mrs Hutch-

---

[46] NGC 3532.

[47] Now known as *Lobelia comosa* L.

[48] James Prinsep (1799–1840), architect and orientalist, an important authority on Indian antiquities, working for the states of Calcutta and Benares. He was a contributor and, later, editor of *Gleanings in Science,* which became the *Journal of the Asiatic Society of Bengal,* of which Prinsep became secretary. He was a brother-in-law of Julia Margaret Cameron, the eminent photographer who took so many sensitive likenesses of the famous Victorians, Sir John included.

[49] This building, designed by Major C. C. Michell, had been opened on February 16, 1834. It occupied the site of the present St. Paul's Church.

inson.—Learnt that *since* Dinner a joint of Beef has been stolen from the *Safe* (now proverbially become the unsafe)

Tuesday, March 3, 1835

Had all the Servants in & Lectured them about the disappearance of the Beef.—The true culprit discovered in the person of Breda's Milk boy.

Wednesday, March 4, 1835

M[r] Hough M[rs] Ellerton M[r] Maclear breakfasted at Feldhausen and after Breakfast Cap[tn] & M[rs] Wauchope,[50] M[rs] Thomson and ? Miss Joined in from Simon's Bay & spent the morning here.

Swept in Ev[en]ing till 12 but bad definition is at last come on & no good could be done

Thursday, March 5, 1835

Read in Lichtenstein's[51] travels—he mentions "De onwetende Fontein aan der Daunis Kloof" in the Hantam District one days journey from Abraham de Wyk's place as abounding in fossil fish 3 feet long in slate stone where the spring rises.

[Opposite page, a sketch of a hand with a pointing finger indicating the above item and the words]   Fossils—locality

[50] Captain Robert Wauchope, in command of H.M.S. *Thalia* and flag captain to Admiral Campbell. He invented the Time Ball for ascertaining the rates of chronometers "in use at Greenwich Observatory and at Portsmouth, St. Helena and the Cape of Good Hope" (*Dictionary of Naval Biography*, 1849). A conspicuous black sphere hoisted up a mast was dropped at noon or other preconcerted time to serve as a time signal. Wauchope's sister married Captain Patrick Campbell, appointed admiral at the Cape in 1834 (see note 104, May 30, 1835). Wauchope was a devout man whose religion stood in the way of his advancement, his refusal to allow prostitutes on board ship being regarded as prudish. At his death he was Admiral Robert Wauchope. The Wauchopes become close friends of the Herschels.

[51] Dr. Henry Lichtenstein, professor of natural history in the University of Berlin, indulged a desire to travel by becoming tutor to a son of Governor Janssens, who was appointed by the Batavian government as governor of the Cape in 1802. This was after the termination of the first British occupation and before the second, both being motivated by strategic considerations connected with the Napoleonic Wars. Lichtenstein's works, published in translation in two volumes in 1812 and 1815, describe his travels, which took him north into the Roggeveld, into the Karoo, and as far east as Graaff Reinet. The Hantam River runs just north of Calvinia. Downes is about eleven miles east of Calvinia, and could be "Daunis."

Sunday, March 8, 1835

Attended Divine Service at Rondebosch NB Caffer Subscription Sermon[52] by Mr Judge—

Collected 19 Bulbs of the Haemanthus Coccineus—

A Desperate Driving of Cows Bulls & Horses out of the Grounds—

Equatorial at Night till 3½ AM.

Discovered π Lupi to be close double $*$

Monday, March 9, 1835

Rose at 11½

Lady Compton & Mrs Grant & Mr Dunlop (Indians) called. Lady C is on her way to England per Victory ? and sails on Wednesday.

Marg^t went over to Observatory

Equatorial at night till 4 AM.

Monograph of χ Crucis[53]

Tuesday, March 10, 1835

Rose Late

Outlined per Camera the great Candelabra[54] bulb for M. A most magnificent flower 2 feet in diameter (i.e. the head or hemisphere which it fills).

Mr Cochrane—Mate of the "Wolf" (? Lion) called on return from China dined with us, Drove out round the back of Newlands in his Buggie after dinner—At Tea Mr Versfeld & brother & Mr . . . [his omission] his friend came by invitation to view the Moon (See opp page)

---

[52] The Arabic word, *Kaffir* or *Kafir*, meaning an infidel, is sometimes used, often with derogatory implications, to refer to members of the Bantu tribes of southern Africa. The "Caffer Subscription Sermon" was in aid of the fund for "relief of the destitute sufferers by the Caffer invasion," and the collection realized £20.2s 7d (about $100 at the prevailing rate).

[53] Surely he means κ, the cluster famous as "Herschel's Jewelbox."

[54] The plant usually known in the Cape as the Candelabra Flower is *Brunsvigia orientalis* (L.) Ait. ex Eckl. Herschel's description is not quite clear; there are, in fact, about twenty flowers, borne on long pedicels in a very large inflorescence. The flowers are dark red, and about two inches long. One can understand Sir John's enthusiasm.

[Opposite page]

#*Tuesday contin^d*

When they were gone, I joined Margt & Mr Cochrane at Mrs Menzies, who had a dance.

NB   Superb Moonlight effect of the tree-grouping in front of their house.—What a country this for landscape gardening

Wednesday, March 11, 1835

Captain Edie 98th called PPC. gave him letters of Introd^n to Baily and Beaufort. Wrote to Johnnie Stewart

[Opposite page]
Wind NW in Mo[rn]ing
    NE in Ev[en]ing, bringing fog
    NW gusts at night

Thursday, March 12, 1835

Sent in to Captn Edie a Note of Introd to Mr Urquhart & to Mrs Lestock Wilson. Packed up & sent in Captn Loyds[55] papers.

Baron M^rs & Miss Ludwig called

Col & Lady C Bell, Mr & Mrs Stewart, Mr & Mrs Menzies, dined here, also Mrs Smith, & Major Macdowall#

Viewed stars & Moon

[Opposite page] # Col Bell assures me he has seen Venus with the naked eye between 9 & 10 AM every morning for 16 successive days

Friday, March 13, 1835

Captn Biden of the Victory called & brought a tribe of his Passengers (perfect strangers) who broke up the whole morning & would (I thought) never go away—One in particular (name unknown) a remarkably forward, chattering person to whom it was hardly possible to be civil   Rode with M to the flats & Gladioluses & Proteas

Saturday, March 14, 1835

A Roasting hot day   Margt went over to the Observatory & brought back the Circle Book. Measured the NW avenue with Skirrows Chain

---

[55] John Augustus Lloyd (1800–1854), civil engineer, surveyor general of Mauritius, 1831–1849. He had served with Bolívar in Latin America.

Sunday, March 15, 1835

Baby Louisa is observed to have cut a tooth, a fact announced by much crying last night.—Attended Dr Phillips's Church[56] in Ev[en]ing with Margt.

Wednesday, March 18, 1835

Dined at Mr Menzie's—Met Mr and Mrs Blair   Col & Lady C Bell, Miss Cato Liesching—Major & Mrs Loughmore[57]

Thursday, March 19, 1835

Mrs. Hutchinson here.

Friday, March 20, 1835

Mrs. Hutchinson went into Town   Worked at Serpentine Walk in Shrubbery

Received 13 letters from England and J.C.S.'s journal which promises to be a high treat.

Began Sweeping for the Lunation.

NB. Rheumatism across shoulders!

Saturday, March 21, 1835

Today being one of the days of hourly Meteorol Obs.[58] began at 11 AM. & continued the whole day & night registering everything registerable

Looked in Equatorial with Ring Eye piece for Halley's Comet—not found in Rumker's place or near it

Mr & Mrs Hutchinson came

Swept from 8^h ST to 11^h30^m interrupting the Sw[eep] for the Meteorol Register After ☽ tried Equatorial but ✳s ill defined

Sunday, March 22, 1835

Kept at home observing the hourly Met Obs^ns   Especially the Actinom[eter] & got a 1st approx^n to the unit of its scale.

Mr & Mrs Hutchinson here

Swept till 11^h30^m—

[56] Dr. Philip's church was the Union Chapel, Church Street.

[57] This is how it is written, but since the latter half of the name is written over, the correct version would seem to be Longmore, as used again later. This then refers to Major George Longmore, special magistrate, who lived in St. John's Street.

[58] See note 179, December 23, 1834.

Monday, March 23, 1835

A Hot N. wind—Therm at 10 AM 92—Maximum 95 after which it
fell so rapidly that at midnight it was 64! Swept

NB Slept the greater part of the day Mrs Hutchinson at Feldhausen

Tuesday, March 24, 1835

Mrs Hutchinson here

Wednesday, March 25, 1835

Major Cloete & Mrs. Hutchinson dined here—

Thursday, March 26, 1835

Mr Schomberg having now finally vacated his house—took possession & chose a set of apartments for myself having a tiring room,
a laboratory & a kitchen to serve for the furnace department of the
Lab[orator]^y &c. NB. The place is swarming with rats & fleas.

Put Stone & Zomai in possession each of better sleeping apartments

Friday, March 27, 1835

Changing rooms & getting up the shelves & transporting Books
papers & Instruments into my new Study in Schönberg's house. Occupied with this all day—At night reduced Actinom^r obs^ns Tormented beyond endurance with fleas. & could get no rest day or night

Saturday, March 28, 1835

Rose late & with headache due to the intolerable torment of the
fleas (caught in Schomberg's house & places) which prevented sleep
till 5 AM & —

Friday, April 3, 1835

Rode over by appointment to Observatory and thence to Town
with Maclear to inspect the Tide Gage which is now fairly got up at
the pier end & going to work under 10 or a dozen Sergeants of the
98th who attend alternately & keep daily watch on the pier. Met
Captn Bance. Maclear rode home with me & spent the day.—

Marg^t went to see Mrs. Schömberg

Saturday, April 4, 1835

Mrs. Ellerton called & Miss Corrie

Sunday, April 5, 1835

Attended Div Serv. at Rodebosch Church Max Therm 71.5

Monday, April 6, 1835

Min Therm in night 52.0 + 3°6 Cl=10

Called with Marg[t] on Mrs Ellerton & Mrs Corrie. Walked with Mrs E. about the grounds of Van Renen's ("The Brewery") which are of vast Extent, Richly wooded, but in a most unkempt State & merely worked as a trading concern to cut & sell the wood.—[Plate 9]

M[rs] Ellerton, Miss Corrie, Mr Hawkins & his Brother (Charles H) dined with us—NB

Loxton all day

Gibbs all day

Young Loxton[59]

[Opposite page]

#NB Mrs Ellerton has resided 50 years in India & preserves the health vigour & complexion of a fresh European and her intellectual constitution seems as healthy as her physical—alert, selfpossessed, and happy. She is a real "pattern old lady"—

Tuesday, April 7, 1835

Gibbs at work

Loxton & Son *not*

Mr Versfeld called about the Peacocks (NB These peacocks)

Stone reported *one hen* left.

Capt[n] Mackenzie and Mr Frazer called on way from India home. In Capt[n] M. Marg[t] recognised an old Dingwall friend "Mackenzie of Drainie" Frazer is one of the Partners in "Frazer—& Co's House [his omission] at Madras & is going home having realised a fortune.— These calls and superintendence of Gibbs Carpenter (who is fitting my new Laboratory with Shelves) absorbed my whole day. Skirrow also called. At night. Obs[ns] of high & low Stars for the 20 ft Meridian. Max Therm 71.0

Min Therm during night 53.0 + 3.5

[Opposite page]

April 7. Brown and Coachman went [to a] Cape Wedding in the Flats & Brought home a great Serpent a Cobra Capello[60]—

---

[59] Gibbs was the carpenter from the Royal Observatory; Loxton and his son, presumably, were also workmen.

[60] This is simply a literary name of Portuguese origin, Cobra de capello, a

Length 6f $3^{in}\frac{3}{4}$
Breadth of head $2^{in}\frac{1}{2}$
Circumference of Body $5^{in}\frac{1}{8}$
Colour Greenish Brown
2 Poison fangs in each upper jaw
Had him skinned & salted.—
NB. His Bile is green and gives a precipitate of picrotoxine[61] with
sulphuric acid

Wednesday, April 8, 1835

Messrs Frazer, Skirrow & 2 Messr^s Hornblower Juniors dined at
Feldhausen. the latter sons of the Captn of the Mary Ann.[62]
Max Therm 74.0 in Evening [?]

Thursday, April 9, 1835

Min Therm at night 57.0 + 3.5

Friday, April 10, 1835

Wrote to Beaufort   Cl=0 & a SE wind
                                                     the whole 24 hours

Getting Laboratory & tool room in order
Dined at the Mess of the 98th on occasion of Captn Eady & Col^l
McGaskin leaving the Cape per Mary Ann. Got between Eady &
Capt Wallace who amused with anecdotes of his residence in N.
America
Maximum Therm during day 74.8

Saturday, April 11, 1835

Min Therm at Night 51.4 + 3.5
Cl = 0 all night. & morning. Calm
Cl = 0[63]

---

hooded snake. The Cape cobra is *Naja flava,* and the description fits this, though
the color is odd, but the colors are very variable.

[61] Picrotoxin is the bitter poisonous principle in seeds of *Cocculus indicus,* but
this does not seem relevant. Although the synthesis of urea from inorganic materials had been achieved in 1828, an event which marked the real beginnings of
organic chemistry, chemical terminology at this time was far from stabilized.
Sir John seems merely to have coined a name for some ill-defined substance from
twin Greek roots referring to the bitterness of the bile and the poison of the
snake.

[62] This was a merchant ship. Fans of C. S. Forrester may like to speculate
whether this was Horatio on a secret mission, or only some real-life figure bearing the same surname.

[63] Units of cloud on a scale of ten.

Cl = 1 Dense line behind Table Hill & a North fog bank rising
Cl = 9. Rain
Max Therm 74.6

Sunday, April 12, 1835

Min. Therm 50.0 + 3.5
Cl = 10. Rain & NW Wind

Monday, April 13, 1835

Sent off Brown & 2 Riding Horses to Farmer Peck's to be ready to start thence tomorrow, with orders to get a 3ᵈ horse for himself at Wynberg.

Tuesday, April 14, 1835

At 9. En route in the Carriage with Maggie for Farmer Peck's.[64]— Then found Brown & the Horses—got coffee & then mounted & rode 1st across Kalk Bay 2nd Fishhook Bay 3rd Else Bay.[65] Then turned off to right quitting the Simon's Town Road & tracing up the Else River to its sources on the hills—Thence partially descended to a flat & desolate Table Land in sight of the Coast South of Hout's Bay & returning descended on Simon's Town by "the Red Hill" & home to Farmer Pecks along the Sea Side. Kalk Bay is a great indent, deeply skirted with wide sand drift which drives up the hills like snow in the SE gales and overwhelms all

[Opposite page]
(Tuesday 14ᵗʰ Contin) vegetation.—

On the Sand are abundant Ribs, Jaws & Vertebrae of whales whitened by weather It is a desolate scene.—Fish hook Bay & the Road between it & Kalk Bay is skirted with houses of the Whale fishers, and a terrific display of Skeleton shapes it exhibits—Ribs Jaws &c form great fences & Enclosures—nay houses—Roofs, Walls &c.—Winding up from Else Bay (where and in Fishhook and Kalk Bays are bad quicksands[66] in the rainy season) skirt up the *invisible* Else River,

[64] Farmer Peck's was a famous hostelry between Kalk Bay and Muizenberg. Established in 1825 by Simon and Jock Peck, farmers from Oxfordshire, it flourished until the eighties. Otherwise called "The Gentle Shepherd of Salisbury Plain," it was famous for its quadrilingual rhyming signboard.

[65] Else, or Elsie's, Bay is just north of Simon's Town, now known as Glencairn.

[66] There are still quicksands at these places. The village of Hottentots and Malays was probably at Brooklands, now a water catchment area beyond Glencairn.

only proved to exist by a stripe of garden ground in a ravine caused by it, and ascend a dreary & rocky waste where however in sheltered nooks Captain Wolse? & Mr . . . & Mr . . . [his omissions] have got "places" i.e. houses with a few stumpy oaks & firs.—Near Mr . . .'s [his omission] noticed immense numbers of beautiful green sugar birds[67] which haunt the rich scarlet Phlomis Leonurus[68] which grows profusely here, & forms a brilliant contrast.—Beyond pass a few wretched huts forming a kind of village of Hottentots & Malays, & then after leaving on left the Descent to Simon's Town & a fine view of the Bay—descend on the West Side into a Table flat the most barren, rugged and desolate it is possible to imagine. Yet the distant Atlantic gives it a grandeur, and 2 lonely houses (at one of which we stopped) recal human associations & there *is* a waggon track. Here saw a regular Bosjesman[69] & Boy. Very short & small. Matted crisp-curled hair coming over eyes—desperately high cheek bones & most ugly face!!—Both were lying only not asleep by the road when we passed.—Both Ditto Ditto when we returned 2 hours after!!—Maggie took rest & some Bread & cheese at the house of . . . [his omission] The Mistress was out & the key not at first to be had, but she returned tired from a walk. Received hospitably, house room & washing basin (& *very* clean Towels) allowed *me,* who had been bulbgathering (Antholyza . . . [his omission]) but she nor her slaves[70] could speak English, nor understand my German—so no communication

---

[67] Until very recently, many Cape naturalists used the terms "sunbird" and "sugarbird" almost interchangeably. It is only in the last twenty or thirty years that the practice of reserving "sunbird" for the Nectariniidae and "sugarbird" for the Promeropidae has really gained ground. The name "sugarbird" derives from the association of these long-beaked birds with the protea, or sugarbush, the flowers of which are probed for insects and nectar. Sir John is certainly referring to the Malachite Sunbird (*Nectarinia famosa*), which has brilliant green plumage and the characteristic long thin curved beak. The distribution area includes that of the sugarbird, but is much wider. It hawks insects in the air, and may hover in front of flowers, or perch on them. It feeds on the nectar of flowers, particularly those of *Aloe, Kniphofia,* and *Leonotis.* Spiders, diptera, and minute Coleoptera and Lepidoptera, are the main creatures eaten. (See Austin Roberts, *Birds of South Africa.*)

[68] *Phlomis leonurus* L., now known as *Leonotis leonurus* R. Br.

[69] Bushman. Herschel's description of the physical type is accurate.

[70] In fact the slaves had been liberated in 1834, but were to be apprenticed to their former masters for four years.

Wednesday, April 15, 1835

Today threatened in the Morning a N[orth] W[es]ter[71] & it came on before night.—At home all day—Planting bulbs & dressing the bulb garden which now begins to make a promising shew—I have at least 100 Sorts of Cape bulbs in progress, & all have rooted & are doing well.

Thursday, April 16, 1835

A Day entirely Raining, one torrent. Went about noting the effects of the rain in obliterating old & making new channels & cutting up the roads, paths &c.—Also examined the Bamboo & Cane layers & cuttings & put in fresh of the latter as I see the former have struck root at the joints—The joints should be cut off 2 in[ches] below the knot & stuck in so as quite to cover the knot in the moist ground. Dined with M[r] & M[rs] Oliphant in Cape Town. After Dinner Dancing commenced *instanter*. Planted 380 Cane joints

Friday, April 17, 1835 (Good Friday)

Went to Church with M. and thence Drove over to the Observatory with M when she took away Caroline & Bella and I staid the night.— A superb night passed chiefly "laborioré nil agens" in *missing* the transits of stars.[72] However took 8 or 9 Mural Obs[us] with the 6 microsc.[opes] N & S of Zenith and deduced the index errors which shew that odd anomaly Airy complains of.

Saturday, April 18, 1835

Rose at 9½ (NB. went to bed at 5½ after viewing the most superb exhibition of the *blue* line of Mount[s] across the flats before Sunrise which is !!! . . . Breakfasted & rode with M. into the low swampy grounds between the Obs[y] & Sea where Maclear presently contrived

[71] Prevailing winter wind in the western Cape, usually associated with rain storms.

[72] In transit instruments, such as the transit and mural circles, star coordinates are determined at the moment when the star is on the meridian. If this is missed, the observation cannot be attempted until the star comes round again, 24 hours of sidereal time, or 23h56m of mean time, later. The altitude to which the telescope must be turned to catch the transit yields the declination of the star. Numerous corrections are applied to the raw measures. One precaution is to read the inclination of the telescope, which is attached to the circle, by six equally spaced microscopes. The mean eliminates error due to wander of the axis. Airy wrote several memoirs on the behavior of the mural circle, without reaching any definite conclusion (see note 55, March 26, 1834, note 83, May 11, 1834, and note 22, February 6, 1835).

to get into a quagmire out of which *he* scrambled but left his horse saddle deep, floundering. It was an ugly affair and some folks passing comforted him by the assurance that had it been 100 yards further N[orth] along the coast, he would *never* have got out. As it was the horse got out &c.

Rode home shooting at hawks & hammer heads.

[Opposite page] "Duke! Save the Nation willy-nilly"
                            were the last words of Silly Billy[73]
A Bad night got no Stargazing.

Sunday, April 19, 1835 (Easter Sunday)
Went to Rondebosch Church.

Waited 1 hour & ¼ Mr Judge came Then he gave an interminable sermon

Then the Sacrament Service

Altogether 4 hours & more! in a Bitter cold wet day—This gave Maggie a vile cold!

Monday, April 20, 1835
Maggie has decided bad cold, is feverish, & sore-throaty, and can't get up. Yesterdays duty was really severe. The cold was worse than any day last "winter" & the rain drenching

Another cold wet morning—Got [entry terminates]

Sowed 600 Fir seeds skirting the Wynberg road

Began a new set of Sweeps from Zenith S[outh]

Swept till 2 AM.

Tuesday, April 21, 1835
Maggie still in bed with good deal of fever.—Dispatched Brown to Hout's Bay to get Bulbs   he took 2 horses, Bags, Spade, &c and brought back 10, says it is too late the flowers are down & the bulbs not to be found.

Wrote to Prinsep[74] (mentioned my having sent home the plan of Collimation for Murals before I got the N° of the Asiatic Journal in w[h] Taylors[75] paper is—NB This reached on the 11th Inst.—Wrote to D S.

[73] The Duke of Wellington, addressed by his sovereign, William IV ("Silly Billy").

[74] Herschel was urging Prinsep to follow his plan of concentrated meteorological observations at the equinoxes and solstices.

[75] Thomas Glanville Taylor (1804–1848), director of Madras Observatory,

[Opposite page]

Enclosed to Prinsep the Met[1] Instructions—

Worked in garden laying bonfires & preparing to sow firs proteas & acacias

Prepared working list—1st Sweep & worked till 2 AM.

Wednesday, April 22, 1835

Rose late. (M. up again)

M[rs] Maclear called shewed her about the shrubberies &c Marked out the cross walks in the Home Plantation & finished getting it cleaned & trimmed.

Swept till 3 AM

[Opposite page]

Oak leaves begin to fall.

Bulbine Bisulcata[76] seeds ripening

Gladioluses begin to shoot anew

Gladiolus Brevifolius[77] in full flower (& seeding fast).

Brunswickiae Minor[78] New leaves up

Disa Aurea[79]

Disa—Pink (rosea ?)

Disa—Chrysostachya[80]   All coming up

Disa—Graminifolia[81]

Thursday, April 23, 1835

Boiled or stewed Seeds. & got in about 100 Castor oil-plant seeds and in evening planted 1070 fir seeds

The Castor oil plant-seeds were first put into water & brought to a lukewarm heat in which when they had been *not more than ten minutes* they had all germinated, the young roots forming a large

---

1830, FRS, 1842. He published *Madras General Catalogue* of 11,015 stars, 1844. He had evidently also devised a method of checking errors of mural circles by the use of collimators (see note 22, February 6, 1835).

[76] *Bulbine bisulcata* Haw.

[77] *Gladiolus brevifolius* Jacq.

[78] Probably *Brunsvigia minor* Lindl.

[79] Dr. Anthony V. Hall says that *Disa aurea* is a previously unpublished name, but from the context of the entry for October 25, 1835, where it is mentioned again, he thinks Sir John is referring to *Disa cornuta* Sw.

[80] *Disa chrysostachya* Sw.

[81] *Herschelia graminifolia* (Ker.) Dur & Sch.

white double protruberance which increased rapidly before I got them into the Earth.—

Swept till 1½ & then Equ¹ till 3

Friday, April 24, 1835

Rose at 11.

Stewed seeds & got the fire-places &c in the laboratory into something like reasonable order—NB The place so dreadfully full of fleas[82] as to be all but purgatory. Called with M⁺ on Lady D'Urban at Wynberg also on Mrs Hare & Mrs Menzies

Swept [erased]

Worked at Equatorial till 12ʰ ST then joined Stone & got sweeping till 3 A.M.—found a planetary nebula.

Planted	60	Coffee plant seeds	
	70	Acacia Lophantha[83]	Seeds
	250	Fir	

Sunday, April 25, 1835

Rose late

Still the same magnificent weather. Summer returned & a clear cloudless expanse of blue!!!

Got my heliostat[84] up (with the oval reflector) & arranged a filter for my optical Expts.—

[Written vertically]

Paid. Gibbs 22: 6

    Christⁿ ⎫
      Boy ⎬ 28: 6

       2:11 : 0

Worked at Equatorial till 12ʰ½ ST then took up the 20 ft & Swept till 16ʰ½ ST. being a splendid night

---

[82] It has to be admitted that even the cleanest of unoccupied houses anywhere near the Cape Flats is speedily invaded by fleas, presumably normally dwelling in the sand.

[83] *Acacia lophantha* Willd., now known as *Albizia distachya* (Vent.) McBride. This species is now one of the worst culprits among the imported Australian *Mimosoideae* in invading and crowding out Cape vegetation. Sir John's note is quite probably the earliest record of *Albizia* being planted at the Cape (see note 181, December 25, 1834).

[84] A mirror rotating at a prescribed rate about a properly selected axis, which reflects a beam of sunlight in a fixed direction.

[Opposite page] NB   The [entry terminates]

Wednesday, April 29, 1835

Captn & Mrs Wauchope came & Dined here Capt & M^{rs} W M^r & Mrs Stewart

Thursday, April 30, 1835

Rain heavy all m[orn]ing
Sowed 150 fir 120 Protea
Whiled out the morning with Shewing Captn W. some optical Expts.

Friday, May 1, 1835

Max Therm 61.3
Min 43.0 + 4.3
Cl = 10. Calm
Cl = 10. Calm
Cl = 10 . . . 0 . . . 5
Cl = 0. SE[85]

[Written vertically] Planted 1400 Protea Argentea[86] Seeds in the Shrubbery.—Also Planted 766 fir from a Mark in Road twds [towards] Devils tooth 109 Paces, 4 deep, up to a deep double mole hole Planted [erased]

Captn & Mrs Wauchope walked into Town & returned at 7 to dinner

Saturday, May 2, 1835

Max of last night 63.2 Min 45.0 + 4.2
Cl = 0. Calm all night and all morning
Not the slightest cloud on T or D.[87]
Captn & Mrs Wauchope left for Simon's Bay[88]

---

[85] The cloud observations indicate that it was for a long time overcast and calm, then cleared, then became half cloudy, after which it cleared completely with the wind in the south east.

[86] Dr. Hall says that in the present context Sir John is probably referring to the Silver Tree, *Leucadendron argenteum*, rather than to the shrubby species for which the name *Protea argentea* has also been used. The leaves of the Silver Tree have hairs which lie flat on the surface in dry weather, giving a silvery appearance.

[87] Table Mountain or Devil's Peak.

[88] The Royal Naval Base at Simon's Town.

Sunday, May 3, 1835

Attended Div Service at Rondebosch

Killed a Black & grey snake said to be very poison—the skin of his neck has 2 bags or enlargements which he *flattens* out in a very singular manner when enraged[89]

Tuesday, May 5, 1835

Sowed 1200 Fir + 85 Oak

Wednesday, May 6, 1835

Wrote to Dr Gird from R. Observatory at 6 PM. about the Pony for Capt Wauchope (NB Trap's down)

Thursday, May 7, 1835

Sent to Capt Wauchope & wrote to Gird again about the Pony

Friday, May 8, 1835

Sowed 500 Protea

Saturday, May 9, 1835

Finished a Large sized Pencil drawing of the Front Garden & Avenue to go with Sir J. Gore[90] or other opportunity with M's flowers, Home. [words erased]

Sunday, May 10, 1835

Wrote a Long letter to Whewell—also to Capt[n] Smyth

M. went in to Dr Phillips's Church in evening

Monday, May 11, 1835

Dispatched (per Thomson & Watson) a regular Power of Attorney (sent out by Drummond from London & received Saturday) for Sale of £4000 Red.[91] Thus Feldhausen is now mine in property—Bought *and Paid for.*

Dined at Col Bell's and met Lord Clare[92] and the officers (Sir J.

---

[89] A ringhals, or spitting cobra.

[90] Vice-Admiral Sir John Gore K.C.B. (1772–1836) was commander-in-chief, East Indies, from December 16, 1831, until May 30, 1834, when he was relieved by Sir Thomas Bladen Capel. Two of the ships on the East India Station at this time were the *Melville* and the *Imogene*.

[91] Four thousand pounds of reduced stock.

[92] The Right Honorable John FitzGibbon, second Earl of Clare, K.P., G.C.H., a privy councillor of Great Britain. Born 1792, he was the elder son of the first

Gore excepted) of the Melville and the Imogen—Lady & Miss D'-
Urban, Major & Mrs. Longmore & Capt[n] Nairn of the Anne Robertson[93]

Tuesday, May 12, 1835
Barometer at 10 AM 29.876 Att. 61.5[94]

Thursday, May 14, 1835
Passed the morning out on the Hill, Taking angles, gathering bulbs
& planting 2000 protea seeds in a circle round the firtree v.[95]

M. worked on the reduction of Bance's obs[n]s for temperature & I
cast them finally up
Cleared off & became very fine   Equatorial work till 3 AM.

Saturday, May 16, 1835
Wrote to Bessel[96]

Sunday, May 17, 1835
Went to Rondebosch Church
Walked home with M[r] & M[rs] Menzies
Swept til 14[h] ST. & went to bed

Earl, Lord High Chancellor of Ireland, to whose titles he succeeded in 1802. A
member of Christ Church, Oxford, he was appointed Governor of Bombay in
1830, and sworn a privy councillor. He left Bombay in 1834, and became
Knight Grand Cross of the Hanoverian Guelphic Order in 1835, and Knight of
St. Patrick in 1845. He married the Honorable Elizabeth Julia Georgiana Bur-
rell, third daughter of first Lord Gwydir, and died without issue at Brighton in
1851.

[93] Not a navy ship.

[94] The reading is in inches. "Att." means "attached thermometer," since a
knowledge of the temperature is necessary for an accurate derivation of barometric pressure.

[95] Probably a reference to a lettered spot on one of Sir John's plans.

[96] Friedrich Wilhelm Bessel (1784–1846), celebrated German astronomer
and mathematician. He worked especially in the fields of positional astronomy,
geodesy, and celestial mechanics. Professor of astronomy at Königsberg, East
Prussia, from 1810 to his death, he was director of the observatory there from its
completion in 1813. He was the inventor of Bessel functions, and secured priority
of publication (over Henderson and others) of the first measurement of a stellar
parallax in 1838 (see note 195, Observations, 1834).

A Letter from Sir John, Feldhausen,
to James Calder Stewart, Canton, China

rec^d 29th Oct./35 [Added in pencil]
Feldhausen near Wynberg C G H.
May 17. 1835

Dear Jamie—I shall begin my letter to you by copying part of one forwarded in Duplicate to Peter which will reply to one part of your late Comm.^ry Then—if room, will proceed to gossip.

"Dear Peter &c . . . &c.
"You said in one of your late letters something about intending in the course of this year to take up one or both of certain Bills of S. E. and Co. left at Drummond's (£5000.) Now if you have not yet done so I gather from a letter of dear Jamies' that there is room for another use for that little capital than simply putting it away to rot in the funds or be wiped away by the National Spunge. He tells me, in short with the Confidence of a Brother that it w^d be of use to him and tend to put him in a better position with any [last word inserted in pencil] his new partners—and in so telling me he does only justice to my real feelings & my earnest desire to be of use to him to alleviate in some little degree the heavy pressure you have long borne and are still bearing with such manly and generous energy. Now therefore If S.E. and Co have *not* yet (1) taken any step about it, but are, as I gather from yours, about to do so—present the enclosed* to Drummond & Co when you take up the Bills and apply the amount according to your own knowledge of Jamie's wishes on the subject to whom I shall write to Canton by the first ship stating what I have now written to you. *If* on *the* other hand (2) the Bills are already taken up, in that case Drummonds, acting on my previous order have of course replaced the money in the 3 per Cent Reduced Stock whence *they* cannot reobtain it, as I have left with them no general power of attorney for the Sale of Stock. In that case then there will be no means of coming at the Principal but the very dilatory and round about one of

*(an order for the amount)

your procuring a Power of Attorney authorizing you to sell the amount of Stock it may have produced and send it out to me here for signature. You must be the judge whether that course is advisable. Should you consider it so, apply to my stock broker Jonathan Harrison of Token House Yard . . . who knows the particulars about names, forms &c and who will get the power prepared for you.

If under all the Circumstances this Capital cannot be of use to James, perhaps it may to Johnnie and if so, have no scruple about so applying it—or in short making it available according to the best of your Judgment for the benefit of "the Concern" according to the true intent and spirit of this present—observing that I require no security beyond any of your notes of hand for the principal and that I will take no interest from a brother, whatever I may have done from the firm of Smith Elder and Co.—

Yours ever faithful

J F W Herschel

So—Now you are in possession of data & you & P. are put in communication on the matter.

Now let me thank you for the interest your Journal has afforded us both. It seemed like travelling over the ground again. Why did you not half scuttle the Gen^1 Kidd & oblige her to put in here. It was a shame to pass us. Cur dextrae jungere dextram non datur—et *vivas* andire ac reddere voces?[97]

Can you procure meteorological observations & those of the tides (especially of the *times* of High & low water) to be made with any degree of care at Canton or any other ports in the Eastern Seas.—I enclose a little brochure of which pages 5–6 contain a recommendation which Maclear & I here have and I hope by this time Lloyd at Mauritius and Prinsep at Calcutta as well as many in Europe, will have begun to act upon. Pray use your endeavours to get us corresponding obs^{ns} on the plan recommended.—I conclude you are provided with Whewell tide-papers and that he has at various times & seasons impressed you with a clear idea of his anxiety to procure such information on that subject as you may perhaps have in your power to *obtain for* him

J.F.W.H. May 17/35

[97] Why is it not permitted to join right hand to right hand and hear and return *living* voices?

We are all well & happy & the Stars and planets continue propitious. Through Dr Phillip we see the Canton Monthly Magazine (or some such title) one N° of which contains an able article on "Free Trade with China"—Adieu

JFWH.

A Letter from Lady Herschel, Feldhausen,

to Caroline Lucretia Herschel, Hanover

N° 7
Rec^d July 21 [In a different hand]
Cape of Good Hope May 19th 1835

My very dear Aunt
          Our last joint letter to you dated Feby 21st has I hope arrived safely & brought up our family history to that time pretty correctly as Herschel himself was the principal writer of it.—And although we have not been blessed with any letter since then from Hannover we will hope that your life has gone on smoothly and happily as ours have done—Every thing here delights us as it did at first, except perhaps a tyrannical Spirt which Colonies are apt to engender among their Inhabitants towards the Natives & which has now involved the Governor in an offensive & defensive war on the frontier of this Cape far enough away from us to do ought but grieve our hearts that such steps should be necessary—Herschel, who is as alive to anything concerning his fellow creatures, as to things above, is much interested in the causes of this war & throws the weight of his name & character into the side which humanity & justice dictate, & which is guarded by a mournful minority at the Cape—As to the stars—Herschel's expression *this moment* is, when coming in from the Telescope with closing eyes—"I cannot positively *keep up* with these fine nights—for *seven* nights I have been up till past four o'clock, & such clear nights were never equalled before—& now I am half dead with sleep"—& so he throw himself on the sofa & enjoys a comfortable nap—His health is *admirable*, & comparing his successful Harvest within the last year to the low expectation he too hastily formed soon after his arrival, he is

more than repaid for his voyage—besides the accumulation of interesting facts respecting the Heat of the Sun's Rays—the Annual
variation of the Barometer which is leading him to suggest new
Theories on this subject—and the Height of the Tides—All which are
more easily observed & amassed here than at home—But the beautiful scenery still tempts us to make many excursions on horseback,
when Herschel always carries his basket & trowel on his arm & robs
the wilds of their lovely flowers—His bulb garden occupies a great
deal of his attention & he keeps me close at work painting the flowers
for Mama—But a still more delightful task claims my attention, that
of teaching my little ones, & this gives me more pleasure than I can
tell any one, while it does not interfere with my enjoying as much of
dear Herschel's society as he can afford me—Caroline who is now
five years old, is getting on pretty well in reading English & French—
in Arithmetic & sewing—& her taste for Music deserves & shall have
the most distinct cultivation—Little Isabella whom I thought rather
stupid has all at once started onwards & almost equals Caroline in
everything except Music—She never learnt reading from any spelling
Book, but insisted on a *lesson* as well as Sister Carry, & nothing can
turn her attention from what she is about—I don't know a fault this
little creature has—energetic, industrious, affectionate & honest as
steel, she rolls about the House, (for she is as round as a Ball) the
delight of everybody—William is already Papa's companion & de-
servedly so, for he is as manly & tractable as possible with an abundant share of intelligence—I found him hammering a nail into a
board, the other day with astonishing correctness of aim—he is only
2 years & 3 months old—My baby is a very dear, healthy little Baby,
with the sweetest temper—& that is all that can be said of a Baby just
8 months old—If London is the centre of civilized Europe, this seems
to be the centre of the rest of the World—for we live in the midst of
accounts & arrivals from India, China, Australia & America—All teem
with interest, & the different Governors & Admirals passing & repass-
ing towards their Governments & Stations, make a point of visiting
us, & some agreeable acquaintances we have made will not be forgotten in England I hope—Admiral Sir John Gore whom perhaps you
may remember at Datchet, has just returned to England from the
Indian Station, & has to mourn the death of his *only son*, who generously threw himself into the sea, off the Cape of Good Hope, to save
the life of a man who had fallen overboard—This has affected us

deeply, for we knew Lady Gore's affectionate disposition—Lord William Bentinck arrived here yesterday with his suite having resigned the Governor Generalship of India on account of his health— we expect him & his Lady here hourly—You have lost your happy & privileged correspondent—my brother James who has gone to China, but to our great grief, could not stop here on his way thither—How much disappointed he was to pass near Hannover & not suffered to enter—His journal of his continental tour has afforded a great treat to Herschel who seemed to tread the ground over again—Of Miss Baldwin & my Mama & Brother you must know better than we do, for arrivals have been few & far between, but we earnestly hope the next will bring news from you—Should you ever be reading any book about the Cape, you will find the 2^d Vol. of the blind Traveller Holman's [98] work, lately published very interesting & tolerably correct. Herschel has just awaked to write with me in most affec^e love to our dear Aunt, & I remain

Your much attached Niece
M.B. *Herschel*

The Diary: May 22 to September 5

Friday, May 22, 1835

At work all day on the Interpolation of Bance's observations—a dull & partly rainy day

Saturday, May 23, 1835

Set Candassa[99] Planting fir seeds.

---

[98] James Holman, author of *A Voyage round the World, including Travels in Africa, Asia, Australasia, America &c &c From MDCCCXXVII to MDCCCXXXII*, in four volumes, published by Smith, Elder & Co., 1834. The author, a lieutenant in the British Navy, lost his sight at the age of twenty-five, but, nothing deterred, embarked on his travels round the world. He arrived at the Cape in 1828, and in spite of his handicap produced a large volume of useful information and even managed a journey to Kaffraria.

[99] A colored washing and mending girl.

Lord & Lady Wm Bentinck[100]—called on their way to Simon's Town for Embarkation.—Sir Stamford . . . [his omission] Dr Turner, and Captn Pakenham accomp^d them   L^d B talked of the Steam Navigation and said it was "impossible that the Euphrates route[101] could succeed" reasons—repairs—want of coal stations—Drunkenness & death of Engineers—hostile & predatory nations &c

Swept till 17^h ST. with interruptions—but over the most monotonous & dull Zone ever seen & in the finest night—NB. It felt dreadfully cold yet the therm was 52!

[Opposite page] L^d B is a man of easy, affable & natural address.

Sunday, May 24, 1835

Attended Divine Service at Rondebosch & heard a very very long sermon from Mr Judge.

Walked out with M & the children to gather the Strumaria Crispa[102] a most beautiful bulb which grows [entry terminates]

Swept

Measured Saturn

Swept till 2 AM

Monday, May 25, 1835

At home all day. Prepared for Polishing. Worked out the Interpolations of Bances's Barom^e obs^{ns} to their mean result

Swept till 12. A strong NW at first then (when I had put up the telesc thinking it useless to proceed & sent Stone to bed) it subsided to a perfect June Calm!!

Wednesday, May 27, 1835

Went into Town to attend Meteorol Committees meeting

Thursday, May 28, 1835

Prepared boxwood cup &c for setting up a zero Barometer

Major Cloete dined here

Col & Lady Catherine Bell came in þ^e Evening to see the stars.

---

[100] Lord William Cavendish Bentinck (1774–1839), second son of third Duke of Portland. He was Governor General of India from 1833 to 1835.

[101] Possibly a discarded alternative to the finally adopted Suez Canal project for a short route to India. The absence of this short cut until several decades later was one of the principal reasons for the continued importance of the Cape.

[102] Strumaria crispa Ker, now known as Periphanes cinnamomea (L'Her.) Leighton.

Shewed him & Cloete Saturn!!! ω Centauri, α Crucis, α Centauri & the nebula about η Argus—the most superb defin[n]

Cloete gave the history of Hintzas[103] death as related to him in a letter from the Governor. See opposite page

[Opposite page blank]

Saturday, May 30, 1835

Rode over with Maclear to Simon's Town to call on Admiral Campbell[104] & Capt[n] Wauchope, also to inspect the Tide Gage

Saw a New Zealand Canoe beautifully carved—which floating on the chrystal-like water of S. Bay looked like a fairy boat or a phantom

Friday, June 5, 1835

Bellas Birthday—the little folks in their garlands—Mrs Hutchinson came over & dined with us

Dr Phillip brought over Adam Kok[105] (Chief or Captain of the Griquas) & his Secretary or son in law—also Mr Campbell Missionary from Madagascar

Wednesday, June 10, 1835

Hauman? the Paarl Farmer called in his waggon & 8 with wine, Dried fruit &c. A curious & primitive specimen.

Capt[n] Beresford called on his return from the frontier. Says Hintza was a man of "princely manner"—fine stature & figure, tall and of good features. # See opposite page

[Opposite page blank]

A most magnificent night but $*$s too ill defined for working.—At Midnight called up M[t] to "take a walk" & "see the Moon". Behold! She was *eclipsed* in the *Zenith* (quite a bite out of her lower limb) & *by her light we read* Mrs Hall's letter cross writing *and all*.

---

[103] Hintsa, paramount chief of the Xosa peoples occupying the land beyond the Kei River. After an invasion of the frontiers of the colony at the end of 1834, the tribes were checked and an additional area of territory annexed. On the following day (May 11, 1835), Hintsa was shot in self-defense by George Southey while trying to escape. (Eric Walker, *A History of Southern Africa*, p. 185.)

[104] Rear-Admiral Patrick Campbell, commander-in-chief of Cape of Good Hope and coast of Africa, May 30, 1834–September 25, 1837. He later became Vice-Admiral and K.C.B.

[105] Third of this name, a younger son of Adam Kok II. He was chief of a Griqua tribe centered on Philippolis (named after the missionary), just beyond the Cape border.

Thursday, June 11, 1835

Rode into Town to Enquire into the matter of the "Non acceptance" of my Bill D for £ 200.—Called on Messr[s] Thomson & Watson & having shewn M[r] W. the documents, took him with me to Mr Ross & had the matter so far explained as the case admits. There has either been a mistake of one Bill for another—or gross negligence on the part of some holder of the Bill—or a forgery. It remains to be seen which—M[r] W. undertook to enquire into the matter *here* of all the parties concerned.

Wrote to Maclear & Deas Thomson[106] of Simon's Bay about Corresp

Tide to Dr Whewells in June [almost illegible pencil scrawl]

Friday, June 12, 1835

A morning and afternoon of Wandering about the Wilds & herbalising & planting and lounging and reading Robertson's Charles V and English letters

Saturday, June 13, 1835

Steady rain all day & night much wanted

Wrote in reply to the Duke of North[ds][107] letter   See Correspondence

Occupied with English Newspapers which have come in in a bag

Sunday, June 14, 1835

Rained Steadily all day [words erased] but desisted about 4 PM.

Read Prayers, with M. at home

Wrote to Hamilton[108] of Dublin

Saturday, June 20, 1835

Rode into Cape Town & Obs[d] the High Water. Joined at 1 by Marg[t] Called on Lady C Bell   Mrs Hutchinson &c

---

[106] John Deas Thomson, naval officer and agent victualler, at the Naval Office, Simon's Town. He was convicted in May, 1845, of embezzlement of £ 10,920.10.5 (about $55,000 at prevailing rates) and sentenced to transportation to Australia for fourteen years.

[107] Duke of Northumberland, the patron who financed the publication of Sir John's observations.

[108] William Rowan Hamilton (1805–1865), infant prodigy and mathematician. Born in Dublin, he was Royal Astronomer of Ireland and director of Dunsink Observatory from 1827. He invented the mathematical method of quaternions, a precursor of vectors and tensors.

Library—Looked over Daniels Coast Scenery[109]—very pretty.

Sunday, June 21, 1835
Rode to Cape Town and took up my pos[n] on the Jetty as before
After High Water Rode to Observ[y] & dined there and then back to
jetty where found Bowler[110] to whom left the Obs[n]—& returned in
time to Sweep

Monday, June 22, 1835
Rose at 6 and commenced the 36 hourly Meteorological Obs[ns]
using Daniel's Watch
At 11 started for the Tide Obs[n] which falls to my share this morn-
ing. Returned about 3 & resumed the Met[l] Obs[ns] which I continued
(Sweeping in the intervals) all night till 5 AM of Tuesday
While Absent Marg[t] took the Meteorol Obs[ns].

Tuesday, June 23, 1835
At 5 AM went to bed. Rose at 11. & resumed the Obs[ns] which M &
I continued till 6 PM when to my dismay I found Daniel's Watch an
hour & 40 minutes slow!!
Occupied the whole of the rest of the night with trying to make out
this error. A most dreadful mass of confusion resulted—a Curiosity
in its way[111]

[109] An illustrated book published at a loss by Smith and Elder.
[110] Thomas William Bowler (1812–1869), artist and illustrator. He came to
South Africa with Maclear as his valet; subsequently he was employed on astronomical work at the Observatory. Dismissed from this, he gave drawing lessons
and gradually acquired an important position as a realistic illustrator of Cape
scenes and architecture. (See Frank and Edna Bradlow, *Thomas Bowler of the
Cape of Good Hope: His Life and Works, with a Catalogue of Extant Paintings;*
with a *Commentary on the Bowler Prints by A. Gordon-Brown.*)
[111] Sir John was much put out by this. In a letter to Maclear he wrote:
"Thursday Afternoon   My dear Sir, I have made so dreadful a *mess* of the Tide
Obs[n] I took on *Monday* as well as of the Meteorological Ob[svg] by using Daniel's
watch—as defies all my power to decipher—i.e. in respect of time. It is something
so compounded of the ludicrous & the melancholy as to be an epitome of the
great tragi-comedy of human life. In truth I am ashamed to shew it to you—yet
as a matter of curiosity and philosophical enquiry it may be worth while. I shall
therefore ride over to the Observatory between the High and Low water tomorrow and 'make a clean bosom' of the whole affair. Tomorrow (Friday) the
stroke of 8 will find me on the jetty. If you will be up, I think I shall take the
Observatory in my way and pick up a time I can depend on. My blunders (if
blunders they were—i.e. if the evil one had not his finger in the pye) went to

Wednesday, June 24, 1835

Occupied the whole of this blessed day in trying to make out [entry erased]

Drew out a synopsis of my Measures of Saturn's Satellites

Received a Parcel of letters from P S.—Grahame—& Aunty & Cousins &c &c—a vast treat after an age of tantalising disappointments

Tuesday, June 30, 1835

Got the Large palm tree under the weeping willow transplanted.

Wednesday, July 1, 1835

Attended the meeting of the S A. Lit & Phil Inst

Thursday, July 2, 1835

A Rainy day.—Marg^t ill & kept all day in bed.

Maclear and Mr Deas Thomson came over at 7 PM dined & staid till 1 AM—Talked about the tide Obs^{ns} and [entry terminates]

Friday, July 3, 1835

M. still very ill with Headaches and sickness—keeps in bed.

In Evening a "dance" at Mrs Menzies' a farewell merrymaking, before they quit Sans Souci.[112]—Capt^{us} Wallace & Cooper came in in auncient coustume as old officers & danced in their Jack boots to the amusement of all present

Saturday, July 4, 1835

Wrote to Schonberg stating that I consider myself out of his debt—but that if any small item has been forgot desiring he will state it in writing immediately. Also asking him & M^{rs} S & family to tea & stars.

---

whole hours—half hours and quarters. It is something enormous—incredible—and to me utterly incomprehensible,

Yours in haste,

J.W.F. Herschel.

P.S. On second thought I will not come via the Observatory. There will be too little time and I shall miss the Observations altogether most likely by so doing."

[The letter bears a pencil note by Maclear]

"A fog. Missed his way and found himself at Tyger B.   T.M."

Tigerberg is about five miles away from the Observatory in the wrong direction.

[112] An estate to the east of Table Mountain between the Observatory and "Feldhausen," now the site of the Girls' High School, for many years the property of Hamilton Ross.

In Ev[enin]g. S. came. Shewed him ω Centauri—Saturn & my blue Planetary nebula near the Cross.[113]

S. said that "I had quite misunderstood him—that he never meant to say that there was anything of consequence unsettled—but there might be some small matter of "interest" (heaven knows what he means   not amounting to £1."

Sunday, July 5, 1835
Attended Divine Service at Rondebosch Mr Hough preached.

Monday, July 6, 1835
Wrote to Whewell about Tides—also to Murchison to accompany specimens of trilobite by Beresford—but lo! Beresford is gone.—

A Strong Steady hot North Wind set in last night or Early this morning very *dry*. Therm^r 78; *Evap.* 23°!!^{114}

Bar. falling   cloudless sky.

Captn & Mrs Bance, D^r & Mrs Adamson M^r & Mrs Hutchinson, Messr^s Polemann,[115] Watermeyer & Baron Ludwig—dined with us— NB. Watermeyer loud & disputacious at and after dinner.

Carry, Bella, & all the children have colds and coughs.
Rich red Sunset

Tuesday, July 7, 1835
Occupied in drawing up a Meteorological report for the S. African Phil Inst.

A burning hot day   Therm. 80. Barom slowly falling. Wind violent & steady NW,   very *arid   Evap^n* 21°!! Cloudless sky—A Rich red Sunset   "Abend Roth"

Friday, July 17, 1835
Rode with Marg^t to call on M^r & Mrs Cumberledge[116] by the "half-way house"

Filled Maclear's Barometer

---

[113] NGC 3918 (see note 37, 1835).

[114] He calls attention to the very large difference between the wet and dry bulb thermometers, indicating a low humidity, very expectional in Cape Town when the wind is north west.

[115] Peter Hendrik Polemann, of Pallas and Polemann, apothecaries, was a member of the South African Literary Society.

[116] Captain Cumberlege was an Indian visitor. The "Halfway House" could have been Merckle's, later Rathfelder's, at the Diep River.

Saturday, July 18, 1835

Polished my two mirrors.

Exp[ts] on the roasting of Colcothar

Reduced D Stars for the 6th Catal[117]

Wrote to Dr Somerville[118] (See Extract).

A rainy day—the whole day

Monday, July 20, 1835

M[r] Vogt called on the part of M[rs] Danford to make a fresh arrangement about Somai the coachman See opposite page—

[Opposite page blank]

Interpolated more of the Table bay tides—a dull & tiresome job owing chiefly to the bad arrangement of the obs[ns]

Rode with Marg[t] out on the flats   Almost [erased] no flowers to be found but the blue early Babianas, & Oxalis & Hypoxis Stellata.[119] No Gladioluses. Slept till—[his omission] & Swept till 23[h] 13[m] S T. A most superb night

Tuesday, July 21, 1835

Rose late

M[rs] Maclear came & spent the day. Interpolated Tides—Observed ♄ [s120] Satell[s].—A superb day & magnificent night. Swept till 21[h] & then went to the Equatorial—and made a night of it.—

M[rs] Menzies called & told of 7 marriages determined on & 9 more reported

Friday, July 24, 1835

Mr Chisholm[121] & Dr Adamson came over to Breakfast & to confer about the Newlands water.

---

[117] That is, entries for the sixth catalogue of double stars.

[118] Mary Somerville (1780–1872), Scottish mathematician and scientific writer. Born Fairfax, she married (for the second time), in 1812, her cousin William Somerville. Moving to London in 1816, she associated with the leading scientists of her day. Her paper to the Royal Society in 1826 ("Magnetic Properties of the Violet Rays in the Solar Spectrum") attracted interest. She authored several books: *The Mechanism of the Heavens* (1831) (a commentary on Laplace's *Mécanique Céleste*), *The Connection of the Physical Sciences* (1834), and *Physical Geography* (1838). Her *Molecular and Microscopic Science* was published in 1869, when she was eighty-nine. Later a resident of Florence, she died in Naples. The Oxford women's college Somerville is named after her.

[119] Hypoxis Stellata = *Spiloxene capensis.*

[120] Sign for Saturn.

[121] John Chisholm, superintendent of the Cape Town Waterworks.

Took a Barometer up to Newlands Spring[122] & also to Munnick's Spring leaving M^t to make corresponding obs^{ns} every 5^m—at home. Thus [?] got ob^{sus} for the level—yet owing to the strange effects of the local winds on this [side] of the Mountain—we were baffled & got no dependable results. NB. The wind here today was SE, moderate or nearly calm but in Cape Town Baron Ludwig Mr Watermeyer & Mr Chase who called this morning told us it was blowing a perfect tempest. Mr & Mrs Cumberledge called. Swept till 3 AM. The most Magnificent of all the Mag^t nights of late

Sunday, July 26, 1835
The Children went to the Observatory in the Carriage & I took the opportunity to send back the two Newman's Barometers of Maclear's & my own for comparison.

Tuesday, July 28, 1835
Walked with Marg^t on Wynberg Hill to the Terrace. Dug out Microlomas[123] and transplanted them. Found *one* Watsonia *Marginata* the only one I have seen in this neighbourhood.—

Wednesday, July 29, 1835
Dined at Mrs Corrie's[124]

Friday, July 31, 1835
Dr Adamson called & had a confab about Chisholm's data for the conveyance of the water from Newlands into Cape Town.
Lady C Bell called. Cold very bad and very stupid
At work making out an effective Equatorial working list
[Previous writing, erased or overwritten, ends with words] Myself very ill with cold

Saturday, August 1, 1835
Occupied in finishing my Equatorial working list—NB about 260 or 270 Double Stars put on the list
[Many erasures follow]

---

[122] Clearly the object of the exercise was to determine relative elevations as a guide to the correct laying of water pipes, a perfectly practicable scheme on a calm day with carefully compared barometers.

[123] A small South African genus of bright-flowered, slender climbers (Asclepiadaceae).

[124] Mrs. Corrie, an Indian visitor, probably the wife of Bishop Corrie (see note 143, September 11, 1835).

Rode with M^t to call on Mrs Hough, & was very glad to get back being shivering with cold in the South Easter [erased] Worked at Equatorial All day. Laid up with a terrible *Catarrh dismally* bad.— Worked at Equatorial from 8 PM to 4 AM. indiff^t N^t

[Opposite page]
Gladiolus Gracilis in fine flower
————Hirsutus D$^{\circ}$[125]
(————Permitatus *long since over*)
Trichonema alba ⎫
                   fully out[126]
————sulphurea! ⎭
Babiana Villosa almost going off
Melanthium secundum [?][127] in flower
Hypoxis Stellata (Large yellow star flower) beginning to be common
Homeria Collina[128] & flexuosa D$^{\circ}$—
Ixia Pendula[129] in flower
Large red aloes going off rapidly
Branching short yellow aloe gone—

Sunday, August 2, 1835
Rose very late (at 1) & two Mr Hawkins's called. Maclear called and after a talk, accompanied him part of the way to the Observatory. Parted at the Camp Ground.

M^t went to Rondebosch Church but was obliged to leave before sermon was over   Found 3 Bassione [?] Ciliatas[130] and had much trouble in cutting them up   Went to Equatorial & measured and reviewed till 5 AM (Morning twilight) being a most *superb* night. Def^n perfect—absolute perfection

[125] Gladiolus Hirsutus: probably *G. brevifolius.*
[126] Trichonema alba, T. sulphurea: previously unpublished names, evidently for plants belonging to *Romulea* (Iridaceae).
[127] *Dipidax ciliata* (Liliaceae).
[128] *Homeria breyniana* (Iridaceae).
[129] This name is generally regarded as being one of the synonyms of *Dierama pendulum,* a well-known eastern Cape and Natal species with pink flowers on a tall inflorescence with thin branches.
[130] It has not been possible to identify these.

Wednesday, August 5, 1835
Attended the Meeting of the SA Lit & Phil. No communications read

Thursday, August 6, 1835
Copied fair the Report from Dr Adamson & self
In Evening went into Town with M[t] and dined & passed the night at Col. Bell's   Met at dinner Mr & Mrs Hutchinson & Mr Kekewich[131]— Handed to Col Bell the report agreed on respecting the Water works

Friday, August 7, 1835
Accompanied Col Bell to see the opening of the main pipe which leads the water to the Reservoir & w[ch] is much corroded though *never full*. Also an experiment of Mr Chisholm on the water delivered by 100 feet of pipe with a fall of 1 inch—the Exp[t] bad & of no use.— Examined the Reservoir
NB   the water is *blue* a decided colour   NB. get to try some Expts on it, a good opportunity. Made calls with M[t] & Returned to Feldhausen after Tiffin with Mrs Hutchinson

Saturday, August 8, 1835
Copying my 6th Catalogue of D Stars, after reducing it at odd times. Wrote to Clift in reply to his request to procure specimens of Elephants Camelopards Hippopotamus skeletons &c

Wednesday, August 12, 1835
D[r] & Mrs Phillip & M[r] & M[rs] Fairbairn called—F. asked me to preside at anniversary of the Penny Subscription Lib[y][132]

Friday, August 14, 1835
M[rs] Hutchinson came in the Morning & M[r] H. to Dinner

Saturday, August 15, 1835
Made up dispatch 2[d] to Whewell enclosing tides in Table Bay & results of interpolation also the curves for the Simon's Bay tides & finished letters. Mr & Mrs Stewart called. Rectified the Eyepiece of

---

[131] Judge George Kekewich, appointed to the Cape bench, 1828. William Kekewich, possibly his son, was a clerk in the Supreme Court.
[132] The "Penny Subscription Library," properly called "The Popular Library," had a maximum charge of a penny per week. Sir John presided over and spoke at the meeting on August 18.

20^{tt} Swept till Moon Rise then worked at the Magnitudes of the stars (Grus, Pavo, Toucan, Hydrus)[133] till 4 AM.

NB   about 5½ AM. M & myself both distinctly heard a loud heavy dull #

[Opposite page]
report like a Great Gun at a distance, or the explosion of a powder mill very distant. [words erased] Query what it was

Saturday, August 22, 1835
Rec^d at Breakfast this morning a Budget of Letters with "lots of Home News"

Monday, August 24, 1835
Barometer fallen suddenly ½ inch a very unusual fall here
Rode out with M. to the place near George's halfway house, beyond Wynberg—where found the richest collection of Flowers. Red Antholyza (Ringens) Pink, & White Babianas—Yellow Lachenalias & a new Microloma [?]
A violent Storm of Thunder Wind & Rain all night.

Tuesday, August 25, 1835
Heavy Rain all day

Mr & Mrs Menzies
Mr & Mrs Cumberledge   } dined here
Mr Halhed—
Occupied all morning with making out Working lists & with a drawing of Cape Town from Baron Ludwig's Garden

Saturday, August 29, 1835
Went Bulbgathering & hawk-shooting behind Col Blair's.
Found superb specimen of the Protea Grandiflora[134]—which took and at night, outlined it—2 Hawks flying round & round & screaming? why

Sunday, August 30, 1835
A splendid day. All the beautiful flowers coming out in such glory that M^t & I in a pure rapture siezed on them and neglecting all other duties & occupations set to work I outlining & she colouring them.

---

[133] Southern constellations.
[134] Usually regarded as one of the synonyms of *Protea arborea Houtt.,* a large grey-leafed species.

When wearied with outlining took a bulb-gathering Ramble up to Letterstedt's woods[135] There under the shade of the Protea Argentea[136] found queen disas[137] &c

Clouded at N^t

Monday, August 31, 1835

Having had a fine Oak 60 years growth & sound to the centre cut off to a proper level, and planed & leveled, got up my portable transit instrument [words erased] & screwed down the stand temporarily, for adjustment

Outlined Ixias

went bulbgathering & hawk [line erased]

Sate up till 5 getting the transit into approximate adjustment.

Stars ill defined & no working with Equatorial

Tuesday, September 1, 1835

Went to bed at 11   Rose at 4 to look for Halleys Comet Maclear having sent word he had got it—Swept well over the place but no Comet found   Maclear probably saw the faint Pl VI. 21 near Mess. 35[138] & took it for a Comet

Wednesday, September 2, 1835

Repaired my drawing board which was broken. Prepared for an Expedition tomorrow. Went into Town to attend a meeting of the SA Lit & Phil Inst. Chisholm produced a pipe lined with Roman Cement on my plan which had been down 8½ months & no deposit was formed in it nor was the cement torn off or softened. He declared it to be a perfectly successful experiment & was ready on the strength of it to lay down the Newlands pipes so lined with full confidence

Brought out Mrs Hutchinson to Feldhausen

---

[135] Dr. A. M. Lewin Robinson supplies the following note: Honorable Jacob Letterstedt, later Swedish consul, of Mariendahl or Mariasburg. The old mill near Ohlsson's Brewery was known as Letterstedt's Mill. Letterstedt Road, Claremont, still survives.

[136] Evidently the Silver Tree, *Leucadendron argenteum*.

[137] Identity untraceable.

[138] The modern designation is M 35, meaning number 35 in the catalogue of nebulous objects of Charles Messier (1730–1817), published in 1771 and 1781. M 35 is the cluster NGC 2168. Pl VI. 21 is the cluster NGC 2266. It is a very rich compact cluster and might easily be mistaken for a comet.

Thursday, September 3, 1835
Started with Marg^t and Mrs Hutchinson for Somerset[139] & Sir
Lowry's Pass[140]—See Travelling Journal—
Slept at Somerset-Village

Friday, September 4, 1835
Somerset to Gordon's Bay & Sir Lowry's Pass—returned to Somerset

Saturday, September 5, 1835
Somerset to Feldhausen

The Travel Diary: A Journey to
Somerset West and Gordon's Bay

Sept 3. 1835. (Somerset)
9^h 50^m AM   en route in a waggon and 8 with Marg^t, Mrs Hutchinson and her maid Macqueen for Somerset-Village & Sir
Lowry's pass.

1   0±   Outspanned ⎫   & gathered
         for ¾ hour ⎬   bulbs & flowers
The rich-scented purple & yellow gladiolus—The scarlet &
other lachenalia—& 1 most exquisite gladiolus with red
streaks on a white ground a wonderful flower

2^h 45^m   Crossed the Eerste River at Lindemann's place, a large farm
house with the only trees to be seen except a row of firs by
another farm on right, of Lawrence Cloete?? The soil &
vegetation changes here it is a stiff clay covered with short
scrubby heath. No sugar bush to be seen—no bulrushes—
no silver trees on the hills which rise abrupt & majestic before us

---

[139] Now called Somerset West, a village to the east of Cape Town named for
Lord Charles Somerset, one of the early British governors of the Cape.
[140] A pass up the Hottentot Holland Mountains named for Sir Lowry Cole,
governor preceding D'Urban.

4ʰ 30ᵐ.   Arrived at Mʳ Morchel's[141] a little beyond Somerset Village,
          a very large farm with an immense Vineyard enclosed in a
          square of tall firs in single rows like a vast colonnade within
          which is a smaller (yet large) square similarly enclosed for
          an orchard. The house stands aside from the squares & is
          like all the Dutch houses one story, thatched & forming 3
          sides of a square on a raised terrace or stoop with the slave
          houses stables &c at a distance forming quite a Town. Ex-
          plained to Mʳ M that we had sent to Mʳˢ Stadler (who
          keeps a lodging house or inn) at Somerset to engage apart-
          ments but owing to Justice Menzies being expected could
          not have them
             on which Mʳ M assured us that we were welcome though
          his was not a lodging house. So we took possession & were
          well lodged and entertained, as part of the family. Walked
          out to look about us but the weather was gloomy & evening
          coming on. So saw little. Somerset stands in a fertile & well
          watered spot, much better soil than on the opposite side of
          the flat being clayey mould.

Sept. 4.   1835.—After a rainy night the prospects of the day looked
          doleful—Waited till noon when the clouds rather cleared
          from the tops of the hills & induced us to proceed.—Drove
          in rough & jumbling stile over Brake & Bush down to the
          sea Strand near the head of Gordon's Bay or Fishhook Bay
          the most North East point of False Bay. There is a good
          road all but the 1ˢᵗ mile from Morchel's house, over a flat
          full of fine Bulbs—But the flowers are as yet not in Bloom
          owing to the wet & want of Sun.—Fishhook Bay (NB.
          There is also a Fishhook Bay near Simons Town) has a few
          rude fishermans huts, & between it & Gordon's Bay is an
          Oil-furnace with Iron Boilers for the Whalers—very rough.
          —The shore is rich with Corallines Spunges & shells & the
          evergreen shrubs & bulbs come down to Waters Edge.
          Took some sandwiches in one of the huts, a very rude place
          into which much rain had poured & transformed the clay
          floor into mud

141 Morkel's farm.

however the owner furnished us with ancient red velvet cushioned scroll chairs! & apologised for other accommodations by the sickness of an inmate. Thence set out again for the "Sir Lowry's Pass" but it came on cloudy & at length rainy, so we could only ascend about a mile beyond the toll bar. & though deprived by clouds of the fine view we had led to expect found much in the Botany to repay us.

In particular a most beautiful white heath or species of Buchu.[142] Descended & got home to Morchel's where arrived about 6 & passed a dull & rainy ev[en]ing in company with our very unamusing and slenderly informed (tho' by no means slenderly personed) landlord & his rather *fine*ish & would be ladylike young 2^d wife who after scolding the slave girls duly seems to think life has no other occupation left. In the Ev[enin]g. our party was increased by a certain young Mr Judge, one of the Circuit attendants of Justice Menzie's troops, who finding no accommodation with the rest came over here. He seemed well informed and proved communicative enough though in one respect his information & assurances that we ought to consider ourselves as on the mere *hospitality* of our host & that a remuneration for our accommodation was not expected & would not be received, proved inaccurate (much to our relief for we had been teezing ourselves with the idea of the trouble we were giving & feeling quite ill at ease about the whole affair as a regular intrusion)

However Morchel himself set that matter right before departing by a [several words erased] distinct & reasonable charge. Surely this is the best way and surely it *is* hospitality or at least good nature in a country where there are no inns or only one here & there, to allow chosen *customers* the benefits of your residence & refreshments taking from them such a reasonable repayment as acquits all sense of pecuniary favor on both sides & leaves both parties gainers in point of fact.

Sept 5   1935
There being no chance of the weather clearing set off

---

[142] A fragrant plant used in the production of buchu tea and buchu brandy.

homewards, making rather a detour in the direction of Stellenbosch & crossing the Eerste Rivier at a point of evil repute for the destruction of passengers in flooded times. The last who suffered was the wife & child of M^r ———— Cloete [his omission] who were carried over the ledge of stones & drowned in his sight owing to the oxen turning round in the water. It is an ugly nasty place but a very trifling bridge would set it right—such a bridge as for example that [entry terminates]

The Diary: September 6 to October 22

Sunday, September 6, 1835

Feldhausen—Being a desperate day of wind & Rain kept home all day   M^r & Mrs Hutchinson dined here (Mrs. H staying over today to return tomorrow).

Found a Box *at length* landed & come to hand per "Fanny". Read some chapters of ————'s [his omission] "Seelen lehre" or Elementary Psychology

Monday, September 7, 1835

A tolerable day. Employed in planting out my bulbs collected from the country—and in reading & arranging the *letter* & packets per Fanny

Dined at 3. Mrs Hutchinson returned to Cape Town

Friday, September 11, 1835

Went into Town to See Bishop Corrie[143] confirm in St George's Church. NB The Church is a handsome room, plain & simple but in good taste the whole area in one with a flat cieling broken by 3 large scroll device [?] ornaments The only highly finished ornament is the pulpit w^h is a beautiful piece of Joinery of Honduras highly polished—The confirmees were very numerous—among them 1 Soldier in uniform. Returning home, *at our gate*, the carriage ran

---

[143] Bishop Daniel Corrie (1777–1837), bishop of Madras (1835–1837), evidently on his way thither to take up office on October 28.

against Capt[n] Knyvett[144] and unhorsed him with a blow of the pole as he attempted to pass *inside*.

Sunday, September 13, 1835

Attended Divine Service in Rondeb. [osch] Ch[ch] where Mr Judge gave a Confirmation Sermon

Friday, September 18, 1835

Pointed 20 feet for Encke's[145] Comet by Maclears azimuths & altitudes & got several Oak trees of the N. Avenue cut away to clear a sight down to the Table Mountain for it.———M[rs] Hawkins called#

[Opposite page]# Friday. Distilled a fresh stock of Alcohol for Chemical uses. Captn & Mrs Wauchope called

Captn W just arrived from St Helena

[Entry resumes]

Dr Liesching called to see Marg[t] who complains a good deal—brought with him an Indian (name unknown) passenger per———[his omission] to Eng[d]

Cloudy till 22[h]½ & ∴.[146] could do nothing with Comet.—Swept till 3 AM & then got round the telescope for Halleys Comet[145] for which searched unsuccessfully per Ephemeris till daylight.#

Saturday, September 19, 1835

Rose late.—Sir Tho[s] Sylvester[147] called A Superb Day. S E and clear blue sky Commenced fitting up the 4 feet Achromatic by Dollond as a Collimator by adapting [word erased] an apparatus to screw on to the Eye End—M [entry terminates]

Sunday, September 20, 1835

Attended divine Service at Rondebosch.

Requested Mr Hawkins to set down my name for the pew No 12

Examined the Cane-ground   found all the Cane cuttings sprouting

At night Cloudy—so Reduced some Sweeps ⎫
                        Commenced letter to Airy.   ⎭

[144] Captain Knyvett was an Indian visitor.

[145] Halley's and Encke's Comets: See note 13, January 29, 1835, and note 157, letter to Caroline, October 24, 1835.

[146] Mathematical sign for "therefore."

[147] Correctly Sir Thomas Sevestre, an Indian visitor, surgeon at the lunatic asylum, Fort St. George, Madras.

ate 1. The twenty-foot reflector as erected at "Feldhausen." Delineation by Sir John;
nograph by G. H. Ford.

ate 2. Camera lucida sketch by Sir John of the sunset behind Table Mountain, done
 the early months of 1834. (By permission of the South African Library)

Plate 3. Miniature of Margaret Brodie Stewart (Lady Herschel).

Plate 4. Engraving by Thomas Brown of Caroline Lucretia

Plate 6. Portrait of Thomas Maclear. (By permission of H. M. Astronomer at the Cape)

Plate 5. Portrait by Lemuel F. Abbott of Sir William Herschel, 1785. (By permission of the National Portrait Gallery, London)

Plate 7. Camera lucida sketch by Sir John of Dorp Street, Stellenbosch. (By permission of the South African Library)

Plate 8. Camera lucida sketch by Sir John of Table Mountain from Mr. Ebden's country house. (By permission of the South African Library)

Plate 9. Camera lucida sketch by Sir John of "The Brewery," Papenboom, New-lands. (By permission of the South African Library)

Plate 10. Camera lucida sketch by Sir John of Diamond Rock and its cave, "The Bower," above Paarl. (By permission of the South African Library)

Plate 11. Sir John's camera lucida panorama from the summit of Paarl Rock. *At to[p]*
Table Mountain and the Cape Peninsula are seen in the distance; *at center,* Macle[od]
observes, against a background of the Drakenstein Mountains; *at bottom,* the Paarl Vall[ey]
lies before the Wemmershoek, Slanghoek, and Du Toit's Mountains. (By permission of t[he]
South African Library)

Plate 12. This camera lucida sketch by Sir John of the Royal Observatory from across t[he]
Salt River swamp is dated June 2, 1837, but must have been begun much earlier, since t[he]
cloud on the mountain shows that the wind was southerly, as it usually is in summer. (B[y]
permission of the South African Library)

ate 13. A highly finished camera lucida sketch by Sir John of Devil's Peak as seen
m the stoep of "Feldhausen." (By permission of the South African Library)

ate 14. Camera lucida sketch by Sir John from the northeast avenue of "Feldhausen"
owing the mountains under snow; birdlike markings identify them, from the Helder-
rg just right of center to the Simonsberg at far left. (By permission of the South
rican Library)

Plate 15. Camera lucida sketch by Sir John of Caroline, Isabella, and Louisa in the northeast avenue of "Feldhausen." (By permission of the South African Library)

Plate 16. Flower drawings by Sir John, colored by Lady Herschel: (ABOVE), *Disa longicornis* and *Disa graminifolia;* (BELOW), *Vieussieuxia glaucopis* and *Erica . . .* (By permission of the South African Library)

Plate 18.   Measurement of the base line on the Grand Parade in Cape Town, December, 1837.
(By permission of the South African Archives, Cape Town)

Plate 19. Camera lucida sketch by Sir John of Cape Town and Table Bay from the foot of Platteklip Gorge on Table Mountain. Blouberg is the hill seen across the bay. (By permission of the South African Library)

Plate 20. On board the *Windsor* on the homeward voyage; undated, but a mountainous coast is visible, and there is a steady moderate wind; a fiddler entertains the passengers (By permission of the South African Library)

1 hour Zone—reviewed at Equatorial in an interval of Clear

Monday, September 21, 1835
[A reference to Captain Wauchope erased]
Another Comet hunt without success after cutting down more trees
Swept till 2½ AM

Tuesday, September 22, 1835
Found to my dismay & annoyance that I have forgotten the hourly Meteorol Obs^{ns} for the Equinox! ! !—
Rose late. Outlined 2 Gladiolus'es for Marg^{t}—Planted 1400 fir seeds to fill up vacuities in my lines—
Examined my Protea Argentea seeds—all come up and look well.—
Proceeded with the tinnery work of the new Collimator Outl [erased]
Swept till 2½ A.M. Definition of Stars tolerably good but not perfect

Wednesday, September 23, 1835
Captn & Mrs Wauchope arrived—they and M^{r} & Mrs Hamilton dined here.—

Thursday, September 24, 1835
Occupied the Morning with shewing Captn & Mrs Wauchope & Mrs. Hawkins Constantia[148] & the Children Experiments on light— &c &c
Sir Jeremiah & Lady Bryant, Sir Thos Sylvester, Captn Fawcett and Captn & Mrs Wauchope & Mrs Hawkins dined here. In evening being fine shewed them the Stars
Set digesting in Alcohol the "Calamus"[149] root
Swept (with Captn W) till 0^{h} Then set a new Sweep & went on till 2^{h}

[148] Either the wine-growing district of that name to the south of Wynberg, or one of its principal ornaments, the historic residence, dating from Dutch Company days, "Groot Constantia."

[149] Dr. Hall remarks: "Calamus is a genus of tropical climbing plants with thin reedy stems, belonging to the Palm family. Many have spiny pinnae at the ends of the leaves which hook on to surrounding plants. Spines (of a different nature), and thin, stiff stems occur in the Cape species of *Asparagus,* which have large tuberous roots, which Sir John would certainly have found a promising subject in his search for Medicinal Principles." This thus appears to be a misidentification by Sir John.

Friday, September 25, 1835

Walked out with M & Captn & Mrs. W up Wynberg Hill after which they left for Simon's Town

Prepared the Woodwork for the New Collimator

Set digesting in Alcohol the root of the "Calamus" [line erased]

Saturday, September 26, 1835

Distilled the Calamus-tincture It leaves a thick heavy acrid oil There remains a Sol^n of Saccharine matter which yields a precip-[itate] to Nitrate of Mercury

Sunday, September 27, 1835

Rode out on the flats & over the Hills & gathered Pepper Plants [this word erased] Satyriums

Swept from $23^h 30^m$ to $3^h. 15^m$

Tuesday, September 29, 1835

Drove out with Marg^t to call on Mr & Mrs G. Bird the new married but found Mrs B at Mrs Hare's where we also called as well as at the Stuarts—Then went on to Constantia-flats & gathered sweet scented yellow Iris bulbs   White, orange-flower scented Satyriums & a new sort of Lachenalia

Wednesday, September 30, 1835

Admiral & Mrs. Campbell called on their way from Simon's Bay into Town

Thursday, October 1, 1835

Passed the day at the Observatory

Friday, October 2, 1835

Dined at Col Bell's to meet Adm^l & Mrs Campbell. Present Col & 2 Misses Bird. Major Gregory. Captn Wallace? & . . . [his omission]

Saturday, October 3, 1835

Col & Lady C Bell.—Admiral and Mrs. Campbell Mr & Mrs Stuart & Mr & Mrs Maclear dined with us & Major Gregory

NB. There being no chance of Captn Wauchope's St Helena seeds growing here, gave the whole lot to Maclear the soil at the Observatory being much richer

Sunday, October 4, 1835
Attended Divine Service at Rondebosch
Outlined the Great sweetscented Satyrium[150] (white) for M.

Wednesday, October 7, 1835
Went into Town. Called on Lady C Bell Mr Harvey's (the new Accountant Gene^1) saw M^{rs} H—Mr Hutchinson—& on Mrs Truter Attended the Meeting of the S. A. lit & Phil Inst.

Thursday, October 8, 1835
Dined at Col. Bells and afterwards went to the Amateur Play— by the 98th

Sunday, October 11, 1835
Occupied in reading a Monstrous budget of European letters.— While so engaged with M. in the Bulbgarden, Maclear brought over young C P Smyth,[151] Captn S's son, his new assistant. with whom held a long entretien & who passed rest of day here—Mr Hawkins just dropped in—After reading letters, walked out with M and CPS. At Night came on suddenly to blow hard from NW & [entry stops]

Monday, October 12, 1835
A vile morning. Wind & rain. James & Stone went on their promised excursion to the Green Kloof.—Lord help them!—Kept Home as siduously all day.

---

[150] Probably *Satyrium candidum* Lindl.; if so, the species had not yet received a Latin name.

[151] Charles Piazzi Smyth (1819–1900), a fascinating eccentric with more than a touch of genius if ever there was one. The second son of W. H. Smyth (see note 84, May 11, 1834), this tall lad was sixteen years old, and had landed the well-paid but thankless job of making good as chief assistant at the Cape, where all his predecessors had failed. Succeed he did, applying himself to observations in the observatory and out in the wilderness on geodetic expeditions. He became Astronomer Royal for Scotland (1845–1888), was elected FRS in 1857 but, pos sibly uniquely, resigned in 1874. He was excellent as a spectroscopist, and much before his time in his project to establish a mountain observatory on Teneriffe. He was awarded a medal by the Royal Society of Edinburgh for his metrical studies of the pyramids of Egypt. This was where the trouble started, for he convinced himself that they contained in their measurements a vast store of prehistoric and recondite knowledge, at least as extensive as modern knowledge. He never forced his views on people, but they caused his retirement from the Royal So ciety, and he is possibly to be regarded as responsible for the foundation of most of the pyramid mystic cults, which the curious will find still flourish today.

Tuesday, October 13, 1835

Dined at Col Bells.—Present Mrs. Blair & Mr Clerke Burton[152]—Master of Supreme Court

Wednesday, October 14, 1835

Dispatched 2 copies of Pickergill's[153] Engraving to care of Mr Withers for D & Jamie at Calcutta. Sent to Observatory for one missing

Thursday, October 15, 1835

Dined in Town at Mr Ebdens[154]—a regular set dinner with a "Route" in Evening. Mr E's house has a drawing room &c upstairs & a dining room below!!

Friday, October 16, 1835

Mrs Hutchinson came in the morning

Skirrow called   Mr Hutchinson & Mr Withers joined at dinner—

Saturday, October 17, 1835

Measured S. and W. Avenues with Skirrow's chain

M^r and Mrs Harvey called

Tuesday, October 20, 1835

Skirrow called.—says a militia or yeomanry is to be organised. The first fire fly of the season seen

Cloudy night   No sweeping

Wednesday, October 21, 1835

Rainy day. Occupied in making working lists, Reducing Sweeps & comparing the Brisbane Catal with AC & with Rumker—N B.

ß is constantly — [minus sign] in RA about 2^s 0 on average in all the ✳'s examined except the Greenwich Standard Stars!!!^{155}

Sent home Skirrows flags—also lent him, a 2^d time, my small theodolite

---

^{152} Clerke Burton, master of the Supreme Court, lived at "Rouwkoop," presumably renting it from William Hawkins (see note 74, April 20, 1834).

^{153} Henry William Pickersgill (1782–1875), English portrait and subject painter. He became a member of the Royal Academy in 1825 (see Frontispiece).

^{154} Mr. Ebden's town house was at No. 29 Heerengracht.

^{155} In this comparison of star catalogues Sir John becomes critical of the work of Dunlop (see note 85, May 11, 1834). The ornate ß means the Brisbane catalogue. For Rümker see note 12, January 29, 1835. The designation AC may mean Airy Catalogue or Argelander Catalogue.

Thursday, October 22, 1835
Rode over to Observatory to confer with Maclear about the dis-
crepancies between Dunlop's and Rumker's Catals

A Letter from Sir John and Lady Herschel, Feldhausen,

to Caroline Lucretia Herschel, Hanover

Feldhausen Oct 24. 1835
Recᵈ Janu 19, 1836 [In a different hand]

Nᵒ 8

[In Sir John's hand]

Dear Aunt—

The last accounts we have of you are that you are elected a
Member of the Astronomical Society—and that to keep you in countenance and prevent your being the only Lady among so many Gentlemen, you have for a Colleague and Sister Member Mʳˢ Somerville.
Now this is well imagined and we were not a little pleased to hear
it.—May you long enjoy your well earned laurels.—

As I presume our news will interest you more than comments upon
what is going on in Europe in the first place be it known to you that
we are all well and thank heaven happy—the children one and all
thrive uncommonly. Not one of them have had a days serious illness
since here we have been. This is a great deal to say & to be thankful
for among 4 of them, Babies and all.—Carry and Bella can now read
very fluently and their Mama is beginning to teach them French and
between us both they will soon pick up a little German—I mean to
make little Willy learn the Black letter[156] as soon as he knows the
English Alphabet, and shall then give him German Books to read and
find out the meaning of, which I am convinced is the easiest way to
to teach children languages. This is the way Carry has got her
French—She really reads it with a very decent pronunciation and
begins to understand what she reads.—As to grammars & dictionaries
they are of no use to those little things

---

[156] Gothic-style type face commonly used for printing in Germany up to
World War II.

The stars go on very well, though for the last 2 months the weather has been chiefly cloudy which has hitherto prevented seeing Halley's Comet Encke's[157] (Yours) escaped me owing to trees and the table Mountain—though I cut away a good gap in our principal Oak avenue to get at it However Maclear at the Observatory succeeded in getting 3 views of it with the 14 foot Newtonian of my Father's (The Glasgow telescope) on the 14th 19th and 24th? of Septr—If you have an opportunity to let this become known to Encke do so—(I shall write to him shortly myself).—It was *in* or *near* the calculated place, but no measure could be got. It looked *"as he saw it in England"*

I have now nearly gone over the whole Southern heavens, and over much of it often. So that after another season of reviewing, verifying, and *making up accounts* (reducing and bringing in order the observations)—we shall be looking homewards.— In short—I have (to use a homely phrase)—broken the neck of the work—and my main object now is to *secure* and perfect what is done—and get all ready to begin printing the moment we arrive in England or if that is not possible, at least to have no more *calculation* to do.

Peter Stewart writes me word of a correspondence relative to "the old grievance" of not being able to persuade you to keep & use what is your own—and I am afraid that his decisive resistance may possibly (most unintentionally on his part) have given you trouble or uneasiness—In future then let this be our final arrangement and understanding on the subject—Messrs Drummond & Co will answer your drafts on them through Cohen, either *for the whole* amount, *or for any part of it, either in one sum, or in separate items,* as it shall suit your convenience to draw it—And if you do not want it, any one year or half year it will stand over on account, at your disposal, till

---

[157] Johann Franz Encke (1791–1865), German astronomer, who became director of the observatory at Seeberg, Switzerland. In 1819 he showed that the comets which had appeared in 1786, 1795, 1805, and 1818 were one and the same, being repeated apparitions of the same comet with the very short period of 3.30 years. Sir John calls the comet Caroline's because she was the discoverer of eight comets between 1786 and 1797, of which one was that now called after Encke. Becoming professor of astronomy and director of the observatory of Berlin University in 1825, Encke received the Royal Medal of the Royal Society in 1828. He deduced a value of 8″.57 (about 3% too small) for the value of the solar parallax from observations of the transits of Venus of 1761 and 1769.

the next, or till you may have occasion for it—Meanwhile, *until so drawn by you*—it will remain with Messrs D. as part of my current account—so that you need be under no apprehension of its disappearing or dropping into Messrs Cohens Strong box for want of a claimant.—I shall leave a little room for Maggie to add a few words before I fold this therefore Turn over the page—and believe me

dear Aunt & *Colleague*
your affectionate Nephew JFW Herschel

Regards and Compts. to all
Hanoverian friends

[In Lady Herschel's hand]

My dearest Aunt—Herschel did well to leave me a corner, for he has forgotten to thank you for the delightful & very long letter you call N° 3 dated April 23^d 1835, which we received with more pleasure than you imagine, nor can I wonder at the fond expressions which my brothers use about you—indeed many young ladies in London are quite jealous of the invisible object of admiration in Han[over] [?] Mr Frances Baily has sent your kind message about the Transactions of the Ast. Soc., to which you are now entitled, & Herschel says he will keep them carefully as an heirloom for Carry & her successors—Your message also from Haup^t Müller about Prof. Gauss's Magnetical Exper^ts arrived safely—and—Herschel has since had the pleasure of hearing from Gauss himself for a constellation of Astronomers were met at Berlin a few months ago when our friend Capt Basil Hall was visiting that place in his travels, & he made them all write by his opportunity to Herschel—There were Encke—Schumacher—Bessel —Gauss—and Olbers[158]—It is on Herschel's list to write to Gauss on

---

[158] Heinrich Wilhelm Matthäus Olbers (1758–1840), German amateur astronomer, a medical man by profession. After studying medicine at Göttingen, he spent all his spare energies on astronomy, having an observatory at the house in Bremen, where, from 1781, he practiced as a physician. As an astronomer he specialized in minor planets and comets, being the inventor of a method of orbital computation for the latter. He discovered the comet named after him in 1815, rediscovered the minor planet Ceres, and was the original discoverer of Pallas. He originated the theory that the minor planets are the debris of a single larger body. He is best remembered today for the paradox named after him, which demonstrates that if the universe is infinite and populated with luminous

the Exp^{ts} I hope that my Brother Patrick will manage to send you safely an Engraving of Pickersgill's admirable likeness of our dear Herschel [Frontispiece]. It pleases me very much—& I hope you will like it—Papa has given you all the news of the children & left me nothing to say, but that I hope you will see Caroline at Hanover before 3 years are gone & now my dear Aunt—God bless you—and preserve you in health till then—Accept the most affec^t love of your attached Niece

Marg^t B. Herschel

I wrote to you last dated May 19th 1835

The Diary: October 25 to November 7

Sunday, October 25, 1835

Read prayers with M. at home—Rode out to the 1st Sand hills on the flats at or near "the Turf Pit" & got an enormous nosegay of Gladiolus Alatus—Disa Chrysostachya[159]! ! Disa Aurea[160] (the blue disa dropped with *gold dust*) Disa graminifolia—[erasure] Satyrium Suaveolens[161]—&c &c &c.

Cloudy—no Halley's Comet nor Stars of any kind.

Wednesday, October 28, 1835

Rode into Cape Town. Called on Lady D'Urban and Dr. Phillip—Rode out—all cloudy & Black S E at Feldhausen

Knocked up a temporary stand for the 7 feet Equatorial telescope —dismantled it & carried it out to the 1st Sand hills on the flats

---

bodies of which the density does not decrease with distance, the sky cannot be dark. The investigation of the conditions which lead to the avoidance of this result goes directly to the foundations of modern cosmological theory.

[159] A species which occurs in more easterly parts of the Cape coast; this is possibly a misidentification.

[160] A previously unpublished name, probably for the Golden Orchid, now called *Disa cornuta* Sw., which has large reflecting cells inside the galea or head which appear golden in sunlight.

[161] A previously unpublished name, possibly used by Sir John for the Great Sweet Scented Satyrium (see note 150, October 4, 1835); suaveolens is an epithet commonly used for scented plants.

there erected it just at Sunset & was rewarded with the 1st glorious sight of Halley's Comet! ! !

Friday, October 30, 1835

Rode over to Observatory & went with M. [Maclear] through all the process of Examining [word written over] the level & Collim of the Transit.

Viewed Halley's Comet in the 14 ft Newtonian refl^r by my Father

Saturday, October 31, 1835

Shewed Halleys Comet to Col & Lady C. Bell—the 2 Messr^s Hawkins Jnr

Sunday, November 1, 1835

Sketched the Blue Disa[162] & the Gladiolus Blandus a superb specimen and finished the Satyrium Chrysostachium[163] for M.

Mrs Hutchinson came. Shewed Halley's Comet to M^r & Mrs Schönberg, Mr S. Jnr. & "the niece"—Old Mr Breda—(the Nestor of the Table M^n)—Eckstein Snr. & Jnr. and divers young dutch ladies of their party. They viewed it for the most part with indifference only Eckstein seemed interested

Monday, November 2, 1835

Left Feldhausen with Marg^t, Mrs. Hutchinson, Caroline & Bella, Hannah and Christine Thomson for Simon's Town.

NB   6 horses of Dixon's[164] to our own Carriage. Set off about 8½ reached Farmer Peck's at 11 where lunched & got into Simon's Town about 3 with Ease. Dined at Adm^l Campbell's—Put up at Clarence's Hotel,[165] a good enough place considering longitude & Latitude

Tuesday, November 3, 1835

Suffering dreadfully with the face-ache or "Sinkings"[166] as they call it here. Staid at home all morning

[162] November is too early for the flowering time of the species to which this is normally applied, *Herschelia graminifolia*. Perhaps this is another species of the same genus.

[163] This name was originated by Sir John, and published by the eminent British botanist, John Lindley. It was subsequently found to be a later synonym of *Satyrium coriifolium*.

[164] Charles Dixon owned a livery stable in Plein Street.

[165] John Clarence was the hotel proprietor.

[166] "Sinkings" (Dutch: Zinkings), neuralgia or rheumatics. Still used in modern Afrikaans.

Wednesday, November 4, 1835
Accompanied Mr & Mrs Hutchinson, Mrs Wauchope, Mrs Barrow &
Miss Osmond to a cottage called Rocklands[167]
Petrif [entry ends]

Basalt dike

Thursday, November 5, 1835
Staid at home

Friday, November 6, 1835
Returned calls
Dined at the Admirals.

Saturday, November 7, 1835
Ox Waggon Excursion to Cape Point

The Travel Diary

[No date: presumably November 7, 1835]

Novr. Saturday—For Cape Point
5. 25. En route from Simons Town in a waggon with 10 Bullocks
for the Extreme Point of S. Africa.[168] Party M. & self, Capt
& Mrs Wauchope, Mrs Hutchinson, Miss Osmond, Mr Deas
Thomson. the Gentᵈ on Horseback Ladies in Waggon.—As-
cended the "Red Hill above Simonstown a rugged & steep
acclivity with this a comparative table land covered with
sand & swampy flats—with occasional looks out to the sea
on West over very desolate rocky ground. Stopped to breakfast on grassy banks beside a poor Cottage of a farmer where
our bullocks were to rest & drink—after an hours rest again

---

[167] Miss Osmond, daughter of "King John" Osmond, Simon's Bay property
owner, whose house was known as "The Palace." "Rocklands" is the farm between Froggy Pond and Miller's Point, on the False Bay coast, now owned by
Mrs. E. W. Lasbrey.

[168] This is untrue; they went only to the southern end of the Cape Peninsula.
The extreme point is Cape Agulhas, considerably farther east.

en route & thus we got on with little diversity of country & now & then wild breaks of Rough arid mountain scenery on left, at the back of the steep hills which overlook False Bay till we were in sight of a wretched sheep & cattle Krall[169] the last humanised point, full in view of the Cape Promontory [sketch of Cape Hangklip]

The Diary: November 8 to November 10

Sunday, November 8, 1835
Div Service at Simon's Town

Monday, November 9, 1835
At 9½ AM. Left the Clarence Hotel Simon's Town (where M[rs] Clarence made us comfortable enough she being a very good kind of well behaved woman, attentive and considerate) in Hugo's Bullock Waggon drawn by 12 oxen for the Salt Pans[170]—& thence to join our Carriage at Farmer Peck's & home—11[h]± Salt Pans—gathered Candelabra[171] bulbs &c—NB a glorious place for Botany—Such Plants.— 1½ PM Farmer Pecks—Lunched there—and then home in time for dinner. M[r] Hutchinson came

Tuesday, November 10, 1835
Outlined for M. a superb specimen of the blue Disa
Mr Poleman[172] called to shew a specimen of the "Aphateia" a most extraordinary Parasite of the Euphorbia roots. Alias Hydnora Africana.[173]

---

[169] Kraal, an enclosure for cattle, or the enclosed village of a native chief.
[170] The Salt Pans are near Noordhoek.
[171] *Brunsvigia orientalis.*
[172] Polemann, the apothecary (see note 115, July 6, 1835).
[173] A brown, fleshy, leafless parasite.

A Letter from Sir John, Feldhausen, to James Calder Stewart, c/o Messrs. R. Turner & Co., Canton, China

received 23ᵈ April 1836 [In a different hand]
Feldhausen near Wynberg—C G H
Nov. 25 1835
ansᵈ 9 June [In a different hand]

Dear Jamie—If I am or appear to be in long arrear to you in point of Correspᶜᵉ it is because you flit about the globe, being in two places or more at once and in taking aim it is necessary to shoot forward of you. Now however you are settled and marked, so in Colonial fashion you may be knocked over deliberately.—

I shall take you up at Turin. Plana's conduct toward you was such as I should have expected from what I knew of him and has given him a lasting claim on my regard. I had lately occasion to transmit him a little memento of that sentiment in the form of an engraving from a certain picture by Pickersgill of which you will ere this reaches you, have also recᵈ a copy.—From Amici[174] I heard last month— Well & expressing himself pleased with his situation His son has recently been the successful competitor for a Mathematical Prize proposed by the Florentine Acad. of Sciences—subject the Equilibrium of Arches and Domes.—Why didn't you marry his daughter who if I recollect right promised to be a pretty pleasing lapin.—

You speak of a paradoxically caught cold in the *Happy Nine Hills*[175]

The Cape Colds will match it—and are equally puzzling.—I never here catch cold except from sudden *heat.* Two or 3 hot cloudless days

---

[174] Giovanni Battista Amici (1786–1863), professor of mathematics at Modena (1815–1825), then studied optics and astronomy. In 1831 he was astronomer of the museum at Florence. After 1859 he devoted himself to microscopy. He was noted for his work on plant fertilization and pathology and his optical work on reflecting telescopes, achromatic microscopes, prisms, and the camera lucida.

[175] The Apennine Mountains; one of the hideous puns in which Sir John occasionally indulged.

coming on after long continued cold & rain gave me not long since a paroxysm of pain &c such as I shall long remember. People here call it the "Sinkings."—

I followed you with Great delight (retracing my own steps, and singing out (O ego! quantus eram[176] &c) all down Italy and into Tyrol to my favorite Bolzano & Munich.—Maggie & I devoured your Journal with our soup & fish—for we used to make it a dinner dish, send James out of the room & read it aloud alternately and now behold— Maggie is wild to go to Italy and has made me down on my knees & swear to take her brats and all.—

The epigrams I sent you were not my own—(I lack that talent) but Grahame's who in the midst of sorrows, which no man feels more keenly than himself—often flings grief to the dogs and disports himself in Such Sallies.

Jones,[177] as perhaps you know, has been made Prof^r at Haileybury in room of Malthus.[178]—My! Why it was from yourself I learnt it—.I don't think it is a situation very fit for him—but anything with duties and regularly recurring call upon his stores of knowledge in his own department, is better than the dreamy half extinct existence in which he was vegetating.

On rec^n of your letters of 9th Decr and 9th Jany. I wrote immediately to Peter desiring him in case of S.E. and Co. proceeding to take up their Bills for the funds of mine in their possession to hold the same disposable for your use, according to his (Peter's) knowledge of your wishes respecting that matter. My letter to him (of date April 26) contained an order to Drummonds, (to be used or not as he should see occasion) superseding my original order to them to reinvest those

---

[176] Ah me! How great I was, etc.

[177] Richard Jones (1790–1855), political economist. He was professor of political economy at King's College, London, 1833–1835, and was at Haileybury, 1835–1855. He was an opponent of David Ricardo.

[178] Thomas Robert Malthus (1766–1834), political economist. After a fellowship at Jesus College, Cambridge, he became curate at Albury, Surrey, in 1798, the year of the publication of his famous *Essay on Population*. In 1805 he became professor of history and political economy at the East India Company's College at Haileybury, near Hertford. This had been founded in 1806, for training officials, and was closed in 1859, when the Company handed over to the Crown. The buildings became a public school. Malthus published, *The Nature and Progress of Rent*, 1815, and was elected F.R.S. in 1819.

monies in the Stocks, and directing them in lieu thereof, to pay over the amount to Peter, taking his personal receipt. Finally by way of placing P. and yourself in Communication upon the subject, I copied what I had written to him, as also the order on Drumm^d—and forwarded the Copies to you at Canton to the address you gave me. There, I presume you will on arrival (barring accidents) have found my letter.

But P. now writes me that (owing to certain circumstances of which you are informed respecting the dishonour of their Indian Agents Bills in London—) they (S.E. and Co) cannot just yet conveniently take up their Bills.—Whenever they do so the Money is at your disposal—and of this, intelligence will reach you from P. as soon as it does me, so that you can at once *act* in the matter without further reference to me about it—and if you are so circumstanced in your new situation that it will be useful to you—why you know me I hope by this time well enough to spare me your thanks being assured that the facility of being of service to you is all the return I look for & a rich one too.

Well now for some of our Cape news. We have been all staring our Eyes out at Halley's Comet—which was so good as to give us a good sight of him in passing to the Sun. By the way, he ought to have been seen in or near his Perihelion, close to the Sun *in the total Eclipse* of Novr 20th which was total in the Indian Seas.—Now if people crossing those Seas ever visit Canton do ask them if they saw this strange phaenomenon—which has never occurred but once before in history (A.D. 60).—

We are all well & flourishing, children & all. Carry is growing tall and learning to write & talk French (but I can't say she makes much progress in *that*). Bella is round and toddling, but sharp as a needle & of a quick sensibility & affectionate disposition—Billy is getting noisy—rides a stick beats his sisters—does mischief—but can hardly articulate a word.—Little Mag has taken a great penchant for me, for which reason I have taken a reciprocal kindness for her, especially as she is the quietest little animal that ever *was* seen and never gives the least trouble to any body.—And ere long I suppose I shall have some more to tell you about! So they fudge me with a pack of blarney[179] about its being a fine thing and a venerable one to be the

[179] Seductive and facile eloquence, supposed to be acquired by kissing an

Father of a fine family and all that sort of thing, at least so Maggie tells me, and so I am fool enough to believe it and let her have everything her own way—by which and which alone peace & quietness is maintained in the family.

The Stars get on well. I have now nearly gone over the whole Southern Sky & over much of it repeatedly.—I don't know that I shall be glad when my task is over. Whether much good—or much evil is in store for us on our return to Europe is uncertain—but this I know that we have been very happy here—and that our residence at the Cape—come what may will always be to me as Malthus somewhere or other beautifully calls some such happiness—the Sunny Spot in my whole life where my imagination will always love to bask. It is a dream—and too sweet a dream not to be dashed with the dread of waking. I wish you *could* have touched here in that confounded Kyd. [The ship, the *General Kidd.*]

Can you procure at Canton regular observations of *Tides* according to the enclosed paper of Whewell's.—Also carefully registered Barometrical and other Meterological Observations according to the "Instructions" also enclosed, and more especially hourly Observations on the days mentioned in pages 5 & 6 as there described. I shall shortly forward you a pronouncing alphabet, and will beg of you to make yourself or get somebody else made familiar with it—then catch a Chinese Mandarin of high Caste (i.e. as high as you *can* catch) tie him down—put *him on the* oracular tripod and make him utter the chinese words for certain things I shall give you a list of—slowly & distinctly, and write them down in my character—And as this is well and truly done so may you thrive.—

All the Brats—Maggie & all howl and scream to your health & happiness and I remain dear Jamie

Ever your truly

JFW Herschel

---

almost inaccessible stone in the tower of Blarney Castle, Ireland—nonsensical, persuasive, Irish speech.

The Diary: November 28 to December 30;

Occasional Memoranda

Saturday, November 28, 1835

Set out with M. to go to Observatory but horse was suddenly lame
& we could not get on—so turned back and occupied the time with
gathering Blue Disa's which are now abundant (Disa graminifolia?)
(Barbata?)[180]

Sunday, December 6, 1835

Attended Div. Serv. at Rondebosch—Mr Smyth called.

Monday, December 7, 1835

Lady Bryant called

Tuesday, December 8, 1835

Mrs Maclear dined here.
Began Sweeping before Moon-rise
After ☽ rose went to Eql & worked a Zone

Wednesday, December 9, 1835

Rose late—at 11½—Dined at 2
Breakfast & Dinner seeming all to run together—
(Gardening &c in interim)
M^t went into Town in the Evening. I staid at home & swept from
1^h 30^m ST. to 4^h 20^m ST. & then went to Equatorial & measured some
of the D. stars of last nights Zone

Thursday, December 10, 1835

Prepared for Polishing by Moulding—Guttering & scarring & colcothar prep 2.

Tuesday, December 15, 1835

Mrs Hawkins spent the day here. Brought a lot of Flower-paintings

---

[180] One of the early flowering Herschelias, a group of ground orchids with attractive blue flowers on a reedy stem, and narrow leaves. The genus is named
after Sir John. In the first citation of the name, the botanist John Lindley refers
to him as the "successful collector of Cape Orchids," and records habitat data
given by him.

of Villet[181]—Mr P. Smyth came to tea & to sweep through the Great Magellan.

Wednesday, December 16, 1835
Mr Smyth left.

Mrs Hutchinson dined here

Thursday, December 17, 1835
Rode early into Town to attend the Examination of the S.A. College as Mr Innes's "Assessor" of the Mathematical class—Called (by appt^mt at 10½) at 10^h 32^m on Lord Auckland[182] the new Govr General of India who is arrived here in the Jupiter—not at home.

Exam^n lasted from 9½ to 4½—thence went & made a *preparatory dinner* at Mr. Hutchinson's where Dr Adamson & Mr Innes had been asked to meet M^t and self.—Mr I came. NB he related the late affair of Gorah as on opp. page—Thence went to Col Bell's to dine where met Col B. Lady Cath B. Lord Auckland 2 Misses Eden L'd A's sisters?—Mrs Wauchope. Mr Menzies—Captn Raynier?—& a Dandy unknown with white teeth sleepy eyes & black curly moustaches.

[Opposite page] # Gorah affair[183] as stated by Mr Innes [entry ends] [Opposite page: reference to December 17?] Got home more dead than alive

Friday, December 18, 1835
A vile cold in head throat chest & all over—The Cape colds caught by a momentary chill or from no apparent cause in the hot weather are really dreadful. Could not go in to attend the Examin^n of the S.A. College as intended.—Inv^d by Sir Jerem^h Bryant[184] to dine & meet Ld Auckland but could not go—In the evening Ld A. & 2 Misses Eden (unless one was Lady A but ??) and a certain Dandy (handsome

---

[181] Charles Mathurin Villet, naturalist and seedsman of Long Street, who had a menagerie and botanical garden in Somerset Road. He was noted for his work as a zoological collector at the Cape in the 1820's.

[182] George Eden (1784–1849), second Baron Auckland, and Earl of Auckland, G.C.B., MA, Oxon. A barrister at law, he was MP for Woodstock, 1810–1814, President of the Board of Trade, 1830–1834, First Lord of the Admiralty, 1834–1835 and 1846–1849, and Governor General of India, 1835–1841. He was created earl in 1839, and was unmarried.

[183] The Gorah affair cannot be traced in the contemporary press.

[184] Lieutenant Colonel Sir Jeremiah Bryant, 64th Native Infantry, Bengal, and Judge Advocate General.

man A No 1) came to see Stars but they would not shine. His Ldshp got one momentary glimpse of the Nebula 30 Dor.

Saturday, December 19, 1835

To bed with cold in the head & chest & all over that I was obliged to write to Prof^r Innes to excuse myself from attending the Prize distribution at the College. M^t also bad with it—& the children.— Stone can hardly speak & in short it seems *universal*. It is evidently the effect of that *heat* which has of late set in strongly. It is surprising how *instantly* these colds come on & in what a tremulous thrilling state of fibre they put one.—All day long I have heard a noise like distant thunder from the friction of the blood in the veins behind the ear. In evening sweeping the Mag^l Cld#

[Opposite page] #Sent Lord Auckland a drawing of the Nebula 30 Doradus—not a *very* correct one but up to the present state of my data—The Stars are not down from micrometer meas^r The Neb. is like.

Sunday, December 20, 1835

Attended divine Service at Rondebosch. After w^h returned & strolled out with M^t to see the Namaqua's[185] (Andreas's) location which progresses but slowly.—Lord Auckland called & joined our walk over Wynberg hill—his object was to get copies of the Meteorological Instructions, being desirous of "giving" as he says a fillip" to that & such like matters in India. He appears to take an interest in promoting scientific objects—seems fond especially of botany & enquired much about the Cape plants &c

In Ev^g Equatorial & Sweeping

Monday, December 21, 1835

At 6 AM began the Hourly Obs^ns for the Winter [partly erased] Solstice. Viz Bar. Ther. Actinom. &c.—Occupied with them the whole day.—at 5 PM went to bed. got up at 9 & sate up all through the night and till 3 [erased] sweeping in the intervals

Tuesday, December 22, 1835

Before Sunrise looked out for Halley's Comet—not seen.— Went on with the Met^l Obs^ns till 4 PM when I went to bed—rose again at about 10 and made a night of it sweeping &c

---

[185] Hottentots.

Thursday, December 24, 1835
Went into Town & paid off the outstanding Mortgage of the Feld-
hausen Estate to the Bank & Dr Neethling[186]—
Began reading Sismondis History of France[187]

Friday, December 25, 1835
Christmas day Attended Div. Service & Sac[r] at Rondebosch
Church Mr Judge

Sunday, December 27, 1835
Wrote to Captn Smyth. and half a letter to Plana.

Monday, December 28, 1835
Finished Letter to Plana—Wrote to Beaufort & to Capt[n] Lloyd
about the Lacaille's[188] Quad[t] Finished Drawing N[o] [omitted] of the
Coming on of a N. Wester.[189] from the Shrubbery.

Tuesday, December 29, 1835
Drawing—Fitting squares into Drawing board.
Identification of Stars
Blowing a fierce gale from NW

Wednesday, December 30, 1835
Worked at Identification of Stars all day. In evening Fair Copied
& rearranged the hourly Meteor[l] Obs[ns]

---

[186] Not a medical man; probably Advocate Johan Henoch Neethling, of 5 St.
George's Street.

[187] Jean-Charles-Lénard Simonde de Sismondi (1773–1842). He wrote
Historie des Français, eighteen volumes, 1821–1834, and completed the work in
thirty-one volumes in 1844. No English translation was available until 1849.

[188] Nicolas Louis de la Caille (1713–1762), an outstanding figure in the history of southern astronomy. He entered the Church but his interest in astronomy
brought him the professorship of mathematics at the Collège Mazarin in Paris in
1739. He undertook survey work in western France, and then under the auspices
of the Académie Royale des Sciences de France went to South Africa from 1751
to 1753. He determined the positions of ten thousand stars, measured a geodetic
base line and delineated the Southern constellations. He also engaged in other
branches of physical science. The establishment of the Cape Observatory is directly traceable to the influence of his work. He published his *Coleum Australe
Stelliferum* in 1763.

[189] This sketch was used to illustrate the article on meteorology in the *Encyclo-
paedia Britannica* of 1857.

This diary (1835) Completely copied by L.G.[190]—Comparison of copy finished with her 19 July 1909

[In another hand]

Occasional Memoranda, November, 1835

Flowers in Bloom
Satyrium Carneum going off
Disa Caerulea? or ? Graminifol. in full flower

Occasional Memoranda, December, 1835

Fir seeds falling in great abundance—Gave Dawes's and Jacks[191] Children 1ᵈ a lb for collecting them—A great quantity thus got (at last got 400 lbs)

Flowers in Bloom. *Watsonia Marginata* (tall scarlet)
Here & there a Gladiolus blandus, *discoloured*
Blue Aristea the last lingering Agapanthus coming in—Dec. 20.
Pale yellow Gladiolus
[Erased; in same hand as previously]
19 July 1909 Finished copying L.G.'s MSS of the year 1835 with her at Bracknold
[Back fly leaf]
[Two rough sketches: 1. The Simonsberg seen from the road into Cape Town, with more distant mountains under snow. 2. Lion's Head seen from some point in the upper part of Cape Town on the lower slopes of Table Mountain.]

---

[190] Probably Louisa Gordon, third daughter (died unmarried, 1929) of Caroline Emilia Mary Herschel, who in 1852 married Sir Alexander Hamilton-Gordon (1817–1890), second son of the fourth earl of Aberdeen. Caroline died January 29, 1909.

[191] Dawes and Jack were laborers employed by the Herschels at "Feldhausen."

1836

The Diary: January 4 to January 12

Monthly Summary, March, 1836
Mar 2. Rec[d] this Volume See entry under that date

Monday, January 4, 1836
Attended Levee at the Governors on his arrival from the Frontier.

Wednesday, January 6, 1836
Attended meeting of the S. A Lit & Phil Inst Presenting there my
Hourly Solstitial Obs[ns] for the Dec.[ember] Solstice

Exhibited Wheatstones Exp[ts] for shewing the short duration of the
Electric light by means of a revolving disc painted in sectors with
Colours—and also by illuminating with the flash the grains of falling
Rice and drops of quicksilver

The Rice a very pretty Experiment. Sir B Urban attended. Introduced by him to Capt[n] Alexander[1]

Tuesday, January 12, 1836
Rose with slight pain in Chest which as the day advanced grew
severe with oppression of Breathing much increased by slight exertion, as walking across garden. Tried to fight it off and worked at

[1] Sir James Edward Alexander (1803–1885), soldier and traveler. Serving first
in the East India Company's forces, he transferred to the British Army in 1825.
He was present at the Russo-Persian War of 1826, the Russo-Turkish War of
1829, the Miguelete War of 1832–1834 in Portugal; served as aide-de-camp to
Sir Benjamin D'Urban in the Kaffir War of 1835; and then went on the exploring
expedition in Africa, of which more later. This brought him a knighthood in
1838. From service in Canada (1841–1855) he went to the Crimea, and thence
to the Maori War in New Zealand. In the arts of peace this fire-eating gentleman
produced a number of books of travel and reminiscence, and was largely responsible for the transfer from Egypt of that notable London landmark, Cleopatra's
Needle. We first meet him here, thwarted in some of his exploring ambitions by
Sir Andrew Smith. Later we shall see the Captain set off, sped on his way by Sir
John and other notables, on the expedition which resulted in *An Expedition of
Discovery into the Interior of Africa, through the hitherto undescribed countries
of the Great Namaquas, Boschmans, and Hill Damaras &c &c.*

Saturn's Satellites but it would not go.—At 2 PM Baron v. Ludwig and Miss L; the American Consul Chase,[2] also Captⁿ Caldwell of the American Brig Levant and 2 other American Captains called——— [his omission] Did the honours very badly being scarce able to crawl then went to bed & *Endured*. This was the worst bodily besetment I have ever undergone it was really terrible. Ended with one of my old fever-flushes and left me next day tempest-tost and half wrecked.

Bodily suffering is everything while it lasts and nothing when over. It fades from memory and one is ready to encounter it again—but the kindness experienced in it lives on and grows fresher & fresher for lapse of time.

The Travel Diary: An Expedition to Paarl

Friday 15 Jan. 1836

$6^h 30^m$ left Feldhausen for Observatory

$9^h 10^m$ En Route with Maclear in Achmet's 6 horse Cart for Paarl &c Big David with Maclear's horse & my Pony   D. carrying my Barometer. Took road across skirts of Tigerberg across a most dreary region of sand & stone.   The hilly part is however of Clay deeply furrowed by the rains to 15 or 20 feet ravines

11 50   Outspan at a halfway house in a dreary corner of the Tigerberg but apparently a rich corn country. The corn is trodden out by droves of horses in circular Kralls & winnowed as in Sicily by the wind   the straw being trodden to powder Gathered plants & bulbs & lunched.

1 21   En Route

2 36± Therm in Waggon = 87

3 35— Therm Dᵒ Dᵒ = 91 well shaded & free from all contact. Leave Klapmuts on Right & as the Paarl Rock gradually opens out country becomes more humanised.

5 53   Paarl. A very long straggling village of neat white houses with fir tree lines & Poplar patches on slope of hill down to

---

[2] Isaac Chase, the United States consul, not to be confused with J. C. Chase.

the Berg River where the Vale is broad & richly spread on the
hither bank with beautiful green vineyards.—Put up at old
Jordaens's which but for one nuisance would be well enough.
They have *newly plaistered* the floor with a mixture of clay
& cow dung the beasts.—Walked up with M. [Maclear] to
reconnoitre the ascent for tomorrows work & came back to
supper & a night rendered sleepless by heat & stench & dogs
howling.

Sat 16 Jan
7 30—En route for Paarl Summit
9 24—Summit of Paarl. On road came to a copious stream 2/3 way
up which is remarkable being drainage of a very small
area. It is said to Run all the Summer.—few or no flowers.
Soil where not granite, very hard baked pot clay of which
they make mud walls w^h last 50 or 60 years in great bricks.—
Climbed the steep granite Block of *Paarl* obliged to take off
shoes & stockings. Again noted the granite dike!! Established
Instruments & began observing

10. 58. Att^3 88.5—Bar. 27.844
11 59—Att 93.0—Bar 27.845
2 55   Att 84.5—Bar 27.765
3 48   Att 82.6—Bar 27.737

Since 2 PM wind SE rising & now strong = 6. or 7.^{4} Maclear
fixed his theodolite on a perfectly horizontal area about 6 feet
across the bottom of a natural Basin on summit. I my Barom
on an exactly similar one 6 feet lower.—NB. also *Such* is
one of the water pools through which the granite dike runs.
Wh is the bottom *a true plane* & *horizontal*. Descended exhausted & dined in a lovely Bower

6^h 5^m Started from our Bower—
Arrived at the brook where long draughts & a plentiful ab-
lution of feet & hands proved most acceptable & refreshing.
This brook is a copious stream & is never dry which is re-

^{3} "Att" is a contraction for "attached thermometer" (see note 94, May 12,
1835).
^{4} The designations 6 or 7 are wind strengths on the Beaufort Scale. They will
seem high only to those who have never experienced the full blast of the southeaster.

markable considering the small basin which it drains & the rocky nature of the higher summits. A little below the "Bower" is a natural house of a single granite flake 28 paces long & 7 internally broad fallen oblique & resting on another [a small sketch, Plate 10] thus.—

The "Bower" is one of the most beautiful combinations of rock & tree I ever saw—Got a rude sketch of it but could not do it justice, my drawing board being quite full of my Panorama from the Summit [Plate 11]. Noticed also another very remarkable specimen of granite exfoliation in a block of vast size almost exactly spherical on a highly inclined slope. Took sketch of it. Situation most romantic.

NB. The view from Paarl Rock is really superb. Only the vast extent of the flats is Ocean-like & monotonous. But the narrow Paarl valley with its river & vineyards & villages & noble background of the Zwartland & Drachenstein mountains, the French Hook Valley, the Donkersberg and Bang Hook—the Prominent Simondsberg &c—are really most striking.—

This proved a dreadfully fatiguing day the wind at last grew strong & blew all the things about and as the Haze thickened over the flats till the Table Mountain could no longer be seen, it appeared hopeless to wait for two Blue lights we had ordered to be fired from the Observatory & from Wynberg House, the more especially as the Obs^y is hidden from this station by the whole mass of Tigerberg. The Bivouack aloft would have been formidable & the descent of that granite slope after dark would have been very hazardous to limbs and insts. So descended & supped & went to bed.

[Sunday January 17, 1836][5]

8 36   En Route from Jordaen's at Paarl. where we found very good treatment & scrupulously clean linen. The impression left by this landlady was this time much more favorable. The old man still lingers on in a very infirm state at 86—After leav-

---

[5] Sir John's heading for this is "Jany 16 — 36." This is incorrect for the civil date, because clearly, a new day is starting, and it is a Sunday. It is just possible he may have been using the astronomical date, which, before 1925, changed at noon.

ing Paarl which we contrived to do in spite of a preconcerted arrangement to the contrary just as all the farmers & their families were coming in to Church exhibiting the gay dresses of their wives & daughters. Traversed a wide plain of thin short vegetation covered with anthills about 2 feet high till we came to the *Col* or Kloof which separates the Simon's berg from Klapmuts Hill, where there are 3, 4 or 5 good farm houses, white, neat & Dutch. Applied for lodging at 2. All family out. Were told that Mr Martin Beer or Bayer lives ¼ hour across the Col who would be sure to take us in   however got 3 blacks to carry up the Instruments to the Klapmuts Peak & [sentence terminates]

11 40	En route from high road for the Klapmuts point		
1^h 48^m	—Therm Att 93.5	Bar 28.200	
2 56	94.5	28.182	
4 22	88.3	28.145	
	Verified lower level being calm & serene	Sun nearly setting[6]	
6^h 54.	Att 71.8———28.144		
7 35	— 70.3———28.150		

The Diary: February 5 to February 7

Friday, February 5, 1836
2^m before 12 at night Alex^r[7] born

[6] One of Maclear's official tasks was to verify Lacaille's survey work, which had produced an anomalous value for the length of a one degree arc of the Earth's meridian in the Cape area, and to extend this by making general geodetic and topographical surveys. The trip to Paarl seems to have been a kind of trial exercise, for at about this time Maclear was just beginning the work which he so successfully prosecuted later. On several occasions, in collaboration with Sir John and Captain Wauchope, marker rockets were to be fired—the "Blue lights" of the entry for January 16. The remark "Sun nearly setting" associated with the last two entries suggests that the times are in local mean time.

[7] Alexander Stewart Herschel (1836–1907), second son of Sir John, M.A., D.C.L., F.R.S., F.R.A.S. He was professor and honorary professor of physics, Durham Armstrong College of Science, Newcastle. He died unmarried.

Sunday, February 7, 1836
Wrote to Maclear about 1 Aquilae[8] being variable. See Sweep 609

A Letter from Caroline Emilia Mary Herschel,

and Sir John, Feldhausen, to

Caroline Lucretia Herschel, Hanover

Nº 10

Feldhausen, Feb. 19th 1836
Recᵈ May 29 [In a different hand]
[In young Caroline's hand]

My dear aunt I am very happy to tell you that I have got a new live doll called Alexander he is very quiet. Mama will write as soon as she is able. Papa has seen the Comet and he looks at it every night. Papa sends his kind love and he will write soon. Mama and Bella send their kind love to you.

I am your loving niece
Caroline E. M. Herschel

[In Sir John's hand]

Dear Aunt—To interpret the enigma little Carry sends you be it known that you have a new Nephew to be called, after his Grandpapa Stewart and his Granduncle Herschel *Alexander*. This was on the 5th inst 1ᵐ 30ˢ before midnight—the Comet then just rising.—Yesterday & today his Mamma took a ride out to inhale the fresh air on Wynberg Hill (i.e. of course not on Horse back—but even in a carriage that is pretty well in a fortnight).—

As Carry hints, I have got a fine series of Comet observations & hope to see it long after everybody else has lost sight of it.

Little Bella would have written to you by this Post but is too busy with a letter to Grandmamma, and two letters in one day was too much for so young a correspondent.—Bella has literally taught her-

---

[8] Now called α Scuti, not listed as variable.

self to write by copying everything she sees written or printed, and writes quite as good a hand as Carry who is so much older.

Pray when you write to or see Gauss, say to him that I have not got the Apparatus for his observations on the Magnet, nor have I a very clear Idea of the method itself, as I have not got Poggendorf[9] here I remember indeed just before leaving England to have read (in a hurry) a description of the method, but I have not a sufficiently distinct remembrance of it to construct an apparatus and I fear my time is now too short to write home for one & receive it in time for any series of Observations.

Your affec[te] Nephew JFWH.

The Diary: March 2 to March 8

Wednesday, March 2, 1836

This afternoon I received this Book in a packet from P. Stewart with the Phil. Magazines for Nov. & Dec. of 1835 and the Supplement. —The Entries made in it previous to this date are either copied from a temporary Journal, or entered from Memory or from a collation of Dates from other Memoranda   Sweeps, &c &c.—

Drew up a notice for the Ast Soc. of the observed "Singleness" of γ Virginis & enclosed it to Beaufort with a letter to Henderson.—to go to Post tomorrow.—

In morning Packed Bulbs for Brown the Nurseryman—for Miss Tunno—for J. H. Nelson & for Dr Jennings.—

In afternoon marked out & cleared of Bush a line of road to lead out into the Wynberg road up the Hill obliquely

M. Rode out to M[rs] Cumberledge and Brought home a lot of her Flower painting

Packets arrived from Sir E. Ryan & from D. Stewart with Enclosures from Jamie.

[9] The reference is to Johann Christian Poggendorff (1796–1877), editor of the journal usually referred to as *Poggendorff's Annalen*, correctly titled (1799–1819) *Annalen der Physik* and (1819–1924) *Annalen der Physik und Physikalischen Chemie*.

Thursday, March 3, 1836

Took horse at 8½ at set off from Feldhausen with David for Houts Bay across the "Kloof" above Constantia on a bulb gathering expedition armed with pick, trowel & Bags, being warned by the full growth of two superb Josephinas in our garden that the time for gathering the Autumn flowers is at hand. The beauty of the Constantia hills has been much impaired within these few days by a great fire which has blackened all the mountain and extending round it over the Kloof has reduced to ashes every green shrub rendering the whole region on the other side (already barren enough) a scorched arid desert.—Its traces Extend for 3 or 4 miles along the Valley but it terminates before we reach the sea.

Hout's Bay itself is sheltered, secluded, but wild, desert & inhospitable. Yet its character seen in the blackness of today's atmosphere, with a strong wind blowing in to shore and a dark background of Cloud is somewhat picturesque with an air of the grand & melancholy. An old blockhouse picturesquely situated on a knoll of the richest and most verdant evergreen under shrubs mingled with toothed & jutting crags gives it an air like an Apennine Sea View of Salvator or Pompeii.—

Found 5 Brunswickia Josephina[10] Bulbs in *deep sand hills* under & among the Shrubs (of Blown Sand from the beach). They are not yet in full flower—mere buds & in consequence very difficult to find as their dried leaves are all torn away & scattered by the Winds. When fully in flower they must be magnificent ornaments of these singular spots.—The scarlet and yellow (? Flame lily? Cyrtanthus Angustifolium[11] is now abundant in swampy places recently burnt, among roots of the Rushes &c. Found superb bulbs of the Satyr^m Carneum!!![12]

Friday, March 4, 1836

Planted out Roots of the Satyrium Carneum and of the Cyrtanthus Angustifolius. A morning of Gardening—

In Ev[enin]g. walked with M. Read Holman's Travels Vol. 4.—

---

[10] *Brunsvigia orientalis,* Candelabra flower.

[11] *Cyrtanthus ventricosus* or Brandlelie (fire lily). Produces a fine umbel of scarlet and yellow flowers only after a veld fire.

[12] *Satyrium carneum,* Rooi Trewwa. A large, fleshy, pink-flowered, Cape ground orchid.

Wrote to Holman and Peter (in M's letters)—Worked at Equatorial. Comet hardly visible for Moon (1 day past full).—Measured D. Stars.—

Saturday, March 5, 1836

Archdeacon Robinson[13] from Madras called. Worked till morning twilight at Equatorial Star Magnitudes—Comet &c. Tried a project for Comparing Stars[14] with Moon by total refl^n at base of a Prism

Sunday, March 6, 1836

At Home all day   Prepared a prism for Star Photometer as a trial Maclear came at 3 & spent ev[en]ing—His adventure with his Horse in the Liesbeek   Mrs. M. came in Ev[enin]g in car[riage] She staid M walked home   Conferred about Reform of Southern Constell^{ns}[15]

[13] Archdeacon Thomas Robinson (1790–1873), formerly chaplain to Bishop Heber. Later he was professor of Arabic at Cambridge (1837), and Master of the Temple (1845).

[14] The first hint of Sir John's invention of the astrometer (as he called it), the first metrical device ever proposed for star photometry. The principle of the instrument depended on the visual comparison of, necessarily, bright stars with an image of the Moon which could be reduced in brightness at will. The instrument was mounted on a vertical support with a long rod carried on a universal joint at its top. A slider could be pulled back and forth along the rod. On the slider was a swiveling cross arm carrying the prism mentioned in the diary. By swinging the rod and swiveling the prism, Sir John could point the rod in the general direction of the star to be observed while catching the moonlight on the prism so that by total internal reflection (involving no light loss), the Moon's rays could be sent parallel to the rod to a short focus lens which then formed a very small, starlike, image of the Moon. By pulling the slider near or far by means of cords, the distance of this image, and hence its apparent brightness, could be varied until it matched the star. Measuring the distance to the lunar image, and applying the inverse-square law gave the apparent brightness of the star. If, while conditions remained the same, a second star was observed, the relative visual brightnesses of the two stars could be compared, using the Moon as an intermediary. Only stars within certain angular limits of distance from the Moon could be compared, and the instrument (illustrated on page 354 of *Results of Astronomical Observations at the Cape of Good Hope*) must have been extremely tricky to use. A comparison of Sir John's results with modern ones suggests that for stars of approximately solar color he may have been able to get mean visual magnitudes good to a tenth of a magnitude or so.

[15] Most of the papers drafted by Sir John on this topic seem to be in the South African Archives in Cape Town. To speak frankly, the southern constellations were a mess. The boundaries were indefinite; many straggled over great arcs of sky; many pairs had closely similar names, leading to ready confusion. Sir John

Monday, March 7, 1836
At home all day   Mrs Maclear returned to Observatory
Equatorial   Comet n[ot] s[een]

Tuesday, March 8, 1836
Sweeping

A Letter from Sir John and Lady Herschel, Feldhausen, to Caroline Lucretia Herschel, Hanover

Cape of Good Hope
March 8th 1836
Rec^d May 17 [In a different hand]

[In Lady Herschel's hand]
N° 9

My dearest Aunt
    I hope little Caroline's letter has preceded this by a fortnight
at least, & gives you some idea how we *thrive* here, & what a nice
little Brother she has got lately—But I had better confess the whole
truth at once, which is that Herschel got a second son on the 5th of
Feby 1836, & thank God, he & his Mama have been behaving themselves pretty well ever since, & I hope not giving very much trouble
to dear Papa, who does not look *very miserable* at this accumulation

---

had a variety of projects, and never seemed firmly to fix on one. In one form the
revised constellations were to be convex polygons of which each vertex was to be
a star. In another the vertices were to lie between stars. A favorite point was to
be the use of "names of recognised Classical Assemblages," such as nymphs and
what not. He tried this all out on Maclear in a series of documents, but spoiled
his, reasonable enough, idea, that all likenesses of animals, birds, and so forth
were to be abandoned, by wanting to retain Orion—but the other way up.
Maclear thought this would bring discredit on the whole idea. An extended account of his proposals is given in a paper published in the *Memoirs of the Royal
Astronomical Society*, 12 (1842), 211. All the constellations have now, of course,
been tidied up, so that, just as the boundaries of the states of the United States
mostly run along meridians and parallels, so those of the constellations run along
meridians of Right Ascension and parallels of declination on the sky.

of honours (or burden?)—In sober truth *I do rejoice* that a Father's &
grandfather's name may now be entrusted to *two* & heartily do I pray
that they both may appreciate & if possible *add* to the value of their
inheritance—Caroline took suddenly to letterwriting while I was in
my room & her first trial was a letter to *me* & then Papa said that the
next should be for Aunt Caroline so he ruled the paper & mended the
pen & left her to herself—& when finished the letter was directed and
sent by Post! I hope its arrival caused a Bottle of Constantia to *dis-
appear* to the health of old friends as well as *strangers*—We have decided that the name of this same Baby Boy is to be Alexander Stewart
after my dear father, & the first name belonging to your family also,
makes a *double union*—The Sponsors are to be Capt & Mrs Wauchope
very dear friends of my parents in Scotland in former times, & who
arrived here most unexpectedly & most agreeably to me.—Herschel
esteems them very much, & Capt Wauchope being a Seaman is alive
on most subjects interesting to Herschel—The other Godpapa is our
very beloved James Grahame whom you know already—I am anxious
now to return to my little Companions Caroline & Bella whose good
habits of regularity have been kept up by a young governess mean-
time, & Bella's ardent desire for knowledge & untiring perseverance
have been gratified by talking with Papa at breakfast time—Thus in
two mornings she learnt by heart the Greek Alphabet & on the third
knew how to write them—on another they both understood the signs
+ and − &c—& this is now their favourite game, to answer Arithmeti-
cal questions on the Slate—However I think Caroline's greatest talent
lays in Music, & her voice daily strengthens & improves but Bella
seems resolved to conquer everything— Herschel says he quite envies
the unwearied industry & the *right headedness* of the little creature,
& indeed she is enough to spoil any Papa & Mama—Little William is
only a Smith & Carpenter as yet, & perhaps would have driven in the
nail in your wall to your satisfaction especially if it was to hang
thereon his Papa's picture which I hope you have received by this
time & approve. A few months ago we were made glad by the rec^t of
N° 4 from your dear hand—dated Augt 6th 1835—

[In Sir John's hand]

Dear Aunt.—Maggie desires me to finish this for her but she has not
left me room to write at length. So I will only devote this space to one

point in your last letter which requires reply.—I have not got Gauss's apparatus and I am not sufficiently acquainted with his method of observing to construct one for myself. Besides which it is now quite out of my power to undertake any extensive series of Observations, being anxious to get home and having still so much to do both in observation & Reduction that I really shall hardly be able to accomplish all I have already in hand.—This Comet has been a Great interruption to my sweeps and I *hope and fear* it will yet be visible another month.—Unluckily when I set off from England, I left all my volumes of Poggendorff and the Nachrichten[16] behind me and none of the former & only a few of the latter have reached me here. I fear it is now too late to send home for anything and I have two series of observations—of the Comparative Brightness of the Southern Stars, and of the Photometric estimation of their magnitudes the former just commencing the latter nor yet begun which I *must* do. Pray explain this to Gauss.—I will answer for *Maclear's* readiness however to make any observations he may desire provided only he is furnished with the necessary apparatus.—Astronomical news I have little but one thing *very* remarkable I must tell you γ Virginis[17] is now *a Single Star* in both the 20 feet and 7 feet Equatorial!!!

Your affec. Nephew
JFW Herschel

[In Lady Herschel's hand]

A kind husband's rich postscript will make up for his poverty struck wife's letter but though very stupid with headache & remaining weakness I am still my dear Aunt's most affect. Niece—Marg^t B. Herschel Herschel is simply K.H.[18] & not a Companion of the Bath, as you have been told.—M.B.H.

---

[16] The *Astronomische Nachrichten* (see note 43, letter to Caroline, November 10, 1833).

[17] The components of this double star were so close as to appear single. Sir John published an account of this star in *Memoirs of the Royal Astronomical Society*, 5 (1833), 171, and wrote to Francis Baily about it in 1836 (*Monthly Notices of the Royal Astronomical Society*, 3 [1836], 197).

[18] Knight of the Hanoverian Guelphic Order.

The Diary: March 9 to May 5

Wednesday, March 9, 1836
At home all day. A night of heavy rain

Thursday, March 10, 1836
At Home all day. Adapted 2. Screws to my 20ft Eyepiece to adjust the wires to Collimator with more convenience than heretofore. NB. Lost much time for want of an efficient set of *Small Drills* NB. This is a want of constant occurrence My set is spoiled and broken.

Marked out the ground for my newly projected apparatus for measuring photometrically the magnitudes of stars.

Worked in Bulb garden after dinner

NB.   Bulbs begin to GROW & call aloud to be planted out.—

Antholyza Ringens[19]   } All sprouting
Antholyza Praealta[20]   } violently
Disa Graminifolia   } Satyriums begin
Ferraria Pavonia[21]   } to grow also
Babiana Angustifolia   }

In short all is alive now among the Bulbs—the Dormant stage is past.

Copied out the Magnitudes of Stars in Bodes[22] Antarctic Map fair— to be emendated & enlarged

[19] An extraordinary Cape member of the family Iridaceae, with peculiar scarlet flowers borne near the ground, and the stem extended beyond to form a rigid "bird-perch."

[20] *Chasmanthe aethiopica,* Suurkanol (Iridaceae).

[21] Generally regarded as a synonym of a Mexican *Tigridia.* Although this could have been brought to the Cape, it is more likely that Herschel was referring to the Cape *Ferraria undulata* (Iridaceae), a plant with dark, putrid-smelling flowers that attract flies.

[22] Johann Elert Bode (1747–1826), German astronomer, who in 1774 founded the *Astronomisches Jahrbuch,* fifty-one volumes of which he compiled and issued. He became director of the Berlin Observatory in 1786. In 1801 he published his *Uranographia,* a set of twenty star maps accompanied by a catalogue of 17,240 stars and nebulae. He is usually associated with the empirical relationship be-

Swept for 1^{h}¼ but it clouded—NB   after heavy rain last night
SUPERB
Proceeded (2 pages) drawing out the list of observed stars from
Ɓ [the Brisbane Catalogue] preparatory to reducing

Friday, March 11, 1836

Lieut Henning RN Captn of Windsor ⎫
                                     ⎬ called.
Mr Sutherland[23]—   ⎭

Went into Town (with Mrs Thompson) to make calls & get cash
Called on Mr. Grainger[24]—Mr. Chase the American Consul—
Baron Ludwig   Worked in garden   Lieut Henning came to dinner.
After dinner swept—Shewed Lieut H. the planetary nebula preceding η Argus.[25]—Then turned 20 ft on Comet which he also saw and
also Margt.   then after H. left went to work at Equatorial being a
superb night for definition   Something quite extraordinary   worked
till twilight.

Saturday, March 12, 1836

At Home all day   Rose very late (12½)
Carpenter at work on a new apparatus for Photometric Comparisons
of stars with the Moon.

                                  ⎧ Babiana Angustifolia
Bulb garden—Planted out   ⎨ Ferraria Undulata
                                  ⎩ Satyrium Cucullatum

No sweeping at Night   Began in night [word illegible] reducing
my sweeps. It is dreadfully slow & tiresome work—having to hunt for
stars in Piazzi[26] &c

---

tween the integers and planetary distances, known as Bode's Law, but his main
contribution in this connection was to give currency to the relationship previously found by Titius of Wittenberg.

[23] Mr. Thomas Sutherland was secretary of the South African Fire and Life Assurance Company.

[24] Robert Granger was a merchant and shipping agent.

[25] NGC 3211.

[26] Giuseppe Piazzi (1746–1826), Italian astronomer (after whom Charles
Piazzi Smyth was named). In 1780 he became professor of higher mathematics
at Palermo, and later helped to found the observatory there, of which he became
director. He produced a catalogue of 7,646 stars, noted for the accuracy of its
positions. On January 1, 1801, for purists the first day of the new century, he discovered the first minor planet, Ceres.

Sunday, March 13, 1836

Attended divine Service at Rondebosch. The 2 Messrs Hawkins Junr. called. the Elder P P C being about to sail for Brazil & England. Walked in Ev[enin]g about with M. about the grounds. Compared Maclear's R.A.'s of my zero stars newly observed by him with Dunlops.—The differences are enormous & irregular—but all one way (with only 2 exceptions).—This further trial decides the fate of the "Brisbane" or rather the Dunlop & Richardson Catalogue. It cannot be trusted in R.A.

Monday, March 14, 1836

Called with M. on Mrs Blair who is on the Eve of departure for England per Carnatic

Swept—among clouds—in intervals very fine definition. Picked up several *exquisite* double Stars in a part of the heavens I had set down as barren & uninteresting!! in my Chart Index

Thursday, March 17, 1836

Set out with Marg^t in Achmet's Cart & 6. with Maggie, Billy, Hannah & Baby on a Flower and Bulb hunt to the Salt Pan between Fishhook Bay and Slangkop-bay.—Passed through Wynberg which every time we see it shews fresh huts or cottages building—and is rapidly growing into an extensive Village.—Beyond the halfway house noticed a Cottage building of *Black bricks* which on Examination proved to be the hard black *Ants-nests* squared. No doubt quite as good as the ordinary *Sunbaked* bricks of the country.

Where the road begins to approach the Muysenberg & the Lake the soil becomes more sandy & I was just remarking to Maggie that there was a good habitat for Josephinas than (the words being hardly uttered) behold a large one and close at hand another Gigantic specimen These we bagged & soon found others in fine flower.—Just opposite Farmer Peck's we found the Haemanthus Tigrinus[27] in flower. —Went on to the Salt Pans but on the Hill Side before Fish-hook Bay beyond the Turnpike observed groups of them 5 or 6 together.— Went on to the Salt pans where found Drimia Elata (enormous bulbs with tall greyish spikes of flowers) and that beautiful Tiger Haemanthus the latter only occurs in the Bush covered hillocks of blown

[27] *Haemanthus tigrinus* is an eastern species. Sir John probably meant *H. rotundifolius*, which is still well known in this area.

sand which skirt the Salt Pans—& there in Profusion among the bushes.—The Salt Pans are now dry & the Salt in Crystals. It is collected for sale.—The whole area of the bottom is covered with what seems white wool but is either a grass or a seaweed—about 4 inches deep—very brittle, and leafless—Took specimens. Having bagged 60 Haemanthus bulbs & as many Drimias as we could stow away returned & in coming back went to the Spot where we had remarked the Josephinas.—Found no less than 10 in one locality A truly magnificent display! On the whole we got 14 of these fine bulbs today. They root very deep & the fibres extend right down at least 18 inches or 2 feet below the Bulb.—Stopped a while at Peck's, returned very tired. & went to bed.

Saturday, March 19, 1836

Museum—

Sunday, March 20, 1836

Rondebosch Church. M. at home
Rencontre with Jansen

Monday, March 21, 1836

Meteorol
Made out report of Speech for Chair [?]

Wednesday, March 23, 1836

Fitting in Fraunhofer's Prism
Bulbgarden
Great N Wester—heavy rain. Mrs Hutchinson sent for did not come [two words illegible]

Friday, March 25, 1836

At home all day. Rain having fallen 2 days ago and the ground being now in a favourable state for planting out bulbs I made a gardening day of it

In Evening got 1 or 2 obs^{ns} of comparative light of Stars by my Star-pole (Astrometer). Many unforeseen difficulties in the mechanical part of it but it cannot help answering—

Set a Sweep to take in Halley's Comet.—It is fading rapidly now but nucleus is Still pretty bright

Swept from 11^h 30^m—to 16^h 3^m ST.

Saturday, March 26, 1836

Went into Town with Marg[t] to see the Exhibition of Dr Smith's Collection from the Interior & Ford's & Bell's drawings[28] which are uncommonly beautiful especially the zoological ones of Snakes, Lizards &c.

Brought back with us Mr & Mrs Hutchinson who remain

Being very tired—went to lie down an hour or two before stargazing but waking found that sleep had got the upper hand so effectively that I was forced to give up all idea of working for the night

Sunday, March 27, 1836

Divine Service in Rondebosch Church Major Dutton called & brought out two Golden Cuckoos[29] a Male & Female (skins stuffed) for Marg[t]s collection. The male is all over brilliant green on the back with yellow breast—The female quite different—brown back with green tips to the feathers and spotted breast.

Identified Stars for Zero Catalogue & copied list to send to Maclear

Mr & M[rs] Hutchinson dined here.

In the Evening the Moon being more than half, tried my new *Astrometer* with a single lens which does much better than a combina-

---

[28] One of Sir Andrew Smith's publications on his expedition is listed by Mendelssohn's South African Bibliography as follows:

Illustrations of the Zoology of South Africa; consisting chiefly of figures and descriptions of the objects of natural history collected during an expedition into the interior of South Africa, in the years 1834, 1835 and 1836; fitted out by "The Cape of Good Hope Association for Exploring Central Africa." By Dr. Andrew Smith, M.D., Deputy-Inspector-General of Army Hospitals; Director of the Expedition. Published under the Authority of the Lords Commissioners of Her Majesty's Treasury. (Five Volumes.) Mammalia—Pisces—Aves—Reptilia—Invertebratae. (Bound in three Volumes). London: Smith, Elder & Co. . . . (No Pagination) Roy. Quarto. 1849:

[The bibliographical note continues] This handsome and valuable work gives full particulars of the natural history subjects of South Africa, collected by Dr. Andrew Smith's expedition in the country lying between 25° and 27°58′ east longitude and 31° and 23°28′ south latitude, embracing part of the Cape Colony, Natal and "Kaffirland." The illustrations, which are of a very high order, were executed by Mr George Ford, who accompanied the expedition, and consist of five plain and 273 coloured plates, while the letterpress is of a most accurate and exhaustive character.

[29] Not a very apt name from the description. Probably the Emerald Cuckoo (*Chrysococcyx cupreus*).

tion [word *telescope* erased] & is far simpler. Got a few Comparisons
—Enough to satisfy myself that the thing will work its work.
　　Got the Comet per Equatorial but it is getting very faint—
Measured a few D[ouble] Stars & went to bed

Monday, March 28, 1836
　　Major Nelson[30] Engineers called with a letter from Olinthus
Gregory[31]
　　Worked at Astrometer till Moon set then went to the Equatorial
& went on (being a superb night) till near daybreak—

Tuesday, March 29, 1836
　　Rose late.
　　Planted out Bulbs
　　Mr Ford called
　　Black S Easter No Stars
　　Went on drawing out Zero List
　　Wrote to Mr Morgan at Bathurst[32] about his Meteorol journal

Wednesday, March 30, 1836
　　Wrote to Messr^s Beer & Mädler[33] thanking them for their present
of Engraving of the Moon and Sending an obs of ♄ with the 6th
Satellite[34] Observed stars with the Astrometer till moon setting—then
got a measure or 2 of D Stars with the Equatorial

---

[30] Probably Lieutenant Richard J. Nelson, Royal Engineers, is meant.

[31] Olinthus Gilbert Gregory (1774–1841), English mathematician. He published, in 1793, *Lessons Astronomical & Philosophical.* A bookseller, he subsequently became a mathematical professor at Woolwich. He was the author of several works of mathematics and biography, and a founder member of the Astronomical Society.

[32] Presumably Dr. Nathaniel Morgan, an 1820 settler from London, who lived at Bathurst in the eastern Cape.

[33] Wilhelm Beer (1797–1850) was a member of a Berlin banking firm, a member of the Prussian Diet, brother of Michael Beer, the dramatist, and of Giacomo Meyerbeer (Jacob Liebmann Beer), the celebrated musician. As an amateur astronomer he worked with his friend Johann H. von Mädler on observations of Mars and the Moon, producing their *Mappa Selenographica* in 1836. He published *Der Mond* in 1837.

[34] Family pride was involved here. On August 28, 1789, William Herschel, at his first use of his new forty-foot telescope of 4 feet aperture, had easily seen the sixth satellite (Enceladus) of Saturn, previously just glimpsed in his 6½ inch. On September 17 of the same year he discovered the seventh satellite, Mimas.

Thursday, March 31, 1836

M^t at Breakfast took a regular lesson in Algebra in my MS. Caroline's Birthday—The little Maclears came over & passed the day with her the whole party being crowned with wreaths of Flowers—the blue—"Cape Lilac?"—

Mr & Mrs Hutchinson took the little folks home

Mr Hutchinson came in evening

Noiseless flight of a Night bird—who came to inspect the vane on top of my Astrometer—which he sate in the air looking at like a humming bird

Astrometer obs begun but a violent South East scud (aloft!—with perfect calm below) cut short the Series.—NB. The scud came on with small flocks forming 30° high to Eastward & then accumulating over Zenith—dissolving to the N as they drifted on—Then it cleared universally for an hour—then came on again thick & fast the T & D hills involved to the very base—& only the ☽ being seen through the swift white veil. Meanwhile a perfect calm on the Surface, though the Cloud seems not above 150 or 200 feet high where it seems to blow *hard*

Sunday, April 3, 1836 (Easter Day)

Div Service & Sac[ra]m[en]t at Rondebosch

Monday, April 4, 1836

Furious winds called by Col Bell the South Easter Holidays
M. called on Mrs Mitchell

Tuesday, April 5, 1836

Ford came

Wednesday, April 6, 1836

Meeting of SA Lit Phil Ins

Friday, April 8, 1836

Wauchopes came. Sir J. B.   Mrs Hawkins &c

Saturday, April 9, 1836

Prepared fulminating Mercury   Went to Mr Ebden's   Capt & Mrs Wauchope.   Col & Lady C Bell   Major & Mrs Mitchell   Mr & Mrs Maclear   Baron Ludwig   Dr Smith—C. Bell   After dinner an alarm of Fire. The Bush beyond the Vineyard all in a blaze—house threatened   Fire

Sunday, April 10, 1836
Alexander Stewart Herschel X[tened]   Prepared

Monday, April 11, 1836
Tried with [erased] Tried Exp[erased] Exped[ition] to Houts Bay with the W[auchopes]

Tuesday, April 12, 1836
Split tree with Fulm[inate] [a small sketch, possibly of an explosion] Ford came Cap[n] & Mrs W. Left. Mr & Mrs Hawkins called Child[n] went in carriage & took W[m] Wauchope to Lady C

Wednesday, April 13, 1836
Rode on Jumbo & told lies about him   called on Schomberg & Breda to thank about fire

Thursday, April 14, 1836
Went into Town   Called on Mrs D Clark—Lady C. B. Mrs Mitchell—Dr Smith—ordered Trowzers at Carfrae's [i.e., Roesch, Carfrae & Co., tailors of St. George's Street]. Got hair cut. Mrs Wauchope met at Mrs Sanders   Bought Caross Jackall at D & J [Deane & Johnson, general dealers, Heerengracht].

Saturday, April 16, 1836
Capt & Mrs Wauchope Lunched on way to Simon's Bay.—

Sunday, April 17, 1836
Went to Observatory & Drank tea   Brought Back Barom   Took M. [Maclear] Chron. Magnet. & Petropol Acad Vol with Struve's Obs[s]   Equatorial work   Furious N Wester came on all night[35]

Monday, April 18, 1836
Bad rainy day
η Argus. Calcul of Micr Meas & Maps [i.e., calculation of micrometer measures and drawing maps based on them]

---

[35] Sir John was obviously bored, the weather being bad. He and Captain Wauchope, about this time, amused themselves by blowing up tree stumps in the avenue of "Feldhausen," and might easily have blown themselves up too, considering how they went on. Jumbo was a horse, owned by the Wauchopes, though perhaps Sir John was trying him out before the Wauchopes bought him. A Caross (now, *kaross*) is an animal skin with the hair on. "Chron. Magnet." must be a journal of magnetism, and "Petropol Acad" is the journal of the St. Petersburg Academy.

Planted Gladiol Watsonias
Sent for Maclears Cart

Tuesday, April 19, 1836

Rainy day
Countermanded waggons.
Mr Ford came
Wrote to Major Dutton
Dist. [? carried out a distillation in the laboratory]
η Argus
Genl History of D S [double stars]
Arranged Stars
Planted Bulbs Glad Angustus
Scarlet Watsonia of Flats

Wednesday, April 20, 1836

Planted Gladiol Blandus
M called on Lady Bryant and on Mrs Schonberg
H at home all day. Swept from 16^h 4^m to—[his omission] & went
on with Equatorial till morning twilight   M. called on [entry termi-
nates]

Friday, April 22, 1836

Left Feldhausen for Houts Bay—with Margt. all the children the
3 maids and in short Bag & Baggage. having taken Mr. Cloete's House
there for a month. Returned to Feldhausen in Evening with Zomai

Saturday, April 23, 1836

A dull & cold day.—At home and hard at work with the preparatory
Catalogue of Fundamental Stars all day

Sunday, April 24, 1836

Rained hard all day—Staid at home—towards Ev[enin]g cleared
a little

Thursday, May 5, 1836

[In another hand, enclosed in square parentheses]
For last sight of Halley on 5th See C.G.H. Observations p 397
W.J.H[36]

---

[36] William James Herschel, Sir John's eldest son (see note 5, April 10, 1833).

A Letter from Sir John, Feldhausen and Hout Bay,
to James Calder Stewart, Canton, China

                                            rec^d 10 Oct. /36 [In pencil]
Dear Jamie—
        (Do pardon this diminutive—I can't write to you as if you
were an alien in spirit and a stranger in feeling or that horridest of
Earth's phantoms a friend with whom one must always stand on the
defensive as if he might one day become an enemy (as some Greek
has it) so again *dear Jamie*—What in the world are you doing in that
land of unaccountables? Making a fortune. Well so be it & godsend
you good-speed in that laudable undertaking—but the day has 24
hours and whatever you may say about 10, 12, 16 of them spent at the
desk—nature will out and I defy Mammon to get full possession of
you so long as you can draw on your stores of memory & imagination
for other higher & richer stores of enjoyment than any w^h his devotes
have any notion of. Be his temptations or his terrors what they may
you have within your own soul many & more potent counter-spells &
may walk unharmed & superior to the gross elements around you "in
maiden meditation—fancy free" where others less richly stored within
would be at the mercy of the circumstances around them. I think
there must be something ennobling and inspiriting in feeling that you
have two spheres of existence *here below*—the one gross & sublunary
in which to be sure you may wonder to see those about you so interested, but in which seeing it is your daily walk, it would be foul
shame not to walk strait and like a man with his eyes open—the other
remote & unsuspected by the sad-souled beings, with whom you exchange the asperity of transient communication full of high associations and sympathies and when time and distance have no office but
to mellow and soften to ripen & blend. Nay but I deny that even
Mammon himself can be worshipped in Canton as he is in London. A
man *there* can give only his time.—his physique In London he can
give his whole soul; & talent even of the highest order would be insufficient to fathom the sea of Enterprise on which a man may
founder in pursuit of gain.—At home the combinations are infinite,

the speculation boundless—with you commerce is confined by stated motive new paths are few and far between, and of more uncertain issue.—*Therefore*—Argal[37] you are not to quarrel very vehemently with the devotion of the plodders whom you find yourself among— *Their* only nostrum is time & Labour—*time* which they would not fill better elsewhere—Labour without which their vis inertio would sink to the soil. And for a man of a mind superior to sordid pursuits I think it must be pleasing to find that by the sacrifice of these ordinary elements he can fulfill these necessities of his state without prophaning the God given strength within him, and debasing the higher powers of his soul in the discovery of new & untried schemes for attaining an end no doubt desireable, but not desireable at such a sacrifice.—The fact is, it is the naked undisguised exposure of a motive which at home men rather shroud in a veil of civilized superiority, which has struck and disgusted you—and besides answer me *this* Why should a man go & spend . . . [his omission] years at Canton—Why to get rich! But is the residence desireable—No—Well then, they wish to shorten it—How? Why by getting rich as fast as they can.—Here then is another & in your eyes a more pardonable motive for ultra-diligence.—Each man earnestly desires *to get home*—he longs to be rich partly that he may command the advantages of wealth—but *more* (for *the immediate cause is always the acting one*) that he may get out of prison.—Therefore Argal, you must not *too* furiously condemn the poor devils whom you represent as your collaborators in the treadmill of Chinese commerce. The wheel has a stated number of revolutions to make & then the trap door is opened—so they bite their lips & set their face as a flint and work like devils to make the wheel spin a little faster. That's all.

Houts Bay April 28. 1836

Thus far had I written when Maggie spirited me off to this most sequestered & unsophisticated place, where there are far more wolf & tiger tracks than horse or cow hoof marks and where the winds sigh & the breakers boom to no ears but those whose owners sleep under the same roof (*a very comfortable one*) with ourselves. You will see by the well meant tho' ill expressed talkee talkee overleaf that the *last* of your letters has come to hand first—that "sent a fortnight ago—a long

---

[37] The Shakespearian gravedigger's ignorant attempt at "Ergo."

business letter" has not yet come to hand.—Now a long time ago I wrote to you via Calcutta at your present address a "business letter." —such a business letter that is as brother should send to brother & friend to friend—authorizing & empowering you to do your wicked will with certain capital (in 2 Bills of S. E. & Co. [Smith Elder & Co.] for 1250 £ each) about which you in a letter written on eve of your departure mentioned that it would be useful to you in your chinese relations when S. Elder & Co had no more occasion for it.—Accordingly I wrote to you & to Peter, [(to each sending copy) scored through] *informing each of what I had written to the other* and authorizing P. in place of paying it into my Banker's (& whenever S.E. & Co might take up their said Bills) to hand it over to you or your orders (taking your note of hand for it & charging no interest— [some lines scored through] according to any directions you might have left with him, or any understanding you might since come to.— Now since then Peter writes word that a delay has occurred on the part of S.E. & Co. (by reason of certain dishonourments in respect of their Indian remittances) in taking up their Bills as intended. But of all that you were aware at starting and were to take such steps at Calcutta as would set all that strait. I presume therefore that ere this time S E & Co. have taken up or are in act of taking up those Bills, and *in that event* the amount is still at your service in the way proposed—if indeed the whole transaction is not by this time—transacted—for to say the truth, communications from Eng^d are growing rarer & rarer, and I begin to perceive the necessity of getting information about what is going on on the surface of the earth by telegraphic communication via the Man in the Moon, as I told P. in the last scrawl I dispatched to him.—One thing only observe in the execution of the arrangement—Your bills from Canton must be drawn not *on me* but on Peter—& that only on full previous understanding *with him.* If this seems to you superfluous to mention you are to attribute it to my utter ignorance of the practical conduct of all these distant commercial trans^{ns} which makes me only anxious to put my meaning distinctly as regards my own position.—

Well now.—A ship from Canton way arrived 2 days ago & brings *no letters.* So I shall send this off.—I am now here (May 9! Good Heaven how time *does* fly.) Bachelorizing—Maggie & all the Brats are still at Houts Bay—all well but I wish they were here—the place does seem so destitute of all comfort without them.

Ere long I shall send you a "Phonetic Alphabet" with a list of words and shall beg of you to get these words distinctly pronounced by some well educated Chinaman in Chinese (i.e. their Chinese homonyms[?]) in your hearing or that of some one you can confide in & written down attending to the exact system of pronounciation explained in my explanation of the said Phonetic Alphabet. If they have any *peculiar* sound you must invent a character to express it & describe as well as you can what the sound is.

No absurdity has ever seemed to me so absurd as the Dogma of the immutability of the laws of Chinese literature.—You might have said the same of the laws of French classical Poetry—yet those laws are now fairly exploded in France by the so-called Romanticist school—and so it must be of everything arbitrary and conventional It is like a dense fog while the atmosphere is still but let a breath of wind rise & all the true horizon is restored—only let the slightest regenerative Political movement take place in China, or let foreigners learn their language and *write boldly in it* according to the dictates of good feeling & good common sense & not adopting their quaintnesses or crotchets—and their old fusty rules and ultracelestial notions will topple down like card houses#.—Let them translate Robinson Crusoe into Chinese or any other book of elementary unconventional ideas (if interesting enough to attract attention for its own sake) The object to be aimed at in the first instance must be to please and amuse them —not on Chinese topics or in chinese phrase, but on general principles and in language of the simplest kind and as little idiomatic as the nature of their tongue admits.—Instruction must follow at a slower pace. They are a vain people, and will long revolt at the idea of being taught by us barbarians.

What you say about my writing a "litel Boke" for Chinese translation must be thought on. It is not easy. I could do it, & I long to do it, if it were only to have it retranslated into English, for there are certain notions on which we English stand quite as much in need of disenchantment as the Celestials—But then the time! The time the evil is sufficient for the day, but the day alas! is far less than sufficient for the evil—No the bother—no—the *claim* duty—call it what you will.

The Map of China & the World came safe to hand. I had seen it before in possession of Baron Ludwig of whom you have perhaps heard Duncan speak. Dear J., don't you go and let yourself be run away with the notion common enough in Canton that every Chinaman is of necessity & par metier a rogue & a cheat. Whenever that nut is broken up, sweet milk will be found within it as well as dirty crust and cohair (or coyar or whatever they call it) outside. Remember they are existing in the extreme of human density, under the worst of governments & social systems. Doubtless every form of human nature, amiable and unamiable is as fully developed in China as in Europe. As to any great prize to be got by cracking the nut, I doubt it however.

Well—another ship with letters the Penelope I shall forward your note about a parcel to our agents in C. Town. "Hamilton Ross & Co" (not *Thomson Watson & Co* as heretofore—remember this in future)—as ill luck will have it your letter came to hand only tonight and at 7 tomorrow m^g I stand engaged to go down to Hout's Bay to a Tiger hunt and to bring Maggie back here (i.e. if I don't get devoured) but with this for Post Stone takes in the note for the Captain, & the encl[osure] will I doubt not be the arrival of your packet with its unknown contents As yet your "Business letter from St Helena" has not turned up though twice within this week letters from India have reached *via* St Helena having being carried on thither & sent back. Well—adieu and believe me ever

Yours faithfully
JFW Herschel

P.S. I hope you have not quarreled with your *Muse*. If you do that (let the claims of business be what they will) you are unpardonable.—

The other day Maggie was twitting me with some sins of that kind of my own in former times—something about arks & doves and adverse winds and as I rode home from Houts Bay "my thoughts" as Mrs Radcliffe's[38] Heroines always say "involuntarily arranged themselves in the following lines" which I send to keep you in countenance—in case your conscience should be ill at ease on that point.—

The Parting Dove
Impatient of constraint, around my Ark
In short and lowly flights my strength I tried
But faint and toil-worn from that Ocean wide

[38] Mrs. Ann Radcliffe (1764–1823), author of *The Mysteries of Udolpho.*

Back to the narrow and overcrowded Bark
　(No home abroad achieved) I sadly hied.
　　There pressed my flagging wing for fresh essay
　And launched anew, to prove, in purer air
　　A wider prospect and a loftier way
　And caught one glimpse, and snatched one trophy rare
　And bore it home and mused for many a day
On Sunny Realms, where grew that bough so fresh & fair
Now fare thee well! thou dim & wavetossed speck
　No more for me fit prison or fit lair!
　No more for me fit cause of dull delay!
Though sore 'twould grieve me yet to know thy wreck.

Your Italian & German Journal was very delightful.—

＃ Will any body pretend to say that a given sentence of English is positively not *translateable* into Chinese? (unless it be one of the sentences of the Comick Annual)—If so there are two human natures & all attempt at approach is hopeless

A Letter from Sir John and Lady Herschel, Feldhausen, to Caroline Lucretia Herschel, Hanover

N⁰ 11
[In a different hand]
Rec^d in
London
Septr 26.
in Hanover
Octr 1. [1836]

[In Lady Herschel's hand; undated]

My very dear Aunt
　　We must not let time slip by too fast without reporting progress to you who are so much in our thoughts & in our affections, al-

though we have had no letters from you later than those already acknowledged in our last joint letter to you dated [omitted] to satisfy our earnest longings to know how you are—We must hope that a kind Providence is still sparing you to us well & hearty, & we can at all events cheer you by the assurance that we are all, master, man & boy of us in good health, & full swing at our winter occupations, for the seasons here are so decided in their characters, that we always expect many days & sometimes weeks together of hard rain in the winter months, & then we all sit down regularly at our Books & work—Papa in one room with his reductions, & in the next room am I with an Infant School round me, for little Willie is now old enough to read & handle his pencil & is very ambitious to "come to his lessons," like his sisters—In the month of May we went for a few weeks to a sweet spot called Hout's Bay, on the Sea Side, about ten miles from this, and enjoyed the beautiful scenery very much, excepting that Herschel could not leave the stars & remain with us the whole time to ramble through the mountains & *Kloofs* with us—but as the distance was so small he often rode backwards & forwards—At last however he found his home so dull & desolate without the noise of his little prattlers that he carried us all home in triumph, & as he & I rode on horseback at the head of our cavalcade, we laughed heartily at the long train of waggons filled with children, nurses & all their appurtenances which followed us, & showed us *very plainly* how much our family importance had increased of late.—Herschel was even guilty of a *bon mot* perverting the sense of that sweet passage of Pope when he speaks of his own early love for poetry—We "lisped in *numbers*, for the *numbers* came"—Herschel is straining every nerve to get through with his Reductions, & as Mr Maclear has undertaken to furnish him with all his Zero Stars, he hopes now to proceed without further interruption—The *prettiest* thing I see going on is his map of η Argus, which is a most wonderful object, but the immense number of stars in it (about 1000 at least) makes the labour very great— Have you seen a very clever piece of imagination in an American Newspaper, giving an account of Herschel's voyage to the Cape with an Instrument [omitted] feet in length, & of his wonderful lunar discoveries[39] Birds, beasts & fishes of strange shape, landscapes of every

[39] This is the famous Moon Hoax perpetrated by R. A. Locke in the *New York Sun* for September, 1835—pretended hot reports of lunar inhabitants and other discoveries made at "Feldhausen" (Plate 17).

colouring, extraordinary scenes of lunar vegetation, & groupes of the reasonable inhabitants of the Moon with wings at their backs, all pass in review before his & his companions' astonished gaze—The whole description is so well *clenched* with minute details of workmanship & names of individuals boldly referred to, that the New Yorkists were not to be blamed for actually believing it as they did for forty eight hours—It is only a great pity that it *is not true*, but if grandsons stride on as grandfathers *have* done, as wonderful things may yet be accomplished—I hear from Mrs Somerville that Dr Ritchie (a Scotchman) has discovered the secret of making achromatic lenses of *any size* & of the most perfect transparency—I often hear [from] my brother James in China, where he must remain for some years yet if his health is good.—But I know that an engraving of Herschel's likeness (such as you have got by this time I hope) & a pencil sketch of his Godson Willie will go far to sweeten his banishment for he loves these both most dearly—I long to place my darling Baby Alexander in Grandmama's arms, for I quite grudge to enjoy his sweet smile by myself, & if I could show them all to *you*, I think I should then be quite happy—But Herschel says that he wants my pen, & as I can't give it into better hands, & can sincerely congratulate you on the change, I will only beg my dear Aunt to believe in the fond affection and respect of her attached Niece

Marg^t B. Herschel

[In Sir John's hand]

Dear Aunt.—The best news I can tell you is that I begin to see to the end of our pilgrimage in this remote corner of the world.—Not that there is not still a good deal to do, but that I am at length in a condition to proceed to *realize* what is done,—to reduce, arrange and classify &c &c.—In this I owe great obligation to Maclear who has undertaken to observe all my Zero Stars & has already furnished me with enough to set me a going.—In short we are beginning now to *look* homewards—though as yet we cannot definitely name our time for starting and right glad (to say truth) shall I be to set foot in Old Europe & Old England once more, and trust yet—through God's Blessing once more to see your face in Hanover.—

We had a fine view of the Comet on his return from the Sun. I followed him till about the 20th day of May—and should have kept him in view longer but for a naughty nebula not down in any Catalogue which came so near his place & looked so like him that he fairly led me off the scent.—To say the truth I am glad he is gone.

I have now satisfactory evidence of the 6th Satellite of Saturn & have had one view at least of what must have been the 7th In short I have at *last* had the pleasure of seeing what only my Father had ever seen before Saturn surrounded by all his seven companions at one view Really a fine family!!—Margaret has told you all about her herself and your little nephews & nieces. Little Bella is a most ready learner and is never easy but when learning some thing. She begins to amuse herself with reading and is always to be seen with her book in her hand poring over it most studiously. Caroline has a pretty Ear for music and perhaps will sing well—Remember me to Mr & Mrs Groskopff—Mrs H. Hausmann & the Beckedorff's[40] and believe me dear Aunt ever your affecte Nephew

JFW Herschel

The Diary: May 6 to October 3

Friday, May 6, 1836

Rose at 8½. Planted Bulbs till 9½. Breakfasted &c and made out receipts &c for men's wages & Meteorol. Mem[oran]da 11h then worked at η Argus Map.—Gave orders to be denied to everybody—but it was a day of ill luck in *that* respect.—1st came Mr Maclear Dr Smith & Dr Adamson on their way to Houts Bay—*they* were of course admitted.—Then came a Major Parlby[41] with a letter introducing certain Mr & Mrs Roberts.—*He* would not be denied but insisted on coming in—& very glad I did see him being Parlby the "Gentleman Farmer" of the Kleine Rivier a remarkable man here & who is said

[40] Mme Beckedorff, who held some position in the Hanoverian court, was a contemporary and close friend of Caroline Herschel's. She had a large family of grandchildren.

[41] Major Samuel Parlby, ex-Indian Army, of Klein River Valley in the Swellendam district, some 120 miles east of Cape Town.

to be doing a great deal of good & setting an excellent example to the Cape Boors which however *they* are said not to have the sense to follow.—Well it was late dinner time & just as sitting down—a man rode up to say Mr Borcherds the Civil Commissioner was waiting to see me at the End of the Avenue about an affair of turning a road to cut *on* a corner which the present road cuts *off*, of Feldhausen.—So obliged to go—There was old Breda & old Schömberg and young Breda & young Schönberg and Munnick the Field Cornet and Eckstein and Watson who has just bought Borcherd's House "Weltervreden" (such a name) & has a road of his own *to turn across my ground*. And Lord knows who beside—& when this was over & I had swallowed a hasty dinner—up rode again Maclear & Dr Smith & Adamson—and they staid till late in Evening & here am I scribbling this record of a day passed as I had determined it should not pass!— One out of 3 or 4 days I have ordered myself to be denied since here I came. Heaven send me grace to save up odd minutes! which are Life!

Saturday, May 7, 1836

At work on Map of η Argus.

Major Parlby

Major Parlby dined here & saw Stars. A Man of evident mentality, active, enlightened, enterprise, & who is forming an Establishment at Klein Rivier but enthusiastic & over sanguine.

Sunday, May 8, 1836

Staid at home all morning waiting for Maclear who did not come— after an early dinner rode over to Houts Bay where found all well— Mr Cloety[42] there

Monday, May 9, 1836

Morning at Hout's Bay.—Took a ride on the Beach trying to shoot sea fowl but they are too wise

Returned to Feldhausen

Cloudy Night. No Comets nor Stars.

Tuesday, May 10, 1836

Feldhausen—

At home all day & more or less at work on fundamental Catalogue. &c.

---

[42] Should be "Cloete," but "Clooty" is how it is pronounced.

Sowed 6 lbs. Firseed Broad Cast.—
David came over from Houts Bay. Carpenter at work cieling the
Nursery   No events   Read Benthams Peines & Recompenses[43] the
Chapters about libre concurrence and Recompenses pour la vertu.—
Wonderful acuteness & closeness of argument—A Book to dissipate
prejudices even where it does not command assent—It makes a
Table rase[44] of the Subject at all events. Sent Maclear another lot of
Stars for Obs^n—Cloudy and rainy night. No star nor Comets

Wednesday, May 11, 1836

At home all day. Commenced operations for Planting the newly
burnt Plot of ground by the North Avenue "the Park". with fir.—Set
Dawes & Jack at it & marked out the lines—
Began a fresh map of η Argus the former being on too small a scale.
Ev[enin]g Fine.—Looked for Comet but found only 2 Nebulae and
one object w^h at the time I felt secure was the Comet but which on
coming to reduce the obs^ns appears to be *most likely* a Nebula Cl II
Observed η Argus.
Made a Sweep.—retired at 3 AM

Tuesday, May 17, 1836

Rode out with David to the Sand hills & after in vain again attempt-
ing to get a Seagull, sketched the "Castle" from an elevated Sand hill
richly covered with Bush & then dismissing David, rode along the
left Bank of the Bay beyond the Castle and caught a highly picturesque view of the Mouth of the bay from the head of a deep gulley
into which the sea has battered a narrow & rocky channel.—Returned
—gathered a basket full of Great Bulbs (? some kind of Squill)—But
my horse got loose & ran away & by the time I hailed the Cart on the
beach with the children & reached it, what with the weight of the
bulbs & the heat was thoroughly tired & couldn't get out in the eve-

---

[43] The reference is to Jeremy Bentham (1748–1832), the celebrated British
utilitarian philosopher, economist, and theoretical jurist. Among the many incidents of his life, and his vast literary output, he included, having been made a
French citizen in 1792, his *Théorie des Peines et des Récompenses*, two volumes,
1811. This subsequently became *The Rationale of Punishment*, 1830. Bentham
had an immense influence on nineteenth-century thought. John Stuart Mill was
among his followers.

[44] *Tabula rasa* (Latin), means a smoothed or blank tablet.

ning except a stroll in w^h shot a few brown birds with yellow under the tails—

Wednesday, May 18, 1836

Houts Bay—All in movement for our return w^h was accomplished Bag and Baggage without accident—Margt & self riding—the children following in the waggon &c &c

Arrived about 3½ having started at 12½ riding easily & were heartily glad to get back.

Got a sweep North of Zenith

Thursday, May 19, 1836

Got a sweep South of Zenith.

Observed a very odd cloud descending from Menzies' Kloof like a stream of smoke to *the very* ground. It increased & filled the Kloof but with an almost unperceptible motion (a calm prevailing). By degrees the Cloud bank over the Bay advanced & enveloped the Devil's hill from Northward.—This seems to forebode rain

Saturday, May 21, 1836

Worked out the theory of a systematic reduction of my Cape Sweeps & prepared preliminary tables for the compound term of the precession

Which all proved abortive & useless

Sunday, May 22, 1836

M kept her bed with a horrid Earache

Maclear & Mr Smyth called

Violent hot N wind came on after Midnight—Therm 70.0 at 2 AM —but sank towards morn[in]g & was 60 at 9 AM Monday with cloudy sky & rain commenced at 10

Tuesday, May 24, 1836

A hot day

Kept steadily at work all day at the reduction of my sweeps.

Wednesday, June 1, 1836

Went in to Town to Meeting of the SA Lit & Sci Asson.

A decisive observation of the 2^d satellite of Saturn (the outer of my Fathers two).—

Tuesday, June 14, 1836

Received letters from England—PS's long letter about Johnnie & his present & future prospects.—

The House on fire—A beam in chimney of the Servants Hall caught & smouldered & was just discovered in time to prevent its catching the thatched roof & extinguished—NB   The beam enters the Chimney & must have gone strait across it from side to side What folly!—

At work on Map of η Argus & working Skeletons

Rainy day with little wind

Wednesday, June 15, 1836

Raining in showers all day & night

Captain Fitzroy,[45] Mr Darwin,[46] Captn Alexander, Mr C Bell & Mr & Mrs Hamilton dined here at 6. Capt F & Mr D came at 4 & we walked together up to Newlands.—Captn F informed me that Dr A. Smith is suddenly this morning (without any previous warning) ordered on permanent duty to Simon's Town [there to superintend

---

[45] Robert Fitzroy (1805–1865), British naval officer, hydrographer, and meteorologist. In 1828, he was appointed to the command of HMS *Beagle* and surveyed the coasts of Patagonia and Tierra del Fuego. Returning in 1830, he commanded a second expedition (1831–1836), on which Charles Darwin was the naturalist. Fitzroy does not mention the visit to Herschel in his account of the voyage. In 1843 Fitzroy became MP for Durham and in the same year became Governor and commander in chief of New Zealand, eventually reaching the rank of Vice-Admiral. Because of his liberal native policy the settlers secured his recall in 1845. In 1854 he became chief of the newly formed Meteorological Department of the Board of Trade, and published his *Weather Book* in 1863. He committed suicide in a depression induced by overwork.

[46] Charles Robert Darwin (1809–1882) is too celebrated to need a biographical footnote. At this time he was naturalist on board the *Beagle*, and would not publish his *Origin of Species* until 1859. He does not mention the visit in his printed account of the voyage, but, on July 9, 1836, he wrote to J. S. Henslow as follows (*Life and Letters of Charles Darwin*, Volume I, Third Edition, 1887):

"At the Cape Captain Fitz-Roy and myself enjoyed a memorable piece of good fortune in meeting Sir J. Herschel. We dined at his house and saw him a few times besides. He was exceedingly good-natured, but his manners at first appeared to me rather awful. He is living in a very comfortable country house, surrounded by fir and oak trees, which alone in so open a country, give a most charming air of seclusion and comfort. He appears to find time for everything; he showed us a pretty garden full of Cape bulbs of his own collecting, and I afterwards understood that everything was the work of his own hands . . ."

Darwin's first publication (with Captain Fitzroy), appeared in the *South African Christian Recorder* for September, 1836.

the health of some 30 or 40 men, all healthy] [his square brackets]
This untoward event must of course be a grievous interruption to
the progress of proceedings respecting the Expedition—Stuffing &
describing specimens & arranging their packing &c besides necessi-
tating their removal from Dr S's house.—In short it is quite a mis-
fortune to Smith & the whole concern.—At Dinner Captn A. broached
the subject at table, but in what we thought an *unfeeling* if not an
unfriendly manner towards Smith—treating it as the most ordinary
occurrence & observing that it was what the Profession renders every
man liable to & that therefore S. could not complain (no mention having been made of his *having* complained, or being *likely* to do so—)
&c &c. All this is very true—but it came not well & grated harshly I
contented myself with observing that "it was a very serious misfor-
tune not only to Dr S. but to the cause in which he was labouring—
that of science"—& no one else taking up the subject the conversation
dropped.—I was surprised at Capt A alluding to it as I had previously
resolved not to make it a subject of remark in his presence

Thursday, June 16, 1836

Raining the whole day.—Reduced a Sweep and wrote to P.S. ap-
proving his arrangements relative to the Printing Concern[47] &c (See
Duplicates Vol 5)

Sorted Birds to go to Plana & [entry terminates]

Sunday, June 19, 1836

Dined at Col. Bells.—

Convers[n] about Dr. A. Smith. Col B. inclines to recommend an application on the part of Smith's Committee to pay the expenses of a
Civilian to do his duty at Simon's Bay. I question the propriety of any
such step

Conv[n] about the management of the Schools[48] & public instruction
in the Colony

---

[47] P.S. and the Printing Concern: It is entries such as this, and earlier remarks
connected with the ordering of stationery, which suggest that Peter Stewart was
an employee at Smith and Elder, and that Sir John, with the generosity which
previous letters show to be characteristic, was anxious to help his brothers-in-law
in their business careers.

[48] The first reference to a subject of intense interest to Sir John, leading in the
end to the establishment in the Cape of a most advanced system of public instruction. The story is best told as a précis of "Sir John Herschel's Contribution

Monday, June 20, 1836

Attended Committee in Cape Town to make the necessary arrangements for Dr Smith's absence at Simons Town.—Maclear offered house room in the observatory for the Collection—accepted.

Ford offered to learn engraving[49] & engrave the drawings—his Expences in England being Paid

Tuesday, June 21, 1836

Hourly Obs[ns] for the Solstice

Wednesday, June 22, 1836

Hourly Obs[ns] for Solstice continued through the night and till 6 PM today.

Friday, June 24, 1836

Went into Town to Execute the Transfer of one "Morgen"[50] of Land

---

to Educational Developments at the Cape of Good Hope," by E. G. Pells, *Quarterly Bulletin of the South African Library,* 12, No. 2 (1957), 58: In 1834 there were 60,000 whites and about 120,000 Hottentots, Coloureds, and Malays scattered over an area twice the size of Britain. Roads were rare, much of the country very difficult, there was little actual money in circulation. Education in rural districts was given, if at all, by vagabond "meesters." With a remarkable display of public spirit, and under the influence of Cousin's report to the French government on "The Condition of Public Instruction in Germany," Sir John Herschel, John Bell, Secretary to the Governor (Sir George Napier, D'Urban's successor) and John Fairbairn, devised a system of education to be supervised by a superintendent-general and executed by a number of adequately paid schoolmasters. The first overt step came in 1837 when Bell, on behalf of the Governor, submitted a memorandum to Herschel. Herschel replied in a letter of February 17, 1838. Both these documents are in the South African Archives (Enclosures to Dispatch No. 11, 1838). With them went Enclosure No. 3, John Fairbairn's "Memorandum or Suggestions, for the Advancement of Education at the Cape," February 19, 1838. There was a second memorandum, dated May 23, 1839, now in the Public Record Office, London (Vol. C.O. 48 (188)), subsequent to Sir John's return to England, and, as his diary shows, he did some lobbying in London on behalf of the proposal. The scheme was adopted by the Colonial Secretary, Lord Glenelg, and put into practice, and still influences Cape education to this day. At its inception the superintendent was to get £560, and the twelve teachers between £300 and £150 each, substantial salaries for the day. The first superintendent, appointed in 1839, was Mr. James Rose Innes, professor of mathematics at the South African College (see note 94, May 19, 1834).

[49] See note 28, March 26, 1836.

[50] A Dutch and South African unit of land measure equal to 2.116 acres.

to H. Watson opposite to his "Place" of Welterfrieden (i.e. devant Borcherds's

Sunday, June 26, 1836

Went to Observatory—took over Carry and Bella.

It is now said that Dr Smith after all will not not go to Simon's Town because Dr Murray has been down to see if *he* cannot arrange with a Mr . . . to do the duty for him of attending to 0 sick out of 25 men the effective force there

Wednesday, June 29, 1836

At Home all day.—Fine day.

Mr & Mrs Steuart called

Captn Rogers H E I C [Honourable East India Company] and Dr Grant called

Copied Met. Obs for March Equinox & June Solstice to send to Mr Robertson[51] Margt has toothache & I have one of my dreadful colds—caught yesterday & rendering existence hardly tolerable.—No working or doing any good—

Splendid calm Ev[enin]g. NB. Moon 1 day after full Wrapped well up & worked at Equatorial and at the Astrometer till 12. Then suddenly came on a furious North Wester—1st with a warm puff or two mixed with Cold—and Stars violently agitated—then fierce gusts & Barometer down 2 or 3 tenths—Still clear sky.—But before morning it clouded and rained—& the wind partly abated

Thursday, June 30, 1836

At home all day. A N Wester & a rainy Even[in]g. Report of Meteorol Committee—

Sunday, July 3, 1836

Attended Div Service at Rondebosch.

Friday, July 8, 1836

At work all day preparing the Meteorological Report for the SAI.

Saturday, July 9, 1836

Anniversary of SA L & S Inst.—Went in. Reelection of officers & report Sir B. presided—Dr. Adamson read the report of Council & my Meteorol. Report

---

[51] Archibald Shaw Robertson, publisher, stationer, and bookseller of Shortmarket Street.

Sunday, July 10, 1836

Walk up Wynberg Hill—Church at Rondebosch. Planting out bulbs

Superb night—Swept from 13^h 45^m—to 21^h 30^m with 2 intervals.

Monday, July 11, 1836

Wrote to Drummond about their deficient Balance—ordering sale of Stock.

Tremendous rain in night

Tuesday, July 12, 1836

Mr P. Smyth called with a batch of Stars from Maclear.—Entered them up

Mr & Mrs Hamilton called.

Packet rec^d from Eng^d Magazines &c &c

Swept till 19^{h}20^m ST. when began to cloud.

Saturday, July 16, 1836

Wind strong and Hot from North & so it continued all day Thermometer 78 and cloudless

Rec^d note from Fairbairn stating I had been nominated for a Commission to take evidence about a proposed Mole in Table Bay.— Begged off on the Score of *want* of time and utter inexperience of such matters.

Rode over to the Camp Ground to see old Mr Hawkins who has had an attack of Palsy found him very ill and apparently very unlikely to recover. He had just taken his passage for self & family in the ship now lying in Simon's Bay.

Swept for 2 or 3 hours, then sent in Stone & went to Equatorial & reviewed an hour's Zone of 5°—then to the Backyard where with the Maps &c worked at the Star magnitudes till 3½ AM

The wind still warm—in early part of night 76° with *excellent definition*

Sunday, July 17, 1836

Blowing Strong *and cold* from North.

Attended Div Service in Rondebosch & then went with Mt to enquire for Mr Hawkins whom contrary to all expectations is much better & apparently in way of recovery

At Night Barom fell to 29.5 with violent storm of wind and heavy

pouring rain. Altogether a tremendous blustery night—The roar of
the wind in the great Fir trees was really awful

Monday, July 18, 1836

A furious day following a rough and outrageous night. Rain in
torrents & occasional lightning the rain *coming down quick after the
flashes*, 1ˢᵗ a little then rapidly increasing to a *smash of water*

Tuesday, July 19, 1836

Another violent & rainy day. At home & at work with the Comet &
Actinometer Paper

Wednesday, July 20, 1836

Major Mitchell brought me some plants from Cafferland
Mr Fairbairn called about the Table bay Mole Enquiry Commission. Though sorry to hold back I felt compelled by absolute want of
time for undertaking a long & laborious enquiry—to adhere to my
request not to be included in the Commission.—

Friday, July 22, 1836

A lovely day—Cloudless & Calm.
Morning occupied with pricking down Stars from Bode's map
preparatory to Charts of the Milky Way—and in giving directions
about divers matters of outdoor arrangement.—In Evening worked
an hour or two at the Astrometer and then went to Equatorial—
where a superb night was rendered nearly unavailing by an immense
formation of dew *inside* the object glass, in spite of the dew cap
which effectively defended the *outside*

Saturday, July 23, 1836

Morning. Mending & improving the Astrometer
Lady C. Bell's. A Juvenile party where the sports consisted of Chˢ
Bell & young Mr Burton dancing in Caffer & Matabele[52] dresses as a

---

[52] Of course, not used precisely by Sir John in this account of party frolics.
Strictly, *Matabele* is a term used by the Sotho-Tswana peoples for invaders of
Zulu stock. There are two unrelated groups, the Transvaal Ndebele, who have
occupied their area for about three centuries, and the Rhodesian Ndebele, who
left Zululand about 1820 and have been in Rhodesia since 1838. In languages of
this group the root is retained as the latter part of the word, and the beginning in-
flected to mean, in the case of a people, the tribe, the individual, their language,
their country, and so on.

Chief & his squaw with a pickaninny—also dipping for apples in flour with the hands behind—grinning with orange peel teeth making orange heads and to crown all Col Bell's *admirable!* representation of our grandmother by blackening his knuckle with burnt cork into a horrid looking face—The Lowrie[53] must not be forgot & Col Bells instruction of him as he lay on his back in his hand

Sunday, July 24, 1836
Too late for Church—Went with M. over to the Observatory. Found Mrs. M. just on the eve of making an additional family arrangement

Monday, July 25, 1836
Went into Town with Margt. to call on the new Lieut. Governor Stockenstrom[54] & his Lady. Found him just returning from swearing in—Capt S. is a native of the Colony & never visited Europe till 1832 when he went to see his relations in Sweden whom he found & liked so well that he was on the point of returning here to sell all his property & go there to settle.

Mrs Maclear Has presented the Astro[n] Royal with a Son & heir—the 1st son after 8 Daughters—

Tuesday, July 26, 1836
Curtains put up in the Back drawing room &c &c

Mr Ford came and after giving M. her lesson of Drawing commenced Willy's picture

As soon as Dark got to work at the Astrometer and the Equatorial and went on to 4 AM. A lovely calm cloudless night

Wednesday, July 27, 1836
A gale of wind from North set in and [words erased] blew hard about noon [words erased] bringing cloud

At Night raining & blowing so worked till 4 AM at Actinometer Paper

Friday, July 29, 1836
Heavy rain and furious wind NW all night, worked till 4 AM Actinometer Paper

---

[53] The louries or lories are parrots or touracos. Presumably Colonel Bell was pretending to talk to a toy parrot.

[54] Sir Andries Stockenström (1792–1864), Lieutenant Governor of the Eastern Division of the Cape (1836–1839).

Saturday, July 30, 1836
Rainy Day. Working at Collimator.
Letters from England and Newspapers arrived—
Occupied all evening in reading them
Heavy steady rain all night. Calm. Worked till 3 at Comet paper

Sunday, July 31, 1836
Attended Div Service in Rondebosch Church a very bad day & hardly half a dozen people there—Cleared a little & drove with M. to see M[rs] Cumberledge at Wynberg whose husband has recently sailed for India leaving her here for 2 years

Monday, August 1, 1836
Fine day.—Drove in to Town with M[t] to attend a Meeting of Exploration Committee to receive a letter from Capt[n] Alexander stating that he is about to start Sep. 1. for the Damarra country, crossing the Orange river near the West Coast & so ad infinitum & asking information
Res[olve]d to furnish him with a Copy of Dr Smith's instructions— & to request him to attend to identification of native pronunciations.

Tuesday, August 2, 1836
Another North Wester coming on. Blowing hard & Clouds on hill— Outlined Antholyza Ringens for Marg[t]. Gathered bulbs.—Found a sweetscented Bulbine.

Monday, August 15, 1836
Showery day—no going out. Dr & Mrs Grant called.

Tuesday, August 16, 1836
Pouring Rain—

Thursday, August 18, 1836
Mr Hutchinson called.
Outlined Massonia[55] for M. & finished up her Antholyza Ringens[56]
Rode out to Halfway House and gathered 45 *Antholyza Ringens* now in superb flower.

---

[55] A genus of South African bulbous plants belonging to the Lily family. The flowers appear in a tightly packed head, close to the two broad opposite leaves, which grow at, or near, ground level.
[56] See note 19, March 10, 1836.

Promised fair for a fine Night but came on again to blow hard and rain and so continued till morning

Worked at Comet paper.

Friday, August 19, 1836

A Pouring day. Barom Rising however, and wind getting round to S, by way of West, It is the conflict of the Cold S with the warm North which brings rain & hail in waterspouts and valleys.—Towards Evening it cleared & promised well but clouded again.—At 2 AM it finally cleared & a decided S wind prevailed—then however too late to Stargaze

Worked all day at the Comet's pictures, dimensions, & places.—

Sunday, August 21, 1836

Being a tolerable day I rode over to the Observatory to take Maclear the Obs[ns] of the Annular Eclipse of May 12 sent me by Beaufort. Found M and Dr Smith in full conclave—Smith amused us with anecdotes of Dr Barry[57] (recently left Cape) who like many very little men is very unstable—in fact a second Dwarf Jeffreys having shot 2 men in Duels & been engaged in many more, the cause being always his extreme over-sensibility to any allusion to *small persons*. This with his excentricities of other kinds (vegetable diet, huge attachment to his dog &c) is constantly getting him into quarrels.

Dr Smith states that he had discovered a mass of Meteoric Iron in Namaqua land[58] beyond the Orange River. It is more than a ton in

---

[57] James Barry (c. 1795–1865), M.D. Edinburgh, 1812, army hospital assistant, 1813, arrived at the Cape about 1814. He resigned and sailed for the United Kingdom in 1825, and rose in the army medical service, retiring with the rank of Inspector General of Hospitals for the Canada Command. The most startling thing about Dr. Barry is that at death he (she) was found to be a woman (Kirby claims a hermaphrodite), who successfully sustained the deception throughout her army career. There are all kinds of legends about him (her) including scandal with Governor Somerset. The diary reference is interesting as a revelation of contemporary gossip. One can only speculate on Sir John's reaction had he known the truth. (Isobel Rae, *The Strange Story of Dr. James Barry, Army Surgeon, Inspector General of Hospitals, Discovered on Death to be a Woman.*)

[58] Andrew Smith had explored parts of Namaqualand in 1828 (see Percival R. Kirby, *Sir Andrew Smith*, p. 70), but had evidently not gone far north of the Orange River. Numerous meteorites have been found in what is now South-West Africa. It is impossible to identify this one with what is possibly the largest known, that at Grootfontein, north of Windhoek, which measures nine feet by

weight. It was seen to fall and was hot and made a great smoke as described by the Natives. They cut it with chisels & use the Iron. Dr Smith cut off a piece, a specimen of which he promised to give me.

Monday, August 22, 1836
Being a Superb night, worked at the Equatorial till day break.

Tuesday, August 23, 1836
Superb. Cloudless day. Calm. everything seems alive at once. The [sentence unfinished]
Bad definition of Stars tonight—worked a little at the Astrometer but could do nothing with Eq[uatoria]l night cloudless & calm till morning

Saturday, August 27, 1836
1st Flea noticed leaping alive.

Saturday, September 3, 1836
At work all day making out a formula for Effect of Parallax and Refraction on the Comet observations.

Sunday, September 4, 1836
Attended divine Service & Sact in Rondebosch Church (NB Captn Alexander communicated).—After Service Maclear's little Boy was Christened, I and Margt standing godfathers & godmothers (M as proxy for Mrs Smyth) and Maclear himself for Captn Smyth.
Heavy rain all the Evening and night.

Monday, September 5, 1836
Worked over again the formulae for the Differ. Refraction & effect of parallax in extra meridian Equatorial Obsns
A Superb Calm night of excellent definition—Began sweeping immediately after dark & went on to $1^h ST = 2^h MT$.—

Tuesday, September 6, 1836
Superb day.—
Drawing outlines of a Pink Babiana and the Babiana Rubro-Cyanea—

---

ten feet, and averages about three feet thick. This contains 17 per cent of nickel, and is extraordinarily tough and hard to cut. See Willem J. Luyten, "The Groot-fontein Meteorite," *South African Journal of Science*, 26 (1929), 19, and Harrison Brown (ed.), *A Bibliography on Meteorites*.

Wednesday, September 7, 1836

Went into Town to attend Meeting of the Lit. & Phil Society. On the way out, went to the Spot behind Rondebosch Church where the Hypoxis Elegans[59] grows and took up 150 roots! Some of the flowers are full 3 inches across and are most beautiful.—At the meeting read Jamie's account of the great fall of Snow in Canton of Feb 8. Also various Meteor[l] communications.—Began to make a push to get the Journals of the Soc[y] regularly published[60] and the Quarterly Meteorological Commun[ns] inserted as they arrive

Thursday, September 8, 1836

M[rs] Wauchope came bringing Willie Wauchope & a variety of pretty & curious matters (Turtle inter alia from Ascension) shells—Waxberry Myrtle & spent the morning—on her way to Lady Cath Bell's.

Saturday, September 10, 1836

Rode over to the Observatory to see Cap[n] Alexander off—Accompanied him with Maclear, Major Mitchel, Dr Murray, and Mr Thomson[61] & another Gent[n] to the first farm house out on the flats where the road leads off to the Kloof by which he is to go—& there took leave. His waggons being in a drain [?] beyond a low sand hill, ½ a mile from the Sea He started lightly equipped in blue short jacket long leather pantaloons—Cary's Barometer *in his hand* a telescope in a leather case on his right side & a Schmahalder's compass on his left—on a vicious beast of a horse with a couple of pistols in his holsters.—Round topped white manilla hat like Mambrino's helmet—His attendant an ugly Hottentot in a blue bag of a cloak bundled on a grey steed looked like a Bundle of dirty linen with a chimneysweepers face peeping out of one of the Clothes Bags—head shoulders & knees all in a heap on the horse & clinging close like John Gilpin by the attraction of Cohesion.

[59] A synonym for the variable species now known as *Spiloxene capensis.*

[60] Sir John's efforts were fruitless. The last issue had been No. 4, Part 3 (July–September, 1834), not published until September, 1835, and the final issue (2nd Series, No. 1), dated December, 1836, was only published early in 1837.

[61] Described by Captain Alexander in his book as "George Thomson, South African traveller." George Thompson, merchant, of the Cape Town firm of Borradailes, Thompson, and Pillans, was the author of *Travels and Adventures in Southern Africa,* 1827.

Returned—Bulbgathering—&c NB. Behind Dekenah's Dekenahs [his repetition] Mill & the Observatory Cottages is one sheet of Sparaxis Grandiflora[62] and Calendula[63]—as if a heavy fall of snow had taken place.—The Scarlet Trichonemas[64] abound on the road Side going up to the Mill

Sunday, September 11, 1836

Attended Divine Service in Rondebosch Church after calling on M[rs] Mitchell

Being a Superb Night, perfect Calm, after dinner Monographed the η [Argus] Nebula with 20 ft & Equatorial & then swept till 3½ AM (3[h] 8[m] ST) Sw[eep] 732.

Monday, October 3, 1836

Wrote to my Aunt.

A Letter from Sir John and Lady Herschel, Feldhausen,
to Caroline Lucretia Herschel, Hanover

> Feldhausen   N° 12
> Cape of Good Hope
> October 3 1836
> Rec[d] Dec[r] 14 [In a different hand]

[In Lady Herschel's hand]

My dearest Aunt

It is so long since we have been blessed with a sight of your handwriting that we begin to long exceedingly for this great treat.— The last we heard of you was from my brother Patrick who sent us some prints of Sir William from you, for which Herschel and even the *children* must thank you in their prettiest way, for they look with great interest at the likeness of Grandpapa which is hung up against

---

[62] A white- to red-flowered cormous plant, sometimes profuse in a small area (Iridaceae).

[63] Perhaps a *Dimorphotheca* (Compositae).

[64] Perhaps a *Romulea* (Iridaceae).

Papa's study wall—another of these Prints was conferred on the Cape Observatory & of course is much prized there—I wish we were as sure of your good health & well being at this moment, as I can assure you of ours—from Papa to Baby the 5th—all rejoicing in the return of the lovely Spring, when we are neither annoyed with the heat of the Sun, nor saddened by the sight of naked trees & drenching rain— Never was Astronomer so petted & spoiled by Dame Nature as *ours* is—Imagine his large Telescope & Equatorial House nestling among a profusion of the richest scented Lupins blue & yellow which cover the enclosure like the thickest grass—and when he sets out on a bulb-gathering walk, his two little girls (who know the botanical names of all the flowers) fly about like pointer fairies to discover the flowers for him while he follows with trowel & basket—Since I last wrote little William has been advanced to the dignity of learning to read, & has begun Latin almost at the same time—he shows no lack of talent in either, but Bella still carries off the day in what ever she undertakes —Willie offered of his own accord to go to Cape Town for some teeth for his little brother, for he said "there was plenty there" & sure enough very soon after two pretty little teeth were found in Baby's mouth—I was positively forbidden to write beyond this half page, & I must obey, hoping that Herschel will well fill the remainder— Our accounts from *all* at home & abroad still continue most cheering, may the next prove equally so from Hanover is the prayer of, my dear Aunt—Your affectionate Niece—Marg⁺ B. Herschel.

[In Sir John's hand]

My Dear Aunt—We must make this a joint stock letter. And first I must say that we have been now for so very long a time without a letter from you that I begin to feel great anxiety lest illness should have been the cause of so long a silence—and it is this kind of anxiety which makes the vast distance that separates us so very painful.—On this, as on many accounts I am now longing to get home—and we have now fully made up our minds to let this year 1837 be our last at the Cape.—Every object which I had in view will then I feel assured be fully accomplished & nothing of importance which I *could have ever expected* to do, left undone. In fact my Catalogue of Southern Nebulae and double stars is nearly as complete as it is at all likely to be, and were it not that the observations of the last 18 months, owing to several great improvements in the mode of observing, are so much superior to the former ones as to make it desireable to go over the old ground afresh, I should consider the work as already done.—

The Comet distracted my attention greatly & absorbed much valuable time. I observed it till the 5th of May and should I doubt not have kept it in view till the middle of the Month had not bad weather for 6 days and an unlucky Nebula mistaken for the Comet, afterwards, thrown me off the scent.—However it was really a most interesting object, and I have a fine series of observations of it from the 25th January till the 5th May both inclusive.—We propose to return home by way of Rio Janeiro, which will not materially lengthen the voyage, and will give us an opportunity to see something of that magnificent country—though of course our stay there will be limited to a very few weeks.

Meanwhile we are really as happy and comfortable here as it is possible to be anywhere absent from our best friends and though the want of intellectual society would prove a terrible drawback on a permanent residence, I can imagine worse things than settlement for life in such a situation as ours. Maggie teaches the children and draws the pretty flowers in a very good and (now indeed) masterly style—She has improved herself greatly in this pleasing art, and surely if ever there was inducement for it it is here. Add to this we keep up a tolerable correspondence,—what with Marg^t's Brother in Canton—Brother in India—Brother in Constantinople—Brother in England—Brothers in Scotland!—and a few good folks who now & then drop me a line from New York—Mauritius—Sicily—Petersburg Geneva and *such like* places—we are by no means at a loss to know how the world wags.—

Although I cannot myself undertake any Magnetic Observations, I lose no opportunity of disseminating a knowledge of Gauss's capital improvements and enforcing their adoption.—There is a new observatory fitting up at Bombay—and having been applied to by the director Professor Orlebar for advice as to the objects to be attended to, I insisted strongly on the advantage of providing the establishment with a magnetic apparatus on Gauss's plan and observing with it in correspondence with others in Europe and I have given the same advice on other occasions. Pray let this be mentioned to Gauss, that he may see I am not unmindful of him. By the way the *principle* of

Gauss's method was mentioned to me by Babbage a great many years ago, as an idea of his own. I think it must have been at least 10 or 12 years.—I do not know however that he ever practised it.—

Well now the paper runs short.—Pray heaven this may find you well and that the very next ship may bring the letter from you we have so long been expecting.

With kind regards to D^r & Mrs Grosskopff—M^rs & Miss Beckedorff —Dr Mühig Captn Müller, Hausmann &c not forgetting Mrs Herschel & the Detmeirs [?]—I am Dear Aunt   Your affect^e Nephew   J.F.W. Herschel

[On front page]

As Maggie says I have disposed of 2 of the prints by keeping one for self & giving the other to the Observatory.—Little Willie shall have N° 3 when he is old enough to value it.

The Diary: October 5 to October 8

Wednesday, October 5, 1836

Rode into Town to the Meeting of the S.A. Phil Inst. chiefly to push the point of printing & publishing.

Saturday, October 8, 1836

Wrote a very long letter to Captn Fitzroy rectifying his ideas upon all the various points in his letter which required it

Mrs Wauchope came.—A calm & delightful day

The Travel Diary: An Expedition to Franschhoek, Genadendal, and Caledon

Oct 20   1836.

10^h 30^m A.M. En route with Marg^t & Mrs Wauchope, William Wauchope, Baby Alexander, Hannah & Julien, M^rs W's maid—

8 in hand. Brilliant morning start nearly calm sky cloud-less—all the flowers in the flats expanded—just a trace of snow lingering on the Double kop[65]

—7.20.   Got into Stellenbosch, much fatigued with a very broiling journey over a sandy heavy track most distressing to the Horses which (being a very bad team) were quite knocked up.—Well accommodated at Konigsburg's[66] a regular Inn.— Took Dinner & got a moonlight walk about the outskirts of the town a great part of which is surrounded with an Oak Avenue bordering a good road.—NB The place is improv-ing—Since I was here last the old stony road through the Town has been broken up & a smooth surface is in progress Called on Oliphant who is going also our way, to arrange so as to avoid interfering with each others sleeping ac-commodations at Holmes's[67] & Gnadenthal

Oct. 21. 1836. Stellenbosch—to French Hook.

Rose and got a very pretty sketch of the hills behind the Town. N⁰ 1.—We were then delayed by endeavours to pro-cure fresh horses & waggons which proved fruitless. So took on the old as far as French Hook to the Field Cornet Hugo's. A heavy sandy road except where up & down the hills which are low *clay* eminences below the level of the mountain ridges of sandstone.—Descending one got a good sketch N⁰ 2 of that noble mountain which I call the Altar since nobody will or can tell me a name for it.—Passed a field of the lovely white gladiolus[68] allied to *Blandus* but wanting the horseshoe marks & having instead a dark pink spot in the throat—Bagged about 50 Again botanised at the stony place where the road begins to turn round the Banghook Corner before the passage of the Berg River—got up the Tritonia? Crispa?[69] which is not yet in flower—here found

[65] "Kop" means a head, or peak. By the "double kop" he probably means the twin Jonkershoek peaks.

[66] Correctly, David Kinneburgh, hotel keeper of Dorp Street, Stellenbosch.

[67] Possibly Holmes' Toll House, of which Sir John made several sketches.

[68] Perhaps a form of *Gladiolus blandus*, which is very variable in the dark pink marks on the perianth.

[69] A member of the family Iridaceae, with cream-colored flowers and leaf margins much puckered and crumpled (crispate).

a Singular low creeping Protea[70] with leaves like those of the
fir tree—flowers *close* to the ground. Rode & Bulbgathered
all the way to Hugo's where arrived the horses quite exhausted—M[r] Bell who rode on had tried to procure fresh—
in vain, all the farmers being at a sale at Haumann's (a great
occurrence like a *fair* or a race, here). Hugo's wife received
us good naturedly & gave us Beds—(we brought our own
stores)

Arrived late but time enough for me to go down to the river
& bathe—& gather a few of the immense abundance of
Amaryllis Belladonna bulbs which had been turned up in
a dyke. Too late to start afresh, secured fresh horses & another waggon—so dismissed the old.

Oct 22   1836.—Hugo's French Hook.
Rose before Daylight. NB by moonlight this retired nook
has a very solemn pleasing effect. But it is barren & wants
trees & water.

Before Sunrise we were fairly en route for the French Hook
(Fransche Hoek) pass up whose indifferently planned &
worse kept road we toiled heavily the ladies in the waggon
Bell and self leading our horses. Half way up occurs the
Black oxide of iron in a state partly pisolitic partly scori-
form, in great beds

After resting awhile on summit commenced descent which
winds beside a very deep & rugged ravine of desolate &
sterile hills—It is much more irregularly laid out than the
ascent & at length comes upon the Toll House (Holmes's)
by a gentle slope beside a pretty picturesque river with a
bad wooden bridge & a tolerable bordering of Brushwood
making a good scene for a sketch w[h] after landing at the
Toll House I returned & drew (N[o] 3) Also got a bathe in
the clear stream below the bridge. Breakfasted at Holmes's
& (the ladies going on per waggon followed with M[r] Bell
on horse

The Valley here opens out into a wide alluvial swamp of
apparently 8 or 10 miles broad & great length bounded by

[70] Evidently of the affinity of *Protea scorzonerifolia* (Salisb. ex Knight) Ry-
croft.

dry barren sandrock—a desolate scene. It is said to be bad
soil, a lake in winter & not drainable, of which I believe
not a word—It might certainly, with industry & enterprise
be cultivated   The river in it is? the Sonder End?[71]
Outspanned after Crossing the River # then made another
stage across the "Eiselyacht" or Eseljagd (? Zebra chase)
a great chain of clay hills which on this as on the Cape side
of the mountains form the lower deposit in the Valleys of the
granite slate & sandstone. It seems to be identical with that
of Tigerberg Sugarberg &c   These clay hills are now dry
& hard as bricks & are covered with bulbous plants among
which are great squills[72] and Watsonias (NB   the Marginata occurs in abundance at the outspan last mentioned)
Here noticed for 1st time a beautiful crimson col[d] Satyrium[73]
also crimson Ixias. Overtook after a hard ride & much fa-
tigue the waggon on the descent to Devilliers' Krall where
we walked down & found a wretched house with mud floors
& fleas and 2 old Hags sitting unmoved in chairs, sisters)
the Genii Loci who spoke no English (nor as far as we could
see any other tongue) Here ascertained the loss of my bulb-
bag with all my treasures of yesterday.—Disconcerted.—
Mounted & rode on in waggon. A very long pull to Gnadenthal through a fine country for rugged mountain scenery.
Roads horrible just before crossing the river before Gnadenthal especially.—Forded the River with some trouble &
with the water in the waggon & reached Gnadenthal just
after dark, where got accommodation in humble but tolerable fashion in the small lodging house established for
chance comers. As soon as arrived good D[r] & M[rs] Lees
called & gave their welcome. Joined by Oliphant & M[rs] O.

Here M[t] saw 6 Caffer Cranes[74] by bridge c 1400

[71] The source of the Sonderend (endless) River, is in these hills, and the party
crossed the Eseljagsberg.

[72] *Scilla* sp. (Liliaceae).

[73] Possibly *Satyrium erectum.*

[74] Probably the Blue Crane (*Tetrapteryx paradisea*), the common crane of
South Africa, a graceful bird some three feet tall, with a wide distribution. The
Crowned Crane (*Balearica regulorum*) is referred to as a "Kaffir Crane" in some

Sunday Oct 23. 1836.

    In morning Divine Service in þᵉ Moravian Church at Gnad-
enthal attended by a large congregation of Hottentots all
very orderly and decent looking & attentive

    [A blank page]

    Caledon[75]—Descending stream at 150 yards above the Bath
house 109.4 runs in a rudely covered channel Near top of
the low flat topped mound still in an artificial channel but
where it comes from a covered place 115.6 Arterial source
117.4 a deep $\perp^r$ [perpendicular] hole cylindrical $\pm$ 13
inches shrubs & roots all covered with fine ochery matter
NO gas bubbles—no smell no taste no sulphureous incrus-
tation NB. Brass scale Therm by Doll[on][d76] requires corrⁿ
of— 0°9 to give true Fahr.[enheit]

    de Koch's Red Disas

Wednesday, Oct. 26.

    5½ AM. Started for my sketch N° 7 about a mile back behind
De Reu's or de Roux's—NB This is the 4th or 5th De Roux's
place we have encountered

8ʰ 30ᵐ En route

9. 15 Summit of 1ˢᵗ ridge

1 30 arrived at Palmiet R[iver] Green sward Bush snake Wolf
      Hair cut

2 35 En route

    [Blank pages: the following odd memoranda are not very
relevant, but show prices paid and the kind of philosophical
musings in which Sir John indulged.]

Holmes's—Waggoner   1 : 6
At Gnadenthal
At Caledon—   8 : 6
At De Kocks *he* paid ∴ [therefore]   0 : 0
At Morkels—3½ DM =   5 : 3 pᵈ
Major Rogers Wynberg Protector of Slaves

---

old writings, but the distribution of this fantastic and beautiful bird does not extend so far southwest. The notation "c 1400" means about 2 P.M.

   [75] He is now at the thermal spring at this little town some eighty miles east of
Cape Town, measuring the temperature of the hot waters as they flow out.

   [76] A London instrument maker.

Major Mitchells house wishes to let } 5 miles
the house or sell it   } 1ˢᵗ house
Mr Brink's house Auditor general rent 180 £
Baker Moore's house

Aphorisms
Rhetoric is the art of telling people the truth in such terms that they should not believe you & lies in such that they shall
Logic is the art of talking unintelligibly of matters of which we are ignorant
Poetry is the affected expression of exaggerated sentiments—
End next Watsons to Center of Ditch = 24 + to mark to line of fence 25 + 2 to mark
        alert—al′ eretta
            or ereata

Hangklip from Somerset peak   33° 26′ 0
            Zero   − 0   3 30
                                    ─────────────
                                     33 22 30
Angle at 9 AM ¼ from Ecksteins Kloof   Max. 96.2 Min 31.2

[End of Travel Diary in this Pocket Book]⁷⁷

The Diary: October 27 to December 31;
Occasional Memoranda

Thursday, October 27, 1836
Returned to Feldhausen from an excursion to Gnadenthal and had the joy of finding all the little ones well (colds excepted).

⁷⁷ We know Sir John well enough now to recognize in these, first, his expenditure in shillings and pence on this latest trip; then some notes made earlier when he was looking for houses; then the aphorisms; then notes he made while pacing out the Feldhausen property; then a thought connected with his phonetic dictionary; a horizontal angle measured from fairly high on Table Mountain; and finally, since the minimum is so low, the readings of the thermometer left on the mountain on his first ascent, when he finally recovered it.

NB.   A sort of influenza has been extremely prevalent—not only in the Cape district but up the Country as far as Gnadenthal. Hardly anyone has escaped. It is strange in the warmer and apparently genial weather which has prevailed for the last week to hear everybody coughing violently.

Wednesday, November 2, 1836

Lay in bed all the morning with a Blister on throat & a great dose of Physic—To drive away a fierce attack of tracheal inflammation

The fire has spread down the steep of the Table mountain and promises to become a great conflagration.

At Night this promise was realised and shortly after Sunset it assumed a grand appearance which continued to increase in Magnificence throughout the night—The whole of the Mountain in all its buttresses, Gorges & Ravines, was crested with lines of Vivid fire—in all sorts of festoons, broken lines & zigzags—sometimes glowing dull red, then bursting out into intense light and shooting up tall masses of flame—It was like the Crater of an Enormous volcano with the lava first peeping above its brim and here & there overflowing. The line extended with interruptions from the middle of the Mainbank to beyond Kirstenbosch Gorge—certainly occupying not less than 2 miles or perhaps 3 of the mountain. At different periods of the night it varied Extremely in appearance according to the State of the clouds which sometimes left the Mountain nearly free and at others partially or entirely concealed the fire. About midnight while viewing it & taking a (most imperfect) sketch of it—a Column of flame rose suddenly on the farther side of Kirstenbosch Gorge so large & vivid that the whole orchard where we stood was brilliantly illuminated every object throwing long shadows. Judging from the apparent height of the Column compared with the distance it could not be less than 100 feet high & 20 or 30 broad—At about 2 AM it was at its maximum of grandeur. Seen from the vineyard where no object intercepted the view—it was a Spectacle of the first class!! fully I should think equalling an Eruption of Vesuvius—but more solemn & composed—for the progress of the flames though evident at ¼ hour or half an hours interval, was not *seen* as a *movement* unless when some great outbreak announced the access of the fire to a nest of inflammable [not Sir John's word, but his totally illegible scrawl begins "infla"] material.—

Once however I saw a vast body of flame stream *down* a Ravine
nearly [Continued in space for November 3]
perpendicular. It descended lambently and with a moderately rapid
gliding motion like the Ball of a Roman candle or a meteor of the
first magnitude—I should think its descent at least 300 feet—in 4 or 5
seconds. I am quite at a loss to account for this—brushwood or dry
fern would not run in a train so rapidly—falling masses would have
broken & *could* not have descended so far owing to the inclination &
irregularity of the ravines Perhaps it might be a great spiracle of inflammable gas from some unseen smouldering masses below which
caught fire at top & burned downwards—but this is a forced supposi-
tion.

Far aloft as if among the Stars, half burned trees continued visible
as glowing specks like great red stars so very high that it seemed at
times impossible to believe them *on* the mountain. It is singular how
permanent these were. One kept its place for many hours with alter-
nate intervals of Extinction & revival.

"Why flame the far summit—Why shoot to the blast
"Those Embers like Stars from the firmament cast
"Tis the fire Shower of ruin &c . . .

Sunday, November 20, 1836
Sir J. & Lady Franklin, Captn & Mrs Macconochie—Col & Lady
Cath. Bell, Dr Smith, Mr Harvey, Mrs Stockenstrom & her Sister Miss
M[78] . . . dined with us

Tuesday, November 22, 1836
Began reducing Sw[eep] 523.

Wednesday, November 23, 1836
Finished reducing Sw[eep] 523   Began Sw[eep] 524   About midnight Magnificent display of the great North West Cloud rolling over
the whole range of Table & Devil Mountains—wind moderate—Piles
of cloud immense & coming down below blockhouse.—About 5 AM
heavy rain—for 2h ±

Thursday, November 24, 1836
Finished reducing Sw[eep] 524.

---

[78] Before her marriage Mrs. Stockenström was Elsabe Maasdorp, daughter of
Gysbert-Hendrik Maasdorp; she had two sisters, Susanna and Margaretha.

Sunday, November 27, 1836

Sir John & Lady Franklin[79] & Captn & Mrs Macconochie came in the evening   Shewed them the great Cluster 47 Toucani[80] & the Magellanic Clouds.

Monday, November 28, 1836

Maclear took his angles from my Boundary Stone by Letterstedts wood—Rode up to witness his proceedings—& made smoke signals from 20 feet to give a rough place of the Instrument. While there Lieut. Worster came up (just arrived from England, to Madras) & offered to take Dr Stewarts mirror—

Went home and got up the polishing apparatus for a 9^{in} mirror— Made a Polisher[81] and left it to cool   Maclear came in to dinner and staid till past midnight.

Being fair measured 6 D stars in Equatorial   M looking on

Tuesday, November 29, 1836

At work at the Mirror for Dr Stewart but could not get it done in time for Mr Worster to take it

Captn & Mrs Wauchope came to stay a few days with their Son Wm Wauchope on his horse jumbo

Mr & Mrs Worster called

---

[79] Sir John Franklin (1786–1847), English Vice-Admiral and explorer, discoverer of the Northwest Passage. After serving at the battles of Copenhagen and Trafalgar, Franklin first commanded the *Trent* on Captain Buchan's expedition of 1818 to the Arctic. He himself commanded two overland expeditions (1819–1822 and 1825–1827) which explored large parts of the North American coastline. He published *Narrative of a Journey to the Shores of the Polar Sea* (1823) and *Narrative of a Second Expedition to the Shores of the Polar Sea* (1828). In 1829 he was knighted, and from 1836 to 1843 was Governor of van Diemen's Land (Tasmania), being evidently en route thither when he called on Sir John. In 1845 in command of an expedition for the discovery of the Northwest Passage, Franklin left England on May 19 with the ships *Erebus* and *Terror*. They were not seen again after July 25 of that year. Several rescue expeditions were sent, and it was eventually established in 1859 that the expedition had made its way through the Passage, and that Franklin had died with all hands in 1847.

[80] Another superb globular cluster in the southern skies very close to the Small Magellanic Cloud on the sky, but at only a fraction of its distance.

[81] See note 153, September 9, 1834.

Wednesday, November 30, 1836
Proved a bad day and a projected party to picnic at "Paradise"—proved abortive
Mr & Mrs Fairbairn & Mrs Phillip dined with us [line erased]
Miss Petrie dined with us.

Thursday, December 1, 1836
Mr & Mrs Fairbairn, Dr Adamson & Mrs Phillip dined with us

Friday, December 2, 1836
Col & Mrs Smith[82] & Mr Hamilton dined with us.—In Evening Mrs Smith sang to her guitar in a very delightful style—Optical expt[s] on Col Smith's eyesight—He has idiochromic vision—The red & green images produced by the tincture of Ianthina[83] appear to him a very good match in colour only by moments he thinks the green image "has rather a bluish look at times." Dalton's[84] book w[h] he has examined & returns he says by day light & candle light agrees well enough with his notions.

Saturday, December 3, 1836
Capt[n] & Mrs. Wauchope left Feldhausen

Sunday, December 4, 1836
Attended Div Service in Rondebosch Church
Swept—Superb night but ill defined stars

Monday, December 5, 1836
A broiling day. Thermom Max = 89½
Calm.—Kept in to work prepared reductions of Sw[eep] 526—
Lady C Bell & Mrs Wauchope called

[82] The famous one, afterward Sir Harry Smith.

[83] It is not clear what this was. Ianthine means "violet colored." Ianthinite is a mineral, $2UO_2.7H_2O$, having violet orthorhombic crystals. This is clearly a colored solution, possibly a plant extract useful in testing color vision.

[84] John Dalton (1766–1844), the celebrated English chemist and physicist, who was one of the founders of the modern atomic theory, was himself color blind. His paper presented to the Manchester Literary and Philosophical Society, "Extraordinary Facts Relating to the Vision of Colours," was one of the first subjective accounts of the type of color blindness characterized by inability to distinguish red and green. He wrote, "Blood looked like bottlegreen and a laurel leaf was a good match for sealing wax." The other aspects of Dalton's career of genius are not germane to the present discussion.

Tuesday, December 6, 1836

A cold cutting day with a black South Easter—At work on Sweep reductions Nos 527.—Wrote to Cousin Mary and finished letter to Demorgan[85] See Duplicates—Got a letter from Biot[86] with the wrong book

Wednesday, December 7, 1836

Attended Meeting of the SA Phil and Lit Inst.

Read them a lecture on Dichromic Vision & shewed them Dalton's book of Colours.—Also Shewed the Dichromatism of the Tincture of Ianthina and [erased] lastly explained how by the action of Electromagnetism a continuous wheel motion might be produced to obviate Hevelly's objection against G . . .'s [his omission, probably Gauss is meant] scheme drawn from the shortness of the *Space* through w^h magnetism acts

Went after meeting with Maclear to see the Barrack Carpenters shop where his measuring Rods are making on Roy's[87] construction [A sketch: the rods as finally used can be seen in the drawing dated July 19, 1837, reproduced from the travel diary]

Posted letters to Demorgan &c &c   No stars tonight

[85] Augustus De Morgan (1806–1871), English mathematician, logician, and bibliographer, who gave his name to De Morgan's theorem in logic. He was professor of mathematics at the newly founded University College, London, from 1828 to 1866, except for a break from 1831 to 1836. He was first president of the newly formed London Mathematical Society in 1866. In mathematics his aim was to introduce complete rigor, and he introduced the method of mathematical induction in 1838. The renaissance of mathematical or symbolic logic, which has applications to computer science, was largely due to him and to Boole. He wrote his whimsical *Budget of Paradoxes* in 1872, and developed a system of book cataloguing adopted by the British Museum. He left University College following a dispute over sectarian freedom.

[86] Jean Baptiste Biot (1774–1862), French physicist specializing in work on the polarization of light. In 1880 he became professor of physics at the Collège de France, and in 1806 made the first scientific balloon ascent with Gay-Lussac. He worked on the refractive properties of gases, and on geodetic measures in Spain and elsewhere. In 1820 he discovered, with Felix Savart, the electromagnetic law which bears their names. He also developed the method of polarimetric analysis for the study of polarizing solutions, such as sugar water.

[87] William Roy (1726–1790), surveyor and cartographer. He engaged in military survey work, becoming Deputy Quartermaster General of South Britain, 1761, and Surveyor General of Great Britain, 1765. He became FRS in 1767, Major General in 1781, and Copley medalist in 1785 for his work in determining the relative positions of the observatories of Paris and Greenwich.

Thursday, December 8, 1836

Feldhausen—No events Maggie went over to Observatory to see Mrs Maclear who has been ill She brought back the 14 feet mirror from the Observatory to be repolished. Prepared Colcothar de novo for this operation & for my own need the Calcutta Mirrors—

Put in order the Laboratory w^h had got dreadfully bad.—

Reduced the P[olar] D[istance]'s of Sweeps 526/7/

Promising fine night began resweeping zone 150—but before an hour clouded thick.

Troubled with an oedematous swelling in right ancle which is probably the consequence of the awkwardness of the Equatorial steps—it has been long & slowly coming on & is now grown considerable & must be abated.

Friday, December 16, 1836

Constructed an apparatus of a Wolfe's Bottle for Actinometer work

Splendid night Strong SE but good definition Measured 12 D stars per Equatorial several of w^h I have not been able to get Eql measures of before

Saturday, December 17, 1836

Being a Superb day.—prosecuted Exp^ts with the new Actinometer apparatus for determining the absolute Expansions

A night of bad definition

Monograph of Neb. Orion

2 or 3 stars per Astrometer—bad

Wednesday, December 21, 1836

The Hourly Obs^ns occupied the whole day leaving little else than snatches for other work—about 9 PM [partly erased] went to bed & rose at 2½ & went on till morning—

Thursday, December 22, 1836

Leaving Marg^t to go on with the Hourly Obs^ns (there being no therm^r for Actinometer) went in to attend the Distribution of prizes in the SA College.—Wylde gave an address on introducing the prize boys to the Governor—w^h I followed [word erased] also Dr Adamson & meeting broke up Mr? Changuia[?] 1st addressing his class in Dutch.

Saturday, December 24, 1836

Rode out with Marg^t. Called on Lady D'Urban & left cards—on Col & Mrs Smith—at home

The Blue Spruce
N [entry ends]

Friday, December 30, 1836

Attended Committee of SA Association to Explore Central Africa, on occasion of Dr Smith's approaching departure.

Saturday, December 31, 1836

Margaret & *all the children* took wing for the Observatory to pass Newyears day there

Occasional Memoranda, March, 1836

In flower

Haemanthus Coccineus—going off about middle of Month

Haemanthus Tigrinus—Profuse March 17

Brunswickia Josephina great Candelabra March 17, abundant

Eriospermum Lanceolatum [? Lanceaefolium; reading very difficult] coming in March 25.

Gladiolus called by Bowie *Brevi*folius Pink & yellow, small, many flowered, coming in March 20—NB leaves very long Numerous in Bud March 25 & flower

The Goat Sucker[88] (March 30 & 31) very loud & frequent in the woods at night [Figure 5]

---

[88] The name of a family of birds (Nightjars or *Caprimulgidae*) based on an incorrect legend about their habits. Leonard Gill, *A First Guide to South African Birds*, says, "A peculiar and extremely interesting family, apparently large relatives of the Swifts and with similar small bill and wide gape for catching insects on the wing; but adapted to a totally different way of life. The Nightjars are specialized for living on the abundant insect life that flies in the dusk and at night; they spend the day in concealment and are specialized for that too, their plumage having the general appearance of dead leaves and lichened bark . . . At night, especially by moonlight, they become vocal: different species have different kinds of notes, some having beautiful whistling calls, others various sorts of churring or trilling sounds." Sir John heard the birds in the autumn with the moon nearly full. However, Mrs. Rowan now takes up the story and says, "From the music for the goatsucker call given in the paragraph on Cape Noises in 'Observations, 1837' one can clearly read the call of the South African Nightjar, *Caprimulgus pectoralis*, also known as the Litany Bird, because it says 'Good Lord, deliver . . . liver . . . liver . . . liver . . . us' on a descending scale. With his musical background, Sir John was quite capable of transcribing this musically with accuracy.

The call given for March 30 and 31, 1836, is certainly not that of *C. pectoralis*,

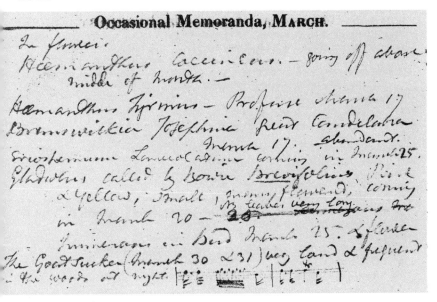

Figure 5. Illustration of the goatsucker's song, from the Diary, Occasional Memoranda, March, 1836.

Occasional Memoranda, June, 1836

June. End. Pink Oxalis (purpureus) and oxalis versicolor in flower.
July
Babiana Villosa (Blue early) full in profuse flower.—July 1 . . . 12
Protea Mellifera[89] in full flower. July 1
Antholyza Praealta coming into flower July 1.
July 12 Antholyza Ringens. in my garden all ready to open—in strong bud. NB.   this is much earlier than usual—quite premature.

---

but it is a fair representation of the calls of two other species, both of which start off with a couple of staccato notes, and finish with a vibrant trill. These are the European Nightjar (*C. europaeus*), and the Rufous-cheeked Nightjar, (*C. rufigena*). The first is a migrant from Europe, not supposed to sing in its 'winter quarters', and, one would think, well on its way back to Europe by the end of March. The second is not known to occur anywhere within a hundred miles of Cape Town. But I would be prepared to bet from the transcribed call that it was one or other of these. Perhaps the European Nightjar is the most likely, since there are several old records of its occurrence in the Cape."

[89] Now known as *Protea repens*, the Sugarbush.

July 6. Saw Gladiolus Watsonius[90] in flower but not yet fully come out. July 26. Fully out

July 1–12.—Disa Longicornis[91] coming into flower

July 27. Protea Mellifera rather going off

About July 9 the spring began to run by the Nursery. July 20 very copious

Occasional Memoranda, September, 1836

Sep: 6.

Babiana. Angustifolia coming into flower

    Rubro-cyanea[92] Full, going off.—

Gladiolus Gracilis going off. Gladiol versicolor full

Sep 11. small brown Morea (? Polystachya) going off. Profuse morea Papilonacea[93] Passing

Occasional Memoranda, December, 1836

Dec 20 ± Cyrtanthus Obliquus (Knysna lily) in flower.

Dec 31. Agapanthus. Tall blue in flower

             Short blue D°

             White   D°

    Crinum Africanum[94] Just flowering

[End Paper of Volume: The end paper contains a few computations connected with the comparison of Prinsep's, Henning's, and Herschel's Barometers, too disconnected to be worth reproducing.]

---

[90] *Homoglossum watsonianum* N.E. Br., a slender plant with showy red flowers formed in a geometrically pleasing way.

[91] ? *Disperis capensis.*

[92] *Babiana rubro-cyanea,* Winecups.

[93] A plant with brick-red or yellow irislike flowers.

[94] This is generally held to be a synonym of *Agapanthus africanus* (L.) Hoffmgg.

1837

The Pocket Book: January 1 to February 17

[Sir John did not receive his regular diary for 1837 until some time in March, and for the first part of the year he kept his diary in the rather scrappy style of his travel journals, in a small pocket book.]

Jan 1. 1837

Maggie & all the child[n] are gone to the Observatory to see the old year out. Midn[t] Clearing off Calm after a long & violent S. wind with vast voluminous masses of cloud.
7 AM Superb morning Calm & cloudless:—
29.536 71.0 [Barometer and temperature readings][1]

Read Miss Baillie's[2] Martyr Obs[d] Actinom[r] at 9
Reduced Sweep 634 Polished thick mirror.
12. Noon—29.518 71.0

Sky covered with cirrous streaks Calm
Rode over to Observatory where found D[r] Smith & Skirrow.—
M[r] & M[rs] Maclear came to dinner—Party Maclear's selves, Smith & 11 children—Suspended Maclear's new 2 feet Bar Magnet with Chisel edges w[h] pointed to Var[iation][3] 8° W in place of 27°— Strange.
Anecdotes of Lions.—The Lion leader of the herd disappointed in

---

[1] The temperatures given are presumably those shown by the thermometer attached to the barometer, which would, of course, usually be in some shaded place indoors.

[2] Joanna Baillie (1762–1851), Scottish dramatist and poetess. Her *Martyr*, published in 1826 but written some time before, relates the martyrdom of Cordenius Maro, an officer in Nero's Imperial Guard who had become a Christian.

[3] At most places on the Earth true and magnetic north differ by a quantity known as the variation. This and other magnetic elements change slowly with time.

his leap, young lion setting example, old raised to greater exertions.—Etiquette in passing a Lion under a bush. Narrow escape of one of Dr. Smiths party who came suddenly on a Lion tho warned by movements of the party Lion made at him—man ran—Lion ready to spring—man turned—Lion clapped tail between legs & made off—Man could not say what made him turn, he was bewildered.

Again suspended magnet—1st Chisel edge ⊥ [perpendicular] then horizontal. Still points about 7 or 8° W of N.—An azimuth compass similarly placed points to 27° W & this all around the room as well as just where the magnet was hung.—*Yet* the poles by trial *are at* the extremities! ! !

Jan. 2, 1837—*Observatory*

Rose at 7.—Took sketch of the building from across the Salt River Swamp [Plate 12].—Lieut. Williams arrived at 8—& was joined by Skirrow. About 9 a Party of 6 soldiers & Stone with 4 of my men joined to help move the Circle.—It was lifted off with perfect ease & without any accident Found the Cradle quite loose & in a most unseemly & in my opinion most unworkmanlike state of wriggling & unsteadiness. It *tilts* from side to side by a very small force applied by the hand and therefore must yield when the circle is turned round, by some appreciable quantity.

Conferred about its remedy. I recommended a differential screw to force it down to its bearing vertically and wedges laterally in place of which latter Skirrow suggested Plaster of Paris—wh is better.—

Conferred also about the placing the Microscopes. It being necessary to cut the arc sent by Jones and he having only sent 2 supports [a small incomprehensible sketch].

Sunday.—1837 Jan. 8

Walked over to Wynberg Church & stood Godfather to Mrs Cumberlege's child christened Beatrice Margaret Herschel (NB This fashion of converting Surnames into Xtian ones is excessively absurd. The Surname is to distinguish families the Xtian individuals) Thence walked with Col¹ King through a labyrinth of Oak woods to Mrs Cumberledges where Margt & I & the children dined & after a stroll in even[in]g to see Mrs Batts's House & Grounds (to be let) returned home

Swept 11 sets of D[ouble] $*$ measures with Equatorial being a night of Superb Def[initio]^n

Mrs Batts was a maid serv^t in Engl[an]d who [entry terminates]

Monday 1837 Jan 9

Rose just in time to get noon Actinom^r Obs for Scale & value of the Actine^4 & Observe Spots in $\odot$ [the Sun] at noon—which are most extraordinary.—

Made a thorough arrangement of Papers, Book Shelves &c Laid out plan of flower beds in front of the drawing room windows. Lieut Williams came to *tea and stars* bringing over Wm Petrie junr.—Shewed them the great Nebula 30 Doradus &c (and several fine D stars in Equatorial & set sweep 647.° [?]—to run over the five D stars & Plan[etary] Neb[ulae] in it

Prepared Platinate of Lime in a state for Actinometer Expt.

Tuesd *Jan 10*

Rose Late having been up sweeping till 3.—

Not Clear Enough for Actin^r standard work—Cl[oud] = 4.

Prepared however [entry terminates]

Spots in $\odot$ projected^5 Wrote to my Aunt.^{6}

Miss Petrie called—brought her Drawing & worked at a flower with M^t—

Dined.—Planned flowerbeds & well in the grove in front of the Dining room window Reduction of Sweep 639.—not being a night for Stars Cloudy & Calm Left off at 12^h 30^m MT

Wedn. Jan 11.—1837

Rose at 11. Spots in $\odot$. Wrote to my Aunt [erased]

Thursd 12 Jan^y

At home all day Noon Spots in $\odot$—not Clear enough for Actinometer A most superb night though Def[inition] not perfect—Worked at map of Visible Stars & Milky Way.—till 1 AM then

---

^{4} Herschel is evidently trying to calibrate the actinometer and to find a radiation unit—the actine.

^{5} By holding a white card behind the telescope eyepiece. A very high sunspot maximum occurred in 1837.2 (i.e., March 14, 1837).

^{6} See below, after end of Pocket Book.

meas^d 3 or 4 Double Stars.—Drew 2 sets Bills per H. Ross 340£
on Drummond.

Frid Jany 13.
Max. Therm. 82 At home all day Reduced sweeps 639, 640.—
Splendid Day tried Act Exp with Platinate of Lime (failed by over
platinising the Mixture). Solar spots—At night a dense black
cloud mass formed on T drifting from S. & at midni[gh]t the whole
sky was covered. till then worked at Equatorial

Sat Jan. 14.—Rose at 10 Cloudless Calm, Superb morning The SE.
however got up—but moderate
NB. A very great series of South Easters (*here* South) Winds have
blown this season and in the beginning of the Month the drift of
cloud was constant & Enormous.
Got obs of Spots & Actinom^r Exp with the Platinate—At home all
day. Read Lacaille's[7] Journal lent by Maclear. It is very meagre

Sund Jan 15/37
Church at Rondebosch afternoon.—AM. Solar spots which shewed
to Lieut Williams—Before that the Actinom^r Expt with Platinate
of Lime NB. Definition of the Sun astonishingly fine.—Rode from
Church to Col Bell, with Lady C & Mrs Smith &c.
Clouded but cleared again, but‡ being a night of bad def^n could
not work the Equatorial so mapped a while & retired at 1.—NB.
The Evenings in the cool lately almost constantly have been passed
in the Grounds with M. Planning new walks & beds &c for the
garden & among the Groves.—It is very pretty pleasant occupation
but a dreadful thief of time
‡ NB. In early part of evening the moon was so admirably defined that I was drawn in to begin at a monograph of part

Mond Jan 16. 1837.
At home all day. Mrs Smith[8] Passed Morning with Marg^t and
brought her as a present a beautiful Panther Skin Caross from
Cafferland.—Reductions of Sweep 643.—Mr Rach called about the
purchase of the land beyond the road. Offered to *sell* for £250 or

---

[7] See note 188, December 28, 1835.
[8] Presumably Mrs. Andrew Smith.

to *let* at 12 guineas a year.—After dinner perambulated with Stone to direct the Cutting of firewood by Schönbergs people
[A portion scored out and illegible]
Towards Evening the clouds drew over from NW gently & quietly & it began to sprinkle increasing to steady rain w^h lasted all night.

Tuesd Jany 17. 1837.
Rose at 10. Rainy morning—Went over to Observatory & with Maclear went through part of the process of Examination of the Circle by the 2 additional microscopes. Then took on Maclear to Cape Town to attend Dr Smiths Expedition Committee w^h was to meet at 12.—The Sec^y Chase did not however attend and Dr Smith himself knew nothing about it   Going into Cape Town were annoyed by a horrid smell of fish on Sunday an enormous quantity of *dead fish* came on shore covering the whole beach at Cape Town & all along the bay as far as Blueberg. Mostly of the smaller sort & none *new*. It is said the same happened the day before the Earthquake of 18 . . . [his omission][9] but this time the quantity is much greater. Government Carts are employed to convey them away from the Town to get rid of the nuisance. One man is said to have died from Eating them but thousands have been eaten *without* producing Death
Dr Smith whom we found at Verreaux's[10] says he never knew the bay so full of large fish & that you cannot thrust a pole from a boat without hitting one, & suggests that the small ones were driven ashore[11] by the great ones or at least into shoal water where they have been killed by the breakers & washed ashore.
Saw at Verreaux's some fossil heads of the Indian Hippopotamus

[9] Probably that of 1809, described by W. L. von Buchenroder in the *South African Quarterly Journal*, No. 1 (1829–1830).

[10] Jules Verreaux, naturalist, who came to the Cape at the age of twelve with his uncle, P. A. Delalande. He kept the Machtenburg Garden in Looyers Plein at this time.

[11] Professor Day remarks: "The small fish driven ashore may have been pilchards. I do not believe they were driven ashore in quantity by large fish—more likely by an occurrence of the 'Red Tide' or algal bloom of peridinians, such as occurred in Gordon's Bay a few years ago. The larger fish may have been snoek which commonly follow pilchards." The phenomenon of the Red Tide, where the sea is colored, and large quantities of dead fish are found, has been recorded on several occasions.

w^h is a decidedly different species from the Cape having 6 strait, projecting teeth from the lower jaw between the side tusks, where ours has but 4

Returned to Observatory and took more readings of the Microscopes & came home to dinner   A cloudy night.

Reduced sweeps. & began letter to Encke. Retired at 2½ AM.

Wedn. 18. Jany 1836. [*sic*]

Rose at 10½. Dull heavy moist warm morning.—Vast dense mountain cloud from North Swept in to bank along D & T but none passing near. Wind rising strong

Thursd 19 Jan 1837.

M^r P Smith [Piazzi Smyth] called, bringing a batch of Zero stars from Observatory   Hard at work all day reducing except 2 hours bulb packing or rather sorting to prepare & labelling

At night a great pile of clouds assembled all along the ridge of T & D. drifting gently from N to S. but *not* pouring over. Wind rising in strong gusts from N but met & checked by the S. Easter which prevents the cloud pouring over as usual.—A vast pile also stationary in S & SE up to alt[itude] 60°.—All Zenith clear   Heavy rain in night [A small sketch plan of the Table Mountain and Devil's Peak area showing the wind directions]

Friday Jan 20 1837

Rose at 9½. Heavy Rain with NW wind. Reduced Sweeps

Packed bulbs for 1½ hours

Sat. Jan 21. 1837

Up to Noon. worked at reductions Then began a series of 24 hourly Meteorol Obs^{ns} which broke into all other occupation for the rest of the day. Occasionally Packed a few bulbs—Finished a letter to Encke. Went to bed at 10 leaving M^t to read off the Inst^s till 3 AM.

Sund. Jan. 22. 1837

Rose at 4 & continued the Met obs^{ns} till noon. Reducing sweeps in a broken & dawdling kind of way in the intervals.—Attended Div Service at Rondebosch   Touched in a Sketch of The Devil Mountain from the Stoop. Violent NW gale set in with Rain in afternoon exceedingly heavy.—Went to bed at 9 very tired & stupid

Mond Jan 23 1837

Rose at 9.—Worked a while at Drawings of Cape Scenery.—
(Devil hill.)—Dr Grant called. Reduction of Sweeps.

Saturday Jan 28 1837

Rose at 10. Packed up Insects for Mr Waterhouse & H Griesbach
& Bulbs for Fox Talbot[12] and dispatched them per Stone to D^r
Smith to carry to England   D^r & Mrs Grant called.
Mr Harvey, Lady Campbell and [entry terminates]

Feb. 7. Completed marking off my nebula Book & began my Register
of Cape Sweeps Vol. 1. Nebulae   This is the beginning of my Cape
harvest Home.

Feb 8

I. Chase Esq^e US Consul & Mr Houghton called in evening. Wrote
to Mad. Lisboa at Rio enclosing Mrs. Graham's letter to her

Feb 9

Sent Mad Lisboa's letter to Mr Houghton to take to Rio

Feb 16

Wrote to Major Dutton thanks for seeds   At Night   Black S.
Easter

Feb 17. Friday

Rec^d letters from India, giving information of Jamie's alarming
illness & being on his way home.—Marg^t went into Town in afternoon   All day a Violent Black South Easter—Cloud down below
Blockhouse[13] & the strongest SE Wind I have known here and quite
cold

[In another hand]   No entries after this.

---

[12] William Henry Fox Talbot (1800–1877), English pioneer of photography,
as Sir John himself was to become on his return to England. He was born and he
died at Lacock Abbey, Wiltshire, where one of the earliest photographs was
made. He anticipated Daguerre in making sun pictures; in 1841 he produced the
calotype or Talbot process, and later he devised a method of instantaneous photography. Also interested in antiquarian and archeological research, he, with
Rawlinson and Hincks, was one of the earliest to decipher the cuneiform in-
scriptions from Nineveh.

[13] See note 146, August 25, 1834.

A Letter from Sir John and Lady Herschel, Feldhausen, to Caroline Lucretia Herschel, Hanover

(Jan[y] 10th 1837)   C. Good Hope

[In another hand]
N[o] 13. Rec[d] Apr[l] 30
[In Sir John's hand]

My dear Aunt

The return of a new year brings to my mind the considerable interval which has elapsed since I began a letter to you on my own account for I don't call postscripts and half pages tagged on to Margaret's letters anything, only just as serving to shew you that your nephew is still in the land of the living, and making progress.—But as this is decidedly the last year of our stay at the Cape and as all our thoughts and wishes are now directed homewards, it seems as if, somehow or other we had made a step nearer to you, and from the accounts my good friend & correspondent J. H. Nelson gave us of his & his Mother's visit to you, and from the sound and healthy tone of mind which pervades your own ever welcome letters, I still look forward with, I trust, a well grounded confidence, to the great delight of recounting to you in person, all our adventures in this corner of the Globe.

As for our news, it is comprised in a brief compass for we live here a very quiet life and *events* are rare. The best intelligence, as you will no doubt think it, is that we are all well, with exception of occasional aches and pains and influenzas, all of a rheumatic nature, the consequent of the sudden changes of this climate, but which, although disagreeable enough while they last, yet as they do not injure the constitution, the sure remedy for is to bear them patiently and wait till they go away. The children thank God are all of them healthy and though not robust (except little Margaret who is such a sturdy rough cast creature that we call her Master Maggie)—not one of them have had three days sickness of any serious nature since our arrival. The change to European Winters will no doubt be trying to them—

but we must hope the best, and as they are not brought up tenderly, they will probably adapt themselves easily to their new climate.

We took an excursion in November to some distance up the country to visit the Moravian institution at Gnadenthal which we found very interesting, and of which (as well as of the execrable roads which lead to it) Margaret shall give you a full account. This is I believe the last specimen of South African travelling I shall experience, and to say the truth, a little is quite sufficient The remainder of our stay here must be devoted to serious hard work & to the reduction and arrangement of the mass of observations accumulated, which is very great, and not a little frightful in prospect, to deal with, there being already between 3 and 400 sweeps registered besides a great mass of observations with the Equatorial and others of a miscellaneous kind. The Comet proved a serious interruption, as it was pleased to remain in view from Jan 25 to May 10 both inclusive during all which time it was observed whenever the weather permitted, & a beautiful series of observations I got of it, as did also Maclear.

I am now at work on the Spots in the Sun, and the general subject of Solar Radiation, which you know occupied a large portion of my Father's attention. The present is an admirable opportunity for studying these things as the Sun is infested now with Spots to a greater degree than I ever knew it, and they are arranged over its surface in a manner singularly interesting and instructive. The sky here is so pure and clean in our Summer that it would be a shame to neglect such an opportunity of making experiments on heat, and accordingly I have been occupied since the december Solstice in determining the constant of Solar Radiation, that is to say, the absolute quantity of heat sent down to the Earth's surface from the Sun at Noon or at a vertical incidence.

I do not think I have ever mentioned to you a remarkable and splendid instance of liberality on the part of His Grace of Northumberland, who without the smallest [these three words erased] has taken upon himself to defray the expenses of publishing my observations at the Cape—and that in a manner the most delicate and considerate imaginable. In consequence "my book" will appear when it does appear—under his auspices, and I hope will do no discredit to his munificence. This is not the only, nor the most remarkable instance however of his attachment to the cause of Science & his disposition to promote and support it.

I shall leave a little room for Maggie to add a PS in return for her good offices in that way to me and conclude by desiring to be remembered to M^r & Mrs Groskopff, Mrs H. M^rs & Miss Beckedorff, Captn Muller & all enquiring friends and to yourself continued health & all happiness & remain your affectionate nephew

JFW Herschel.

P.S. I have been pestered from all quarters with that ridiculous hoax about the Moon—in English French Italian & German! !

[In Lady Herschel's hand]

My very dear Aunt—I am glad even of a corner to tell you how sincerely I wish you a happy new year, & a comfortable existence through it considering the ordinary ills of life, & then may you be farther spared until you welcome us all in Europe, & some of us I hope in Hannover *next year*— It seems so strange to be keeping our New Year's revels under a broiling Sun, or rather clustering under the shade of a magnificent Willow from St. Helena in full leaf, while the children bring baskets of flowers to tie on the branches—This New years day however we all assembled at the Royal Observatory & the youngest of us got up to see the glorious Sun rise over the distant Hills & Carry and Bella have promised never to forget it—Willie was four years old the other day, & this morning I heard of Papa having introduced him to the Spots on the Sun—My other little pet boy is never tired of walking *on his own two feet* though scarcely 11 months old & is the plaything of everybody in the House—My dear Brother James in China desired his most affec^t respects to you when I sh^d write—Believe me my dear Aunt to be your much attached Niece

—M. B. Herschel

The Diary: March 19 to April 6

Sunday, March 19, 1837

Attended Divine Service at Rondebosch. Mr Hough

Actinometer & Sun Spots.—

Continued letter to Fr Baily.—

Morn. No sweeping & bad definition no measuring

Tried Experiments with pierced card-board aperture[14]—Woven Cane (back of an Indian Chair) & a wire sieve, placed respectively before the mouth of the 20-feet turned on Canopus. Nothing can surpass the splendor & regularity of the Phaenomenon.

Monday, March 20, 1837

Rode by appointment over to Observatory to confer about the measuring Rods for the Base line.[15]

[14] If a screen is placed over the objective of a telescope and a small number of apertures made in it, then light from stars will produce striking diffraction patterns due to the wave nature of light. Sir John was much exercised by the striking beauty of the phenomena but does not seem to have carried the analysis further. He was, of course, familiar with the wave theory of light so far as it had then been developed.

[15] It has already been remarked (note 6, January 16, 1836) that one of Maclear's official duties was to undertake survey work, and in particular to check the anomalous result found by Lacaille (see note 188, December 28, 1835). Also, on December 7, 1836, he had a design for measuring rods to be used in this work. It is now necessary to detail Maclear's work in chronological order. He began with an historical investigation to discover exactly what Lacaille had done. He first tried in 1836 to measure a base line on the low-lying ground near the Observatory, but floods came and spoiled the line. Toward the end of 1836 Maclear laid out and measured a base line on the Grand Parade in Cape Town (Plate 18). With improved means a base line was successfully measured near the Observatory in 1837, and concerning this we hear much in the diary. Finally, in December, 1837, in the height of summer, Maclear completed a new base line on the Parade, marking the ends with old guns sunk vertically into the ground. The site of these guns was built over by the construction of the Cape Town suburban station, but, as this is now being demolished, there is some chance of their still being *in situ* and recovered. After these preliminaries Maclear, followed by Sir David Gill, went on to make systematic surveys all over the Western Cape, extending over the whole of southern Africa. After Union in 1910 the work was taken over by the South African Trigonometrical Survey Office, and, quite

In the way, behind Ebden's home came on a Spot where I noticed no less than 18 great Falcon-hawks in the air soaring at once & looking about down pounced a huge bird like a Turkey in size. Thinking

---

apart from topographical and cadastral surveys, formed a major part of the geodetic survey of the thirtieth meridian, extending all the way from South Africa to the North Cape of Norway.

During Maclear's operations he had the benefit of Sir John's valuable, ingenious, but sometimes rather mercurial advice. Maclear also had Lieutenant Williams of the Sappers, Piazzi Smyth, and various other helpers. In 1837 they got on fairly well, with a long interruption due to floods, and a short one due to the wind blowing over one of the measuring rods. When this happened Sir John wrote to Maclear:

"My dear Sir,

I am sorry for the accident but I confess I was at no time sanguine as to the completion of the base without some interruption of the kind. The rods are top heavy and in case of any new operations commenced here with such instruments I should urge on the attention of those concerned a thing which if you recollect I did at one time mention as no imprudent precaution . . ."

Then comes a description of the rods on Roy's pattern, and the gloomy

"P.S. the first 700 feet of the Irish base were obliged to be remeasured by an accident similar to yours"

A second undated note must refer to the same incident:

"I feel today so great an increase of certain very unpleasant sensations which were creeping over me the whole of yesterday, that I hold it very doubtful whether I shall be in a condition to come over tomorrow or at all events for more than an hour or two in the warm part of the day—indeed I begin to perceive that another entire day's duty in your swamp will go nigh to lay me up for the winter, and I cannot help feeling that there are many days' work before that measurement will be completed—to say nothing of some unlucky jog which may render it necessary to remeasure the whole"

It says much for the high mutual esteem of the two that this kind of letter did not lead to the rupture of their friendship. Finally, when in December, 1837, Maclear had remeasured his base line on the Parade, Sir John scrawled the following:

"Your expedition in matters of business puts to shame my tardiness & proves to me that I am getting hardly fit to inhabit a bustling world like this. While I speculated on how long it is to take to get the guns—how long to get them sunk—how long to take up and sink again in consequence of detected errors—and how long to get all in order—behold your Parade base is measured—as it is I am rejoiced that I have *escaped* being present. That *broil* would have *done* me thoroughly. The next thing I expect to hear of is Messrs Maclear, Smith Williams & Co—ill of brain fevers . . ."

The foregoing three letters are in the South African Archives in the Maclear Papers.

it a Secretary Bird[16] I rode up & behold a dead horse fresh skinned & this bird & his Companion with many crows on him. They were enormous Vultures of a light Brown Colour with full ruffs round their necks at the shoulders like feather tippets, & so large & heavy they could hardly fly. Seeing them stupid I rode back for my gun—but before I could get it every possible contretem's occurred.—1st James was out shooting with the rifle—2dly my 2-barreled gun was foul & one cap plugged—2dly [his repetition] no powder flask to be found & only 1 bullet & that too large forced this down into wrong barrel with no powder—obliged to unscrew cap & put in powder—then lost the bullet when gently blown out—then no caps to be got—at last got Stone to hunt & we found some great shot & a few Pistol bullets—cleaned the plugged gun &c &c en route—but not a bird to be seen!

Reached Obsy at 2½ PM. conferred & in part executed (i.e. fitted up a beam compass with 2 Microscopes) & returned—

Much fatigued—no Sweeping or other work—went to bed.

Thursday, March 23, 1837

Occupied the whole morning writing to Mr Baily

Mt kept to her bed all day

Mr Hutchinson called & took home Mrs H in the Evening—

Miss Petrie & Miss Watt Called

Tried some further experiments on splitting oak trees with Gunpowder—The action is that of a violent *blow* as of a great hammer given from within—hence a strong confinement of the powder by a plug is unnecessary.—A screw was tried with a notch cut longitudinally to receive a priming but the effect was not materially greater than when a mere plug of sandy earth was used gently rammed [sketch of the wood screw].

Friday, March 24, 1837

Executed a Diagonal Scale for MaClear's Base meast—on Mica for the constant separation of the ends of his measg Rods—a tiresome & fidgetty job requiring great care & much time & after all could not do it to my satisfaction

---

[16] The Secretary Bird, *Sagittarius serpentarius*, a very large bird, whose legs look as if clad in eighteenth-century knee breeches, while the head quills resemble quill pens stuck behind an ear, has a, possibly exaggerated, reputation as an outstanding killer of snakes. The Cape Vulture, *Gyps coprotheres*, is now seldom seen in the western Cape.

Rec^d letters from Hodgkin & Gemellaco. [?]

Meas^d 3 D Stars near η Argus but Def^n Bad

Saturday, March 25, 1837

A Day spent at Observatory in "deciding upon" & partly executing the fixing of 2 microscopes on a Bar for a Beam Compass—

Sunday, March 26, 1837

Attended Div Service at Rondebosch

Tuesday, March 28, 1837

A lovely calm day with no events but reducing Sweeps, superintending Christian planting our bulbs in the Enclosed yard, a sweep of 3^h at night and a set of Equatorial measurements till 3 AM.

Wednesday, March 29, 1837

Anecdote of Major Johnson of Paarl (now dead) later Magistrate (? Slave Protector). Authority M^{rs} W——ch——e [his omission]

Mr Williams saw a horse in extremis horribly tormented by certain wretches—he called bystanders to witness & go to Mr Johnson but some had a headache—some were bound elsewhere &c so he went himself & complained—his reply was. Sir I see you are a stranger in these parts—when you have been here a little longer you will learn not to busy yourself about what does not concern you

Min of last night $= 60°0 + 2$

Clear till 1 AM & Calm   then came on very singular crossing lines of feeble cirrus Vis[ible] in Moonlight like much extended Nebulae thus [a sketch of a network of two sets of inclined parallel lines of cloud]

Max of day 89.1

Beautiful day Cl $= 0$ till 6½ PM when it began to drift from S along D & T & over Zenith but at 9 it Cleared & was fine till 1 Swept till Moon when cirrous streaks were forming. Def^n bad.

Thursday, March 30, 1837

Min of Last night 60.3

Max of today 76

10 30 AM overcast Dense Cl on T.—S wind gentle

Cleared in Evening but soon clouded over again with a half-black S. Easter

Friday, March 31, 1837

Min of last Night 70.0
Splendid morning but clouded

Sunday, April 2, 1837

Div. Service at Rondebosch

Monday, April 3, 1837

In the Evening M^t and I rode over to the Observatory where we remain till Wednesday—in order to be present at the laying off the measures on the Rods for the Base line in progress. Arrived about Dark—& after tea Maclear & I set to work making divers preparations for the morrows work which occupied till Midnight—

Clear day but Clouds collected at usual hour (1^h after Sunset) for some days past & soon entirely covered the whole horizon (at Observatory) and so remained all night

Tuesday, April 4, 1837

Rose at 7.—Observatory
At work with Maclear preparing for the laying off his scale.—
Breakfast.—Col. Bell. Major Mitchell—Col Lewis[17]—Mr & Mrs Cooper[17]—in addition to the families.—
Proceeded to 1st repeat the 4 foot scale of Dollond—by Microscope Beam Compasses on top of a long edge-bar of Deal trussed—for a 20-feet measure making cross lines on Brass pins fixed in filed down & Burnished.—By means of a T square & a lancet cutter.

2^d. Transferred the final dots from the surface of the edge-Bar to a higher level i.e to the tops of 2 Brass uprights laterally attached, & cut to the height of the Summits of two Brass pins on the ends of the trussed measuring rods—

This was done by a stepping apparatus of my contrivance [Figure 6]. A a block cut to a true strait edge below—B a cylindrical vertical hole—x a plate of mica fixed (let in) below with a cross x drawn on it with a steel needle—B a lens.—This cross being bisected (as seen in lens) by the division on the lower pin, a lancet line is drawn by the square strait edge D across the upper edge of the brass upright—& being done alike at both ends no error is committed in the *interval*.

---

[17] Lieutenant Colonel G. G. Lewis commanded the Royal Engineers; Mr. Cooper was the barrackmaster.

Rose at 7. — Observatory

At work with Maclear preparing for the laying off his scale. —

Breakfast. — Col Bell. Major Mitchell — Col Lewis — Mr & Mrs Cooper — in addition to the families. —

Proceeded to 1st repeat the 4 foot Scale by Dollond — by microscopic Beam Compasses 5 times on top of a long edge — bar of Deal trussed — for a 20-feet measure making Cross lines on brass provisional in field done & burnished. — By means of a T square & a lancet cutter.

2d. Transferred the final dots from the surface of the edge bar to a higher level i.e. to the tops of 2 brass uprights laterally attached, & cut to the height of the summits of these brass pins on the ends of the trussed measuring rods —

This was done by a stepping apparatus of my contrivance A a black cut to a true straight edge below — B a cylinder drilled vertical hole — x a plate of mica fixed (let in) below with cross X drawn on it with a steel needle. — b a lens. — This cross being bisected (as seen in lens) by the division on the lower pin, a lancet line is drawn by the square straight edge D across the upper edge of the brass upright — & being done alike at both ends no error is committed in the interval. —

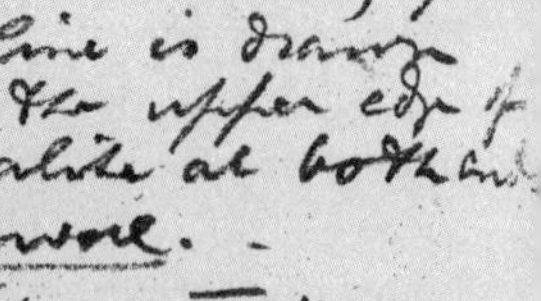

3d — There were then transferred (by a T square straight edge & lancet cutter as before) to the trussed Rods.

Finely fair all day. At Feldhausen it was also raining but not hard. Strong calm

Figure 6. Sketch of Sir John's stepping apparatus, from the Diary, April 4, 1837.

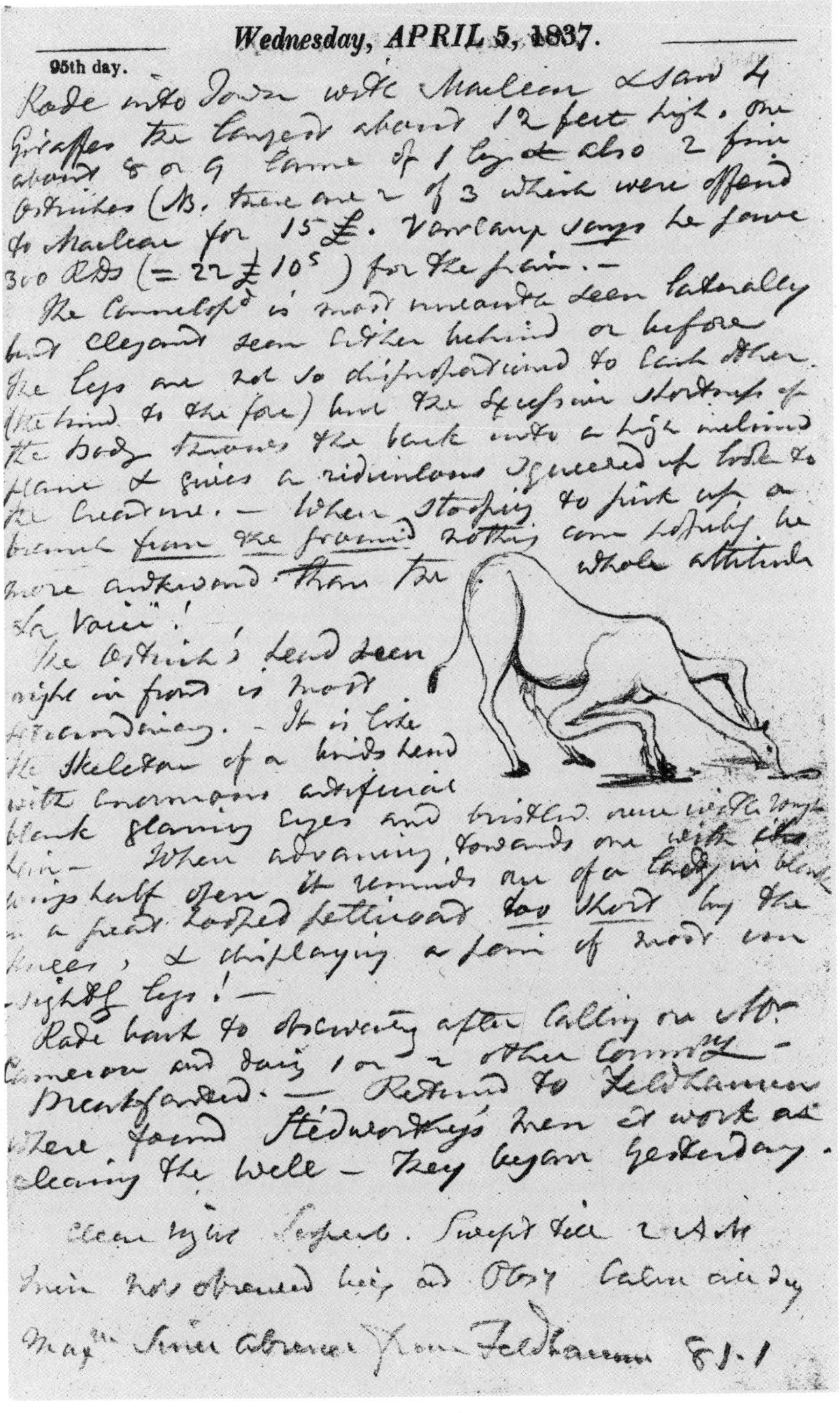

Rode into Town with Maclear & saw 4 Giraffes the longest about 12 feet high. One about 8 or 9 some of 1 leg & also 2 fine Ostriches (B. there are 2 of 3 which were offered to Maclear for 15£. Vancamp says he gave 300 Rds (= 22£ 10s) for the pair. —

The Camelopd is most ungainly seen laterally but elegant seen either behind or before. The legs are not so disproportioned to each other (the hind to the fore) but the stiffness & shortness of the body throws the back into a high inclined plane & gives a ridiculous squeezed up look to the creature. — When stooping to pick up a branch from the ground nothing can possibly be more awkward than the whole attitude or voice!

The Ostrich's head seen right in front is most extraordinary. — It is like the skeleton of a birds head with enormous artificial black glaring eyes and bristles over the top — When advancing towards one with the wings half open, it reminds one of a lady in black in a great looped petticoat too short by the knees, & displaying a pair of most unsightly legs! —

Rode back to observatory after calling on Mr Cameron and ... one or two other ... — Breakfasted. — Returned to Feldhausen where found Stedworthy's men at work at cleaning the Well — They began yesterday.

Figure 7. Sketch of a giraffe, from the Diary, April 5, 1837.

3$^{\underline{d}}$ These were then transferred (by a T square strait edge & lancet cutter as before) to the trussed Rods

Mainly rain all day. At Feld^{hn} it was also raining but not hard Nearly calm.

Wednesday, April 5, 1837

Rode into Town with Maclear & saw 4 Giraffes the largest about 12 feet high. One about 8 or 9 lame of 1 leg & also 2 fine Ostriches (NB. these are 2 of 3 which were offered to Maclear for 15 £. Von laup [?] *says* he gave 300 RDs (= 22 £ 10s) for the pair.—

The Camelop^{d18} is most uncouth seen laterally but elegant seen either behind or before. The legs are not so disproportioned to each other (the hind to the fore) but the excessive shortness of the Body throws the back into a high inclined plane & gives a ridiculous squeezed up look to the Creature.—When stooping to pick up a branch *from the ground* nothing can possibly be more awkward than the whole attitude. La voici! [Figure 7]

The Ostrich's head seen right in front is most extraordinary.—It is like the Skeleton of a birds head with enormous artificial blank glaring eyes and bristled over with rough hair.—When advancing towards one with its wings half open it reminds one of a lady in black in a great hooped petticoat *too short* by the knees, & displaying a pair of most unsightly legs!—

Rode back to observatory after calling on Mr Cameron and doing 1 or 2 other Comm^{ns}—

Breakfasted.—Returned to Feldhausen where found Stedworthy's men[19] at work at clearing the Well—They began yesterday.

Clear night. Superb. Swept till 2 AM   Min not observed being at Obs^y Calm all day   Max^m since absence from Feldhausen 81.1

Thursday, April 6, 1837

Wrote to Uncle Jamie

Therm Max 78.5

Min of last night 52.0

Calm Serene morning.

Calm. Cirrous Cloud in patches   Made an ineffectual attempt at

---

[18] Camelopard = giraffe.

[19] Joseph Stidworthy, builder, mason, and ornamental plasterer, of Sir Lowry Road.

sweeping in the Zenith but cirrous cloud baffled it & continuing worse, gave up. Therm Min 57.1

A Letter from Sir John, Feldhausen, to James Calder Stewart

Thursday April 6. 1837—

Dear Jamie—It was you may be sure a joyful day that relieved us from the cruel anxiety in which the account of your illness rec^d via India had placed us—relieved that is so far as your own accounts of yourself up to St. Helena & those of good Mr Macintosh, from personal intercourse *could* do—and which indeed would go far to place one at ease about your final prospect of recovery were it not that the line remained to be crossed and all the horrors of it damp calm *stewing*. But we will hope all things—you have not been thus preserved in such extremity, be sure, *for nothing* & I cannot but believe that there is yet a long life of vitality & usefulness in store for you.

All other things apart—I am glad you are out of that horrid prison. I had no conception till reading your details what a sacrifice was involved in a mercantile residence there. To one of your temperament it must have been most dreadful.—Depend on it you are *recalled*, as from a sphere unfit for the exercise of your faculties.

I wish you *could* have touched here—but I see now that it would have been *right* to keep you—and to part would have been doubly & trebly painful. But I am sure you must have heard us talking about you as you passed.—If you saw *any* house at all in Hout's Bay it must have been our ci-devant home for the best of reasons—there is no other—

A gentleman of the name of Whiteman sent by a friend (he could not call himself) your message relative to Meteorological & Phonetical Matters at Canton.—I greatly fear I cannot now devote time enough to the latter point to do any good—but in one N^o of the Canton Register I perceive that the *one essential* thing, *the careful discrimination of sounds really different*—is fairly brought into its due degree of prominence. My own impression is that a *distinct character* would be best rather than any adaptation of existing alphabets—but I fear there is no prospect that any linguist would conform to its use. —They are for the most part a class of erudite if remarkably limited views beyond the immediate circle of their own studies, and very difficult to turn aside from their own preconceived notions & habitual conventions. Still less do I think that any body of Linguists would *ever* agree by meeting—correspondence—or any other mode of discussion—upon a character based upon any rational principle. The general tendency is to refine too far—to attempt a *perfect* representation of every minute inflexion & intonation in language.—To take up the matter thus is like studying Meteorology by making minutely detailed maps of the forms & distribution of the clouds hour by hour over every country in the Globe. We must interest ourselves with infinitely broader features.

You are quite right—I am sworn to silence about all that passes in the Sun Moon & Planets—Oh I could a tale unfold! that you may rely on. I am delighted however to see that your Muse keeps her plumage in trim—though for her comfort's sake I could wish her a supply of better subjects than physical science can afford.—Here are a few contributions to your thesaurus of versions of M. Angelo's Quatrain though if you seriously expected me to finish a round dozen Why Lord help you![20]

1.

> To be of Stone is happier than to Sleep
> To dream of hope & joy, & wake to weep
> On crime and shame. Then rouse me not to woe
> Break not my senseless State!—Pass on! Speak low!

[20] The original of these poems is a quatrain by the celebrated sculptor Michelangelo:

> Caro m'è'l sonno, e più l'esser di sasso,
> mentre che'l danno e la vergogna dura;
> non veder, non sentir m'è gran ventura;
> però non mi destar, deh, parla basso.

Dated 1545–1546, this was translated into English by Creighton Gilbert, *Poems and Letters of Michelangelo*, edited by Robert N. Linscott, as:

> I prized my sleep, and more my being stone,
> As long as hurt and shamefulness endure.
> I call it lucky not to see or hear;
> So do not waken me, keep your voice down!

2.

Happier than happiest Sleep, which wakes to pain
Is this Dull, dark unconscious marble state
Refuge from wrongs & woes deplored in vain
Recall me not!—be still—A word might animate

3.

O Sweet is Sleep; but sweeter still this veil
Of dreamless Stone, which Shuts out pain & care
Which wrongs & griefs alike in vain assail.—
Spare me! and speak not—Save in Whispered prayer.

4.

Toil maketh Slumber sweet.—but to the heart
O'erborne with witnessed misery and shame
What Sleep save that which locks this marble frame
In sightless, senseless gloom, can peace impart?
Stranger who dreadst this awful sleep to break
Tread softly and in solemn whisper speak.

Well now I must write more at length about a thousand points of interest that crowd upon me when writing to you, when I have a little more time Though really when that is to be I know not for my Reductions! are pressing so awfully upon me that I have let *all* my correspondence get into a degree of arrear which it is now no longer possible for me to hope to make up. Only this I fling at you that you may not be kept in anxiety lest your packet from St. Helena should have missed.

Your Italian Lion Book [?] is most interesting—It is easy to see that you have been reading history con amore—But what makes you so fond of tracing coincidences purely accidental (of Days—places—birthdays &c I mean). I like your parallel of Coriolanus with Piramo —however being a parallel of realities—of causes & effects.—Such parallels lead us to recognise the common laws of our nature acting uniformly at distant times & places. We have been much delighted lately with Drinkwaters translation of Schillers[21] Maid of Orleans.

---

[21] Johann Christoph Friedrich von Schiller (1759–1805), the celebrated German poet, dramatist, and philosopher, wrote *Die Jungfrau von Orleans* in 1801. Sir John so admired him that he produced translations of Schiller's works, notably

As a translation its freedom of versification is admirable. Schiller seems to have been aiming at shadowing forth a beau ideal of female purity & enthusiasm & truly a Sublime one he has created—

Well now I must end & with the prayer that Heaven which has preserved you so wonderfully will still watch over & protect you to the end   believe me ever yours   JFWH.—Maggie will tell you all our domestic affairs.—

The Diary: April 7 to May 7

Friday, April 7, 1837

Wrote to F. Baily—also some notes (to Mr Stewart & Miss Pattle) Prepared Curves and Tables for Sw[eep] 489   Therm. Max 69.5 Mrs Cumberledge called PPC. being about to sail tomorrow per Abercorn for Calcutta   Swept 2 Sweeps—beginning at 8.30 ST and going on to 14 30 ST.—with an interval of an hours sleep—Then went & measured Saturn & α Centauri   Therm Min 57.5

Saturday, April 8, 1837

Margt went into Town to have Caroline's teeth drawn.—Violent South Easter. Clear. Actinometer & Sun Spots. Bar 29.920   At-t[ached thermometer] 66—

Mrs Cumberledge called en route for Cape Town in a "peck of troubles"—Dispatched James to help her out of one or 2 minor ones   Wrote to Arago about the Meteors & Solar Spots.

Bella volunteered to say by heart 2 whole pages of Peter Parly[22] which had taken her fancy—It was the speech of the Indian "Page 406—"Then the red men were rich and happy: now they are poor and wretched &c &c"

Dined—Mrs. Cumberledge returned having been unable to Embark—

Der Spaziergang, both into English and Latin verse. (*English Hexameter Translations from Schiller, Göthe etc*, 1847, "The Walk," translated by Sir J.F.W.H., ? 1844, and Schiller's *Spaziergang*, translated into Latin Verse, 1867.)

[22] "Peter Parley," pseudonym of Samuel Griswold Goodrich (1793–1860), author of more than one hundred books for juvenile instruction.

Dined again.

Reduced R.A.'s of Sweep 289

Measures of Saturn's Satellites but Definition very bad—A violent South Easter & lovely clear sky all night. Stars on the full fret, *but not* dilating & contracting suddenly in puffs.—

Sunday, April 9, 1837

Escorted Mrs Cumberledge into Town and saw her embarked on board the Abercorn

NB. A Strong South Easter, but less than yesterday—only early this morning the Capt[n] said it blew a perfect hurricane—

Saw the process of Embarking horses at the Jetty—a very rough clumsy way

Min (Sat night—Sund Mng) 57.0

Max of day 73.5

Saturns Satellites meas[d] (this takes nearly 1 hour) Star-mapping till 4 A.M. being a night of very imperfect Definition

Monday, April 10, 1837

Rose at 10½. Superb Day Such a sky! Calm

Transferred Double Stars to register

Gave Xtian orders about stopping leaks in the roof &c—Sun Spots. —Actinometer.

Pricked off one of Bodes Stereographic hemispheres & entered up all the triangles already examined in it as an index map.

This took (Dinner included) till fully dark At intervals put mirror No 3 on polisher *cold* to lie & mould it—moving it occasionally.—

Being a very bad night for defining though superbly clear—went on with Mapping & taking Magnitudes breaking off to get tea and to Measure the Satellites of Saturn—till 3 AM when gave in being tired though the night was still superb.—

NB. Much time was lost today over the Actinometer which took a freak & gave an unaccountable set of results

Spots in ☉ are distributed very [illegible] over Surface & one large one is *nearly on Equator*

Min Therm out of order & could not be permanently rectified

Tuesday, April 11, 1837

Rose at 10. Most glorious calm & serene morning—not a vestige of cloud any where

Entered up D. Stars in Register.

Registered Obs[ns] of Magnitudes

Reduced the Elements of Sw[eep] 488 & projected the Curves.—

Abstracted Clarke Abels[23] Chapter on Geology of the Cape.

Washed over Colcothar for Polishing and cut and scarred the mirror (a long and dirty process)

Measured Saturn and two double Stars [erased] Projected the Sun which is very remarkable, but the day was not favorable for Actinometer as at noon Cirri formed in every part—They partially indeed almost wholly cleared before Evening & the night was fine though with [word erased] a slight cirrous veil w[h] affected the Magnitudes also *Def[n]* Bad.

Shewed Carry Bella &c Jupiter & Mars in Equatorial

Measured Saturn & 2 D Stars

Friday, April 14, 1837

Came on a violent N.W. gale about 1 AM quite suddenly while measuring Saturn

Saturday, April 15, 1837

Planted 3 lbs Fir seeds on the strip of ground above the terrace road on Wynberg Hill.

Sunday, April 16, 1837

Attended Church at Rondebosch

Entered up Neb & D Stars of last weeks reducing

rewrote letter to Arago

Monday, April 17, 1837

A Calm Cloudy gloomy day with sprinklings of rain & much threatening—

Kept hard at work the whole day at the registering & drawing out in order of all the data for the places of the summits of triangles in the map of η Argus—a day of hard work & little *apparent* progress.—

---

[23] Clarke Abel (1780–1826), botanist. He was the naturalist in Lord Macartney's mission to China (1816–1817), but his collection was lost in a shipwreck. Later, he was physician to Lord Amherst, Governor General of India. His observations of Cape geography occur in his *Narrative of a Journey in the Interior of China, 1816–1817,"* London, 1818.

Towards Evening sowed some lots of garden seeds in the new flower beds opposite the drawing room windows

At night being perfectly cloudy—worked the Curves of a Sweep—& worked over colcothar for repolishing another Mirror

Wednesday, April 19, 1837

Reduced Sweeps

Spots in Sun verified [erased] resumed though the day was unfavorable

Thursday, April 20, 1837

Reduction of Sweeps proceeded

Wrote to P. Stewart in reply to letter rec^d this morning via Mauritius per Atalanta steamer

Evaporated juice of Hottentot figs.

A most oppressive hot day. Therm 81 Calm—thick heavy haze in atmosphere & a veil of cirro stratus with signs of a hot wind—

Nevertheless these as all other late strong symptoms of rain have done passed away—the heat declined & it was a cool pleasant evening with a high Mackarel drift

Observed end of an Eclipse of the Moon

Measured Saturn's 6th [?] & 7th Satells among Mackarel cloud after w^h it clouded thick. NB. It took from ———34^m to———54^m after the limb of the Moon was *complete* for that side where the Eclipse was to recover its light equal to the opposite limb. The completion of the limb might be observed to 5 seconds.

Friday, April 21, 1837

[Several exceedingly faint lines erased]

Rose at 5 [remainder illegible—? entered for wrong day and subsequently rubbed out?]

At Noon began the 24 hourly Obs^n wh continued till 10 PM when I went to bed leaving Marg^t to go on

Saturday, April 22, 1837

Rose at 5 and proceeded with the Meteorological Obs^{ns} which continued till Noon—Having got hardly a wink of sleep, what with feverish symptoms & the barking of the Dogs—was much fatigued & harassed & fit for nothing all the rest of the day—

Sunday, April 23, 1837

Piazzi Smyth called with a [sentence unfinished]

Monday, April 24, 1837

Mr & Mrs Maclear came over to Feldhausen Inter alia M mentioned that Major Gregory & Mr Deas Thomson have lately been to "Worcester" shooting & one morning before breakfast (i.e. Dinner) had killed 80 brace of Partridges with 2 guns

Sallust. Bellum Catil.[inae][24] by a passage about Scipio it appears that the Roman nobles kept the images of their Ancestors *in Wax* in their houses.

Tuesday, April 25, 1837

Wrote to Mr Eliot—Captn Highat who sent (of the Jardine) some Chinese maps celestial & Terrestrial &c. & to Col Bell

Sun Spots.—

Evaporated Juice of Hottentot fig

Experiments on the aqueous infusion of Black Ebony wood. It contains a Colorific acid analogous to the Rocellic Acid only Brown

Clouded over—No Sweeping or other work out of Doors.

Finished reading Sallusts Catiline war with Portugueze translation

Wednesday, April 26, 1837

Heavy rain all night & morning till 8.AM

Occupied reducing Sweeps, & Meteorological Obs[ns]

Today at noon Stedworthy's man had finished bricking in the well & went off the premises

Entered in Register Reduced Neb and D Stars

Heard of arrival of Thalia from Simon's Bay—Arrangements consequently in Progress for Marg[t] & children to go down on visit to Mrs Wauchope

Clouded over soon after dark   No Astron[l] work.—

Reducing Meteorolog[l] Obs[ns] &c

---

[24] Gaius Sallustius Crispus (86–c. 34 B.C.), Roman historian and great literary stylist. He was a soldier who entered public life, and, after a profitable governorship of North Africa, retired to devote himself to literary pursuits. He wrote *Catilinae Coniuratio* (*Bellum Catilinae*), which deals with Catiline's attempt at revolution in 63 B.C. It appeared in 44 B.C. and was followed by *Bellum Iugurthinum,* dealing with the war with Jugurtha of 111–106 B.C.

Thursday, April 27, 1837
A very rainy morning. Calm

Worked at reduction of Sweeps &c & other work—inter alia a general clearing & arrangement of Books, Apparatus &c w[h] had got into sad disorder

A wet cold day Kept home

Rec[d] a letter from Mr Dawes via Madras where it had been carried on

Began Sallusts Jugurthine war in Latin & Portugueze.

Clouded & Cleared alternately till about 10 PM when it grew clear enough for Obs[n] & continued with occasional drifts of huge black masses *from the South East* growing finer & finer at last perfectly cloudless—worked at Equatorial till 3 AM.—

Friday, April 28, 1837
A Superb Morning.—All idea of rain gone & everything glowing & sparkling calm & serene. This is Cape Sunshine in perfection.

Stedworthy came—Paid him for his work at the well. He says that No fossils have been found in any wells dug within his knowledge at the Cape—

In this of mine they worked till they came to quicksand and then stopped

Thursday, May 4, 1837
Observed η Argus for the Map   Rediscovered (decisively) the 6th (II[d]) Satellite of Saturn[25] & obtained distinct proof of its being a Satellite & no star by its own & the Planets Motions

Friday, May 5, 1837
Marg[t] Carry & Bella left Feldhausen on a visit to Mrs. Wauchope at Simon's Town

Saturday, May 6, 1837
Got another Obs[n] of Saturn's 6th Satell[ite] on opposite side of its body

Sunday, May 7, 1837
Maclear came over & spent the day

---

[25] See note 34, March 30, 1836. The double numbering refers to the order of discovery (6th) but the second in order outward from the planet.

A Letter from Sir John, Feldhausen, to
Caroline Lucretia Herschel, Hanover

N⁰ 14   Feldhausen
near Wynberg   C.G.H.
May 7, 1837
Received July 22 [In a different hand]

Dearest Aunt

    Your note to P. Stewart dated January last was forwarded here by the Berenice Steamer which made the passage from England in 52 days, 36 of which only were spent *at Sea*. But it reached my hands only yesterday, having been detained in London waiting opportunity for Enclosure.—The last direct dispatch we have of you is dated Oct 20th 1836. and with it Dr. Heeren's Meteorological observations for the September Equinox. Peter Stewart also enclosed the same Dr. H's observations for the December Solstice. For both these very valuable communications pray offer my best thanks to Dr. H. to whom I shall myself write as soon as I can find time, though at present I am dreadfully pressed with the burden of my Astronomical Reductions.—

    I grieve to find by your note to P.S. that your health has been so indifferent and when we read the accounts of the dreadful winter you have had in Europe and that horrid influenza, it makes us long, yet fear to receive further notices from you as it must have been a severe trial to your weakened constitution. However we will trust that you have escaped its attacks and in that hope look forward to our meeting next year.—*Here* neither influenzas nor Choleras have yet made their appearance and our report is all well—no sick list—i.e. abating Rheumatisms which are very severe.

    Marg^t and our two eldest little ones Caroline and Bella are gone for a few days to Simon's Town on a visit to Mrs Wauchope the Captn of the Flag-ship's lady and on Wednesday I am going to fetch them home therefore I write this without prejudice to a long letter which she will write to you soon after her return in which she will detail all our domestic matters.—Meanwhile I will try to entertain you with some celestial affairs in which it it [his repetition] is delightful to find you still taking so much interest.—As you allude to Saturn's Satellites in your letter of October 20 I must tell you that I have *at last* got decisive observations of the 6th Satellite (the farthest of my Father's new ones) I had all but given up the search in despair, when no longer ago than *last Thursday* (May 4th inst) being occupied in taking measures of the angles of position of the 5 old Satellites with the 20 feet & a new polished mirror—behold there stood M^r 6th as here represented a little short of its greatest preceding[26] elongation.—I kept it well in sight from 14^h 26^m Sid T till 16^h 35^m in which time it had advanced visibly in its orbit from *below* the line of the Ansae (as in figure [Figure 8]) to *above* thus—In this interval the planet had moved over fully one diameter of the body towards the preceding[26] side—and therefore had it been a star, must have passed over it whereas it preserved the same apparent distance all the while from the Edge of the ring.—(NB.   Saturn not very far from the Zenith on Merid[ian])

Next night *Friday May 5.* Saturn most gloriously seen—quite as sharp as any copper plate engraving with power 240 and full aperture—All the 5 old Satellites seen & measured—No other seen.

May 6th (last night). The 6th Satellite again well seen & measured —being now on the opposite side thus (fig. 2.) Now considerably short of its greatest *following*[26] elongation.—Distance just as before.— and as on Thursday it was kept in view long enough for Saturn to have left it behind by its own motion had it been a Star.—The change of situation agrees perfectly with the period 1^{day} 9^h which is also the reason why it was not seen May 5, being on that night near its inferior conjunction.[26]—So this is *at last* a thing made out. As for N° 7 I have no hope of ever seeing it.—If your eye sight will suffer you to write to Bessel, I am sure he will be interested by this observation—as he is the only Astronomer who troubles himself about the System of

---

[26] Going ahead in the general direction of orbital and rotational motion in the solar system, that is, to the west of a given position, having a smaller Right Ascension. Following = to the east, or larger Right Ascension. Inferior conjunction = having the same Right Ascension but on the nearer side, that is, in the case of the Saturnian satellite, practically in the same direction as the planet, but nearer than it, so that it would be invisible against the planetary disc and not visible against the sky.

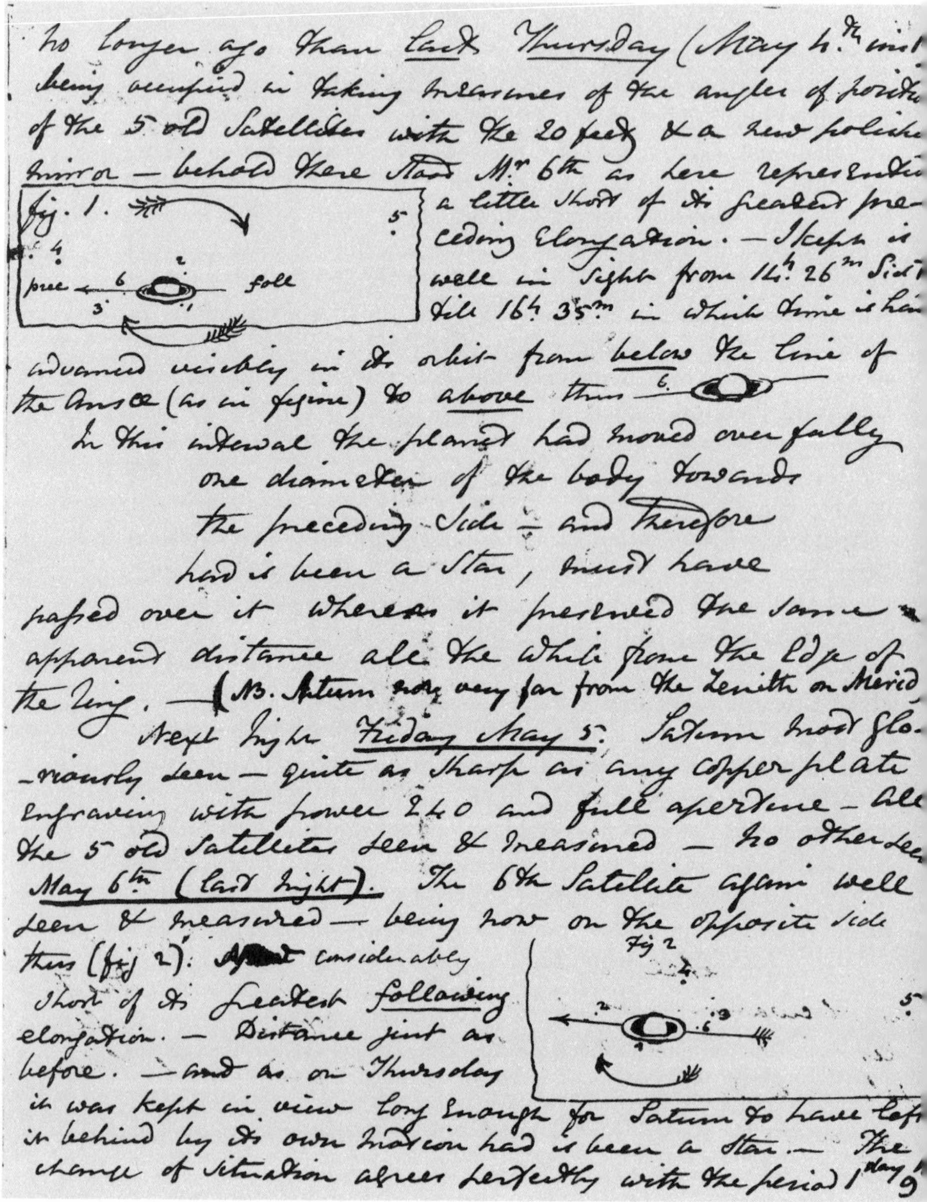

no longer ago than last Thursday (May 4th inst) being occupied in taking measures of the angles of position of the 5 old Satellites with the 20 feet & a new polished mirror — behold there stand Mr 6th as here represented a little short of its greatest preceding elongation. — I kept it well in sight from 14h 26m Sid till 16h 35m in which time it has advanced visibly in its orbit from below the line of the Ansæ (as in figure) to above them — In this interval the planet had moved over fully one diameter of the body towards the preceding side — and therefore had it been a Star, must have passed over it whereas it preserved the same apparent distance all the while from the edge of the ring. — (NB. Saturn not very far from the Zenith on Merid)

Next night Friday May 5. Saturn most glo-riously seen — quite as sharp as any copperplate engraving with power 240 and full aperture — all the 5 old Satellites seen & measured — no other seen

May 6th (last night). The 6th Satellite again well seen & measured — being now on the opposite side thus (fig 2). A considerably short of its greatest following elongation. — Distance just as before. — and as on Thursday it was kept in view long enough for Saturn to have left it behind by its own motion had it been a Star — The change of situation agrees perfectly with the period 1day 9

Figure 8. Sketch of the satellites of Saturn, from Sir John's letter to Caroline Herschel, May 7, 1837.

Saturn.—I shall myself write to him shortly about it but should like to have a few more observations.—So now farewell once more & with many kind remembrances to all Hanoverian friends   believe me Your affecte Nephew   J F W Herschel

[Half of the final paragraph written on the first page]

The Diary: May 10 to August 30

Wednesday, May 10, 1837

Rode over to Simon's Town to bring back Margt & children—At Farmer Pecks met her & Captn Wauchope & the Admiral—Dined at the Admirals & returned [erased] passed night at Captn Wauchope's —Passed Mr Deas Thomson in a hobble about his horse having £1200 in his gig—NB. A great increase of Small Cottages i.e. "Pondhoks" or "Ponducks"[27] going on in the flat tract between Wynberg & the Muysenberg corner—a wide amphitheatre which is now in active process of "Settling"

Col. Blake's anecdotes of his "Boy" the best boy he has—personally attached—would not leave for 40 Dollars per month offered while he (Blake) gives but 10—Only has one or two peccadillos on his character   he cut his own Father's throat & afterwards was tried for murdering his Mother!—

(A dead silence   no comment on this anecdote by any of the party present)—NB. his Father recovered—his Mor died—he flung a chest against her wh struck her on the heart—she staggered out of the hut—vomited blood & in a short time (a day or 2) died—[a sentence erased]—Found not guilty because the Doctor swore she might have died of something else—Also his story of the Malay poisoning pins, hair, nail parings vomited in a ball like a [subsequent words illegible] —(all revenge spite &c)—

Thursday, May 11, 1837

Rose at 7. Took Sketch of Simon's Bay. Breakfasted with Mrs Wauchope Captn W being obliged to keep his bed with a cold—

---

[27] Pondok = shack.

Walked down to jetty at Admiralty to see the yards of the Pelican manned to receive the Admiral (Campbell) who goes on board to inspect previous to Pelican's Departure.—Called at Mr Thomson's—Left Card at Col Blakes in reply to one left for me at Capt Wauchopes

Purchasing & Packing Shells & other curiosities in front [illegible] Some Snails Eggs![28] as large as Pigeons & some great Termite Queen Ants with nest of one (Royal cell) given by the Admiral & a fish with regular paws[29]

Monday, May 15, 1837

Coming on of a NW Cloud—the most beautiful phaenomenon—it came slowly pouring over T & D dissolving so rapidly as to melt into a delicate gauzy filmy mesh like veil floating, with the most exquisite grace and rounded adaptation to all the greater masses of the Mountain—far down almost to the level of the Brewery. At intervals its skirt wreathed up into unearthly aspiring fibrous lines and tangled webs [a rough sketch of the cloud]

Tuesday, May 16, 1837

Marg^t went over to Observatory & brought back all the little Maclears, Mr & Mrs M being gone to Riebecks Castel[30] & Lacaille's North point.

At home and at work about η Argus all the Morning.—

Wednesday, May 17, 1837

A calm densely clouded day with occasional small rain—Country very wet   At home and at work all day on the arrangement & reduction of my D Star Equatorial Measures.—

My Barometer thrown down & broken by Hannah though placed as I thought in so secure a situation that not even the children could upset it! ! With it went to wreck almost my last thermometer—

Marg^t went in to Town where it seems they have had torrents of rain

---

[28] Professor Day thinks these are probably lizard eggs.

[29] Possibly the mud skipper (Periophthalmus).

[30] Riebeek Kasteel: A mountain peak to the north of Cape Town used by Lacaille as a station in his survey work. Named for the commander, Jan van Riebeek, of the first group of European settlers at the Cape. Lacaille's north point was still farther on, at Klipfontein.

Cleared somewhat towards Evening with exquisite definition but it did not last.—Enormously voluminous cumular Masses (one *a globe* fully as large as Devil M^n) rolling up from NW & W—In intervals worked on Data for η Argus & reduced Double Star Measures

Thursday, May 18, 1837

At work on D Star Papers.

Planned & put in hand with the Carpenter a "Panorama Drawing board (a pair) to hold many papers ready for Camera [lucida] work.—

At night Cleared off as it has done for several nights from about ½^h after Sunset to 8½ or 9 PM in w^h interval Def^n was more perfect for a while than I ever saw it—T Argus was • . [two unequal dots representing separated star images] in Equatorial bearing strong illum [ination]

Got a *pretty sure* elongation of γ Virginis

Friday, May 19, 1837

At work all day at D. Star papers

Also Drawing & carpentering up [word erased] the Panorama Board w^h seems likely to answer.—

A dull Cloudy Calm day with occasional Sprinklings of rain—In Cape Town it has rained hard—

No observing at Night though for half an hour it looked promising

Saturday, May 20, 1837

[A circle, presumably to indicate full Moon, at heading]

Men finished planting with fir seeds the whole area of the Burnt grove which is all cleared.—

Rained heavily in Night *though full* ☽

Dismissed Carpenter

Maclear called just returned from Klip Fonteyn

Superb night full ☽.—S wind very strong after near a fortnight calm & cloudy—

Measured 16 D Stars the greatest N° in any night (Eql) since I have been at Cape

Monday, May 22, 1837

At Earliest dawn En Route for Versfields [Versveld's] in [illegible] to ascend the Table Mountain under his guidance.—Marg^t self James & David on horseb^k sending 4 men to carry a chair & 2 with provisions.

Morning of all Mornings most Superb.—Moon & Dawn contending.—
Reached Verfields just before Sunrise—found him ready Started im-
mediately.— at [entry terminates]
   Thermometer up on TM
     Min 31.2 Max 96.2
     Left registering
   Therm left on TM examined
     found Min 31.2 Max 96.2

Wednesday, May 24, 1837

Colours Fire at Protea
Rode round Kloof a geological examination
Ball
Night Superb

Thursday, May 25, 1837

At home
N W
N W Cirri

Friday, May 26, 1837

N W Violent Great fire at Wynberg Sky Clouded Up.
   Rode with Margt to see the fire among the Forests of Silver Proteas
which was very sublime & grand.—
Called on Mrs Hamilton &c
Clouds came slowly on from NW at 2 increasing & lowering till
they poured over T & D and as night came on heavy rain with much
thunder

Saturday, May 27, 1837

Heavy and continued rain the whole night and day—At work reducing the RAs of the η Argus Stars—and reducing & Booking Double
$*$ s
No Events—
In the 25 hours from 9 PM last night to 10 PM tonight have fallen
$3^{in}05$ of Rain. Rain-tub taken in—

Sunday, May 28, 1837

A rainy Morning—i.e. frequent & sudden showers with deceptive
gleams—at home all day—no Events

Monday, May 29, 1837

A Splendid day.—Calm & serene. No cloud—

At work all Morning at the Sun & the Actinometer—Laid down Stars in η Argus map—Reduced and Booked D Star Obs[ns] & took final (I trust) diff[es] of declin of leading Stars in η Argus map.

Then got the 20 feet & examined 4 of Dunlops D Stars θ Centauri, ι Cent[i] ζ Circini, and v^2 Centauri—They are all single

NB. Deep Snow on the Hills on the Stellenbosch side covering the rugged slope of the Mount[s] [words erased] of Donkers'hook & De Toit's Kloof[31]

Tuesday, May 30, 1837

This morning Mr Fry[32] brought to Mr Hough from Vygekrall[33] on the flats a piece of Ice in a blanket

Refitted a tube to my Barometer and filled it but the tube is too small and the fitting is done with soft Cement

At night got a few Measures of D Stars—but after a fine Sunny day it grew thick & at last the haze became too bad for observing

Reduced measures of $\triangle$ decl of η Argus & proceeded with the Catal to accompany the Chart of that $*$

Wednesday, May 31, 1837

A decided Stormy day—During all forenoon blowing a tempest of wind & rain—towards Evening Lulled but grew worse again at Night.—

Finished reducing & Booking in the rough my Microm[l] Measures of D Stars up to the present time (396 in N° of w[h] 58 are of 1st Class.)—Thus this work is so far *secured*. NB. The Angles are excellent the Distances on the contrary very bad

Capt[n] Redman of the Hindostan called with a letter from Peter— Re [entry terminates]

Thursday, June 1, 1837

Into Town with Marg[t]

Wrote to Mr Rutherford[34] in answer to his letter about the Subscription for F. Enclosed a draft for ——— [his omission].

---

[31] Jonkershoek and Du Toit's Kloof.

[32] The Reverend John Larkin Fry R.N. (Ret.) was second chaplain of Rondebosch, 1840–1861. He had come out as a naval chaplain in 1831, but was not attached anywhere officially.

[33] Vygekraal, now in the Athlone district.

[34] Howson Edward Rutherford, merchant, formerly president of the Society

Friday, June 2, 1837

About 9 A.M. the weather looking better after dispatching David into Town with Rutherfoords letter & one sent here for T. Bowler of Robben Island—set off with Marg^t & Carry Bella & Willy for Fishhook bay to see a whale[35] which is lately caught & lying in act of being cut up.—The day proving pretty good with even Sun glimpses we got a very good sight of it.—Close to shore under a rack on which were mounted a winch & tackle to turn it over [sketches of whales]

Saturday, June 3, 1837

A tremendous rain the whole of last night and today—a steady pour without ceasing

At work all day at home

1. Letter to Col^l Smith[36] about Hintsas[37] death
2. Reductions of η Argus Stars
3. Reduced the constants for 10 Sweeps

Never saw such a rain at the Cape The roar of the cataracts from the Table Mountain is heard as I write ($2^h\ 45^m$ AM Sunday morning) with all the doors & windows, & shutters closed like a steady wind—but it is quite calm—or rather like the continued roll of breakers on a beach.

At $2^h\ 15^m$ Emptied the raingage tub which was nearly full. It had in it $7^{in}.8$ of Rain which had collected since 9 AM on Thursday—Left it out being still raining as Steadily as if it never meant to leave off

Monday, June 5, 1837

The gage had in it [sentence terminates]

Friday, June 16, 1837

[Entry mistakenly made in space for July 14, restored to proper position; Herschel's phrases, such as "entered here by mistake," suppressed for clarity]

---

for Relief of Distressed Settlers. In 1837 he was on the committee of the South African Infant School and a commissioner for the Guardianship of Juvenile Emigrants.

[35] Probably a sperm whale, from the sketch.
[36] See note 108, June 24, 1834.
[37] See note 103, May 28, 1835.

Off before Sunrise to Observatory to assist Maclear at his Base measurements. Much valuable time lost at his Breakfast table, where NB. Mr Williams stated on Authority of one Harris just returned from the Tropics that 4000 Boors have assembled & occupied Massalikatzis country with their train of Cattle, Waggons &c and are going to Establish themselves there! !—Rode into Town to make enquiries at Transfer office . . . about mode of proceeding.

M. had reckoned on finishing the Base in one day!—I gave him 4— but from experience of today I doubt if a week will do it   When the day was Ended—Not one Rod had been laid off! The whole day was consumed in trials & adjustments

In afternoon (at 4½) Mr Grey[38] the printer came by appointment to conclude his bargain with M. about the printing of the Cape Obs^ns I was present by Maclear's express desire—Agreed 1st to print similar to Cambridge Obs^ns at £7: 12 per sheet—2^d Grey to be responsible for loss by fire on his premises—3^d Grey not to charge extra for corrections of Errors of his own making—but deviations from the MS to be charged extra—at *reasonable price*—

Marg^t & Caroline & Bella came & we went with 2 of the little Maclears to see Mr Powell's Exhibition of the Oxyhydrogen Microscope in Cape Town at Commercial hall—a most wretched bungling affair. Night Superb but Definition of Stars not good—Went Early to bed

Saturday, June 17, 1837
[Entered by mistake in place for July 15]
The moment I rose off to Observatory & the Base Measurement at which Worked the whole day till Sunset—returned immediately Dined & commenced as soon as Dark with the Equatorial, remeasuring Double Stars. NB. The Time *wasted* in small adjustments in this Base work is beyond all imagination & is enough to tire the patience of any body. After all today's & yesterdays work not above somewhere about 25 or 30 Rods have been laid off. At Breakfast with Col Lewis, Captn Wauchope, Mr Skirrow, Lieut Williams   An immensity of time was lost at Breakfast—
[Further note calling attention to mistaken original place of entry]

[38] It is tempting to identify Grey, through misreading, with the well-known Cape printer, George Greig, but this does not seem possible, as the latter withdrew from the trade at the end of June, 1835.

Monday, June 19, 1837

En route at 9 AM for observatory to the Base—met Mr C Bell who informed us of Mrs Robertsons sudden death—arrived on the ground

Wednesday, June 21, 1837

Attended Mrs Robertson's funeral in Cape town—

Thursday, June 29, 1837

Wrote to Prinsep—to the Secretary of the Albany Institute suggesting Meteorological registers to be Kept in Schools

Friday, June 30, 1837

Went into Town & completed Transfers of allotments of Land to Whilhelmina Walker and Samhaai the free Black

Heavy rain all day & Night with few & brief intervals—
At work on Meteorological report

Saturday, July 1, 1837

Snow on Table Mountain (or ? hail)

Monday, July 3, 1837

Heavy & continued rain from morn to night and all night—
At work on Meteorological report the whole day.

Wrote to Mr Deas Thomson & to Captn Wauchope to organize a set of Obs^us of tides for Whewell

Tuesday, July 4, 1837

Heavy rain all night—at Morn found that the tub set out at 3 PM yesterday had got . . . of Rain—

Rain ceased about 10 AM all but drizzle After that a most delightful soft balmy day with an *universal roar* of waterfalls from the table mountain filling the air like the rushing of the breakers on a shingly sea beach.—A Solemn and Grand sound indeed! The Mountain itself concealed in Cloud & falling rain.

At work all day at the Met. Report, and reduction of Sweeps except a walk in Morning round by Abrahams Cottage &c.

Wednesday, July 5, 1837

A Day of Violent wind & pouring rain at home & at work the whole day

At night Wind Rain & Thunder

1 AM. Bar 29.583 + zero. Lightning, Rain, Wind & all that is fierce & outrageous.

Thursday, July 6, 1837

Noon Rain in tub $= \dfrac{4.9 + 3.4}{2} = 4^{in}$ 4 since 3 PM. on Wednesday

Saturday, July 8, 1837

Went into Town to attend the Anniversary of the Lit & Phil Inst— but found it put off by reason of "Expected bad weather."

Called at Mr Hutchinson's office to arrange for M^{rs} H coming out on Tuesday & to enquire about Grand Jury for w^h I have a Summons for Saturday next.—Returned Outlined Lipania Sphaerica—as long as daylight lasted.

In Evening worked at reductions—they go dreadfully slow.—D [terminates]

Dipped occasionally for variety & to keep from falling asleep over Sir J. Mackintosh's life[39]

Sunday, July 9, 1837

Examined the rain tub.—Since 3 PM on Tuesday 5^{in}2 of rain have fallen

Friday, July 14, 1837

[Originally entered, somewhat unsystematically, in space provided for June 16]

Mrs Hutchinson here Violent Rheumatic Pains across þ^e shoulders —caused as usual by the Sudden heat & dry wind—which seems to suck one's soul out—Especial[l]y after the long wet weather—Towards Evening cirrous streaks began to appear w^h increased all night up to a half cloudy sky

Saturday, July 15, 1837

A hot violent day *July 15*. with Thermom at 79 Dry wet 58 (Diff ——— [his omission]) and N Wind Strong & gusty—at home all day. Actinometer—Sun Spots—Reductions—Preservation of Flowers in Sand—Reading of Evid^{ce} before Aborigines Committee[40] &c.—

[39] Biography of Sir James Mackintosh (1765–1832), philosopher, written by his son, R. J. Mackintosh, 1836.

[40] Report of Select Committee on Aborigines, August, 1836; a further report made in 1837.

Sat July 15 Attended the deferred anniversary of the Lit & Phil Inst [he emphasizes the date since this, too, is written in the wrong place]

Sunday, August 20, 1837
[In a childish hand written on ruled lines]
There has been a violent storm of rain & wind all last night & today. Two ships were blown on shore last—night, & one was a total wreck
[In Sir John's hand]
After one tremendous gust about noon which partly unroofed my study & one or 2 more like ones, it fell Calm and continued so all Evening & night. Barom at noon 29.52 at 1 AM it had risen to 30 this is the greatest change in one day I have seen at the Cape

Monday, August 21, 1837
[In the child's hand]
Admiral Sir Patrick Campbell & Lady Campbell came here today on their way to Simons Town

Tuesday, August 22, 1837
[In the child's hand]
Poured of rain all day

Wednesday, August 23, 1837
[In the child's hand]
Captn Wauchope & Mrs Wauchope lunched here today on their way to Simons Town & left Jumbo the pony. In the evening we heard the news of the death of king William IV.

Thursday, August 24, 1837
[In the child's hand]
The new groom George Lancy came today, Mama went into Town & engaged a governess Miss Dixie who will come to us on the 4th September

Friday, August 25, 1837
[In the child's hand]
Sent to the Observatory. George brought back some proofsheets and a fine bunch of flowers. Dr Grant called. He told about the Ichnumon[41] who fights snakes & the bird who lights his nest with fireflies & many things more.[42] He staid a long while. Today Papa finished

---

[41] An archaic name for a mongoose.

[42] Possibly the Australian Bower Bird, which decorates its bower (not a nest) with anything shiny or bright.

copying out his measures of double stars to send home by Captn Wauchope#

[In Sir John's hand]

Tried to get Some Equatorial measures but it clouded after much time lost.—Tormented with headaches as Mamma was with Rheumatisms—

#Took a set of angles from Station F on Wynberg hill—all the flowers now coming out Superbly—a fine clear Calm day

Sunday, August 27, 1837

[In the child's hand]

We walked through Mr Hamilton Ross's[43] to Chur^ch, to hear Mr Hough preach. Mr Maclear came back with us, & dined & slept at Feldhausen. Dr Gra^nt gave Papa a stuffed mole, & some very curious specimens of sand stone, found at Wynberg.

Thursday, August 29, 1837

[In the child's hand]

A rainy morning. D. Gran[t] came, & Papa went with him to see the Sand petrifactions at Wynberg, but it rained so hard, they could not farther than Capt^in Clarks. D. Grant came home & staid to dinner.

Wednesday, August 30, 1837

[In the child's hand]

Papa reduced three sweeps & booked them

A Letter from Sir John and Lady Herschel, Feldhausen, to Caroline Lucretia Herschel, Hanover

N° 15

[In Caroline's hand]   Rec^d November. 12.   [Probably Sept 7]
[In Sir John's hand]

My dear Aunt

A few days ago I received your most welcome letter of the 30th March with Dr Heeren's Observations respecting which I shall write

[43] Hamilton Ross lived at "Sans Souci," Newlands.

to him by the same post which takes this—and indeed I cannot do better than enclose my letter to him in this* so as to make you my interpreter of the high thanks I owe to him for the trouble he has been so good as to take in this cause.—

I need hardly say how much we are rejoiced to see your handwriting once more though that joy is damped by your complaints of winter indisposition and such a winter! by all accounts—May this prove a better!—and may we hope to find you in no worse health & spirits when we come to see you *next Summer* in Hanover—For so—if it please God to lead us safe home according to our present *altered* plans—we most assuredly propose to do.—

I say our *altered* plans. For you know our intention was to have embarked next March for Rio Janeiro and there to have spent two or three months, after which to have taken passage in the Brazilian Packet for England which would have probably detained us till October and have rendered a visit to Hanover that season impracticable —But by striking off this Brazilian trip and taking our course directly homewards so much time will be saved and all the rest of our domestic arrangements become so much simplified that it seems like finding a treasure, as a fund of time will thereby be placed at our disposal—the first fruits of which, as in all love and duty bound we have determined nem. con. to devote to you. Or rather I should say that when on talking over with Margaret all the *pro's* and *con's* of the question whether to return home direct or viâ Brazil?—*this* consideration at once decided it in favour of the direct course, her desire to see you outweighing every consideration of amusement or temporary gratification which a visit to Rio could offer. So now be sure dear Aunty and keep yourself well and let us find you in your best looks and spirits.—And although what you say respecting our good Mrs Beckedorff's health is somewhat deplorable, yet I will indulge the hope that she too will perform a part in the Dramatis Personae of the happy meeting.—Meanwhile as the time of our departure hence approaches we shall take care and apprize you of all our Movements, respecting which it is impossible at present to speak more precisely.

We are now, thank God all well though both M^t and myself have

* On second thoughts I will send it separately as I do not see why I should give you any trouble about it.

been suffering much from Rheumatism—which has in some measure interfered with my observations during several months past. However I consider my Sweeping work now as completed—and so that is of little consequence. The children are all quite well and as merry as the day is long especially now that the winter rains are abating & the flowers are bursting forth in their usual splendid profusion.—

I have got reduced and fair copied (to send home by the "Thalia" (the Admiral's ship which will leave this station ere long) the first six hours in R.A. of my "Southern Catalogues of Nebulae and double Stars"—which contain 654 Nebulae and Clusters, and 475 Double Stars—and am now hard at work proceeding with the reduction & arrangement of the rest; hoping to find other good conveyances for them in duplicate, and thus to secure the results obtained against accidents by sea &c.—I shall also send home by the Thalia a copy of the Results of my micrometric measures of about 400 Southern Double stars with the Equatorial. (I think I mentioned in my last having got several good and unequivocal observations of the 6th Satellite of Saturn (my Father's 2^d) and traced it round many revolutions, in the period assigned by him)

—The following may serve as specimens of the way in which observations of the same nebula on different sweeps come out on reduction to the common epoch 1830. Jan 1^{st44}

Neb No	RA.	1830.0	NPD.	1830.0	Sweep	Neb. No	RA.	1830.0	NPD.	1830.0	Sweep
593	5^h 39^m	51^s.3	159	11 7	522		5^h 52^m	42^s.6	157° 21′53″		653
		52.8		11 21	508			42.8		21 58	658
		53.4		11 22	653			43.9		21 56	522
		53.7		11 14	656			44.5		21 53	760
		53.9		11 6	508			46.4		21 35	508
		53.9		11 9	509			46.9		21 47	538
		54.0		11 2	513			48.0		22 23	512

There is just arrived a King's Ship in Simons Bay on her voyage to England which offers an unexpected opportunity of dispatching this so I will conclude as Maggie will add a few words just to promise a longer account of our internal arrangements ere long.—Pray remem-

---

44 "Neb No 593" is 30 Doradus. The object in the right-hand column (NGC 2130, also in the Large Magellanic Cloud) had no number assigned at the date of the letter.

ber me very kindly to D^r and Mrs Groskopff Mrs Herschel Mrs Becke-doff Hausmann Dr Mühig and all [hurried: partly illegible] Hano-verian friends & believe me all   Your affect^e Nephew

JFW Herschel

PS. be so kind as to *seal* the enclosed to Dr Heeren

[On front in Lady Herschel's hand]
My dear Aunt
        Herschel has stolen from me the pleasure of telling you our hopes of seeing you next Summer if we be all spared, which Heaven grant we may & do you yourself up in all your fur cloaks & defy the winter's cold & frost, until a few months more bring Summer & us to warm & cheer you. But I won't be able to talk to you in German for a *teacher* I have been at the Cape, & our accounts of the little ones will prove it to you—Our next letter may bring you *more family* news, in the mean time the older ones join me in affectionate love to dear Aunt Herschel & I remain her attached Niece

Marg^t B. Herschel

A Letter from Sir John, Feldhausen,
to James Calder Stewart, London

                                    Sep/37 [Pencil in a different hand]
[In Sir John's hand]

Dear Jamie—I won't pretend to tell you the delight it gave me to see your handwriting from England after all the alarm and disquiet we have suffered since hearing of your altered health & destiny. Priva-tion—or the imminent danger of privation is the only sure test of value—and though I hardly think it is in the nature of things you should be loved better, yet I do think you are valued the more—or rather with more of the consciousness of estimation—for having a touch of the lost sheep about you.

The ratiocinations, hints & suggestions in your letter about the children were all most proper and prudent and dutiful both in the filial
and avuncular sense and I assure you were well deliberated & debated over.—However for some time past I had felt misgivings about
that Janu Rio[45] expedition and a glimmering of what might be called
the Common Sense view of the matter had begun to dawn upon my
mind. On the one side Nature in all her Charms—Rock Sea & Forest
and River—Birds Beasts Flowers Serpents & Carapatoos[46]—Human
nature in a new aspect (none of the pleasantest that) &c &c—and all
to be enjoyed only by a certain detour &c. On the other a tremendous
sacrifice of absolute time—inevitable separation of family (for not
only Willy but two of the other little ones at the least *must* have been
sent home—that were *quite inevitable*—and in fact I had at one
time determined to send them all 3 to Slough—)—and a sort of un-
definable Sense of risk—not personal risk but that sort of risk which
a man runs who carries a dispatch and wanders out of the road on his
own vagaries. The letters from *home* brought matters to a Crisis—
Maggie & I held solemn argument for 3 days & 3 nights without eating drinking or sleeping or for one instant changing the subject—
No debate in Congress was Ever more lengthy or more wearisome I
guess. At last a trip to Hanover to see Aunty gave a Casting vote—
and so as I doubt not she has written in an epistle she has been all
the evening penning the upshot was that we are to get home as the
crow flies by the India Ships in Feb March April as the case may be—
(I think Feb out of the question—March possible April probable)—
Granting we start April 15—we shall arrive June 30 I calculate.—
Allow 6 weeks in London and (not to throw our visit to Hanover too
late) we shall get en route by August 15. Now here is a bit of an entre
nous—I should most gladly have some good reason for avoiding the
abominable speechifying & flummery of the September Meeting of
the British Association. and I think our Hanoverian trip well managed may carry us nicely on the safe side of that *treacly* affair. Not
that I mean to abuse the Institution—but that really there has crept
into their meetings a style of mutual be-buttering the reverse of good

[45] A Herschelian play on words. Rio de Janeiro is "river of January." "Janu"
an abbreviation for January.
[46] A curapato is any South American tick of the genus *Amblyomma*, troublesome to man and beast.

taste,—and independent altogether of that which is merely a matter
of personal annoyance—I don't want to be drawn into any of their
work for the next year at least, having quite as much on my hands as
I can possible accomplish without taking any extra duty—Argal it is
not unlikely that we may stay abroad till somewhere beyond the
middle of Sept^r—Then comes the point of winter Residence—Slough
is unoccupiable except by myself Bachelor fashion or with some of
the brats to "climb on Father's meditative knee" as the poet hath it.
And house hunting in the county is out of the question in the bad
months when especially for Cape seasoned frames warmth is everything & everything is warmth & where can one be be [his repetition]
warm in the winter but in London or the Collieries? So *that* (the
former) must be our winter destination

Now will you (& Pat) though I write this to you as having probably
most leisure take these data into the sunny side of your noddles and
make some preliminary enquiries as to Residences in or about Regent
Park where there may be easy access to Grandmamma & to fresh air
and try and send us something more than general ideas on the subject
of situation—furniture and tho last not least important Rent (always
a formidable word in London!). What we cannot now determine but
might determine on your report—in time to give you notice before
our own arrival so as to enable you to act for us is whether to regard
our first residence in London as merely temporary & employ it in
looking out [illegible word] for a more permanent one—or or [his
repetition] at once to take a house for a year certain—In a word will
you get us & within a reasonable time send all the information you
can pick up on the subject of houses vacant or likely to become so—

I have been employed all this evening in writing Meteorological
letters till my head is like a cloud & so I must pretermit many a grave
gay lively & severe thing I should like to say till a more sunshiny
moment—at present the dews of Morpheus begin to steep my brow
in poppies and I must remain ever your affectionate though sleepy

                                                                JFWH

[In Lady Herschel's hand]

The only thing I have to add is that the *size* of any House taken for
a year should be the same as Nottingham Place or R. Smalls in York
Terrace—Certainly not so small as those in Nott. *Street*—The *great*

requisites being a situation on the same side of the New Road as Grandmama, & *as near as possible* to a garden, convinced as I am, that our little ones require *much fresh* air, & the activity of their minds depends on the activity of their bodies—Perhaps an *un*furnished House supplied with furniture (as I think is common enough) by an Upholsterer at so much p. Cent on the value per annum would be less expensive than a Furnished House—& I can't help thinking that a House taken for 1 year from June w^d be the best on all acc^{ts} for us—I remember so well the accommodations & situations of all the Terraces near Mama that but short descriptions will suffice on these points, & it is on the important one of relative expenses that we ask advice & this we will look for anxiously—Your affec^t Sister C.G.H. Septr 8th all quite well May all be found so by this————

The Diary: September 14 to November 15

Thursday, September 14, 1837

Full Moon day.—This time the full Moon was a Cloudy one—quite a case in opposition to the rule of Clear nights on full Moon nights— [D° yesterday [& also Friday   Days before & after] [his brackets]

Sunday, September 17, 1837

Mr Hawkins called & asked us to his wedding w^h we were however obliged to decline

Church at Rondebosch—Mr Hough's Sermon on Death of King William[47] & Accession of Victoria

Monday, September 18, 1837

The first decidedly fine day after a long interval of Cloudy & misted weather. And most delightful indeed it is—At work at home all day. preparing the 20 feet to recommence Sweeping after the rains & bad weather it requires a thorough setting to rights—

Set Sweep & began as soon as dark with every promise—but had hardly begun when it clouded with A MACKAREL drift slowly rising

---

[47] William IV died on June 20, 1837.

from sw   NB—and this increased till the whole sky became totally overcast—

This morning M[rs] Oliphant called at 2 AM (Sep 18–19) Calm. Rain.

Tuesday, September 26, 1837

Capt[n] Richardson of the Boyne (to Bombay) called with his wife & Miss . . .

Wednesday, September 27, 1837

Mr & M[rs] Grey & Lushington[48] & Mr & M[rs] Maclear dined here

Thursday, September 28, 1837

Went in To Town—Called on D[r] & M[rs] Murray.—Met Grey & Lushington by appointment at Baron Ludwigs garden

Sunday, October 1, 1837

Dined at Col Bell's where met M[r] & M[rs] Hamilton & Miss Fane
NB.   This proved the only fine night for a very long time before & after.

Monday, October 2, 1837

Capt[n] Henning & D[r] Grant called.

Tuesday, October 3, 1837

Agreed by Agency of Messrs H. Ross & Co with Messrs T[n] Watson & Co Agents for the Windsor (Capt Henning) for our passage to England in March next for £500.—
Rec[d] Buckland's Bridgewater treatise

Wednesday, October 4, 1837

Definitely agreed to H. Ross's arrangement with Henning
M[r] Ebden called
Capt[n] Wickham of the Beagle and his 1st Lieut M[r] Stokes called on their way to Simons Town
Wrote to Buckland
The Manifesto of the Kirk Session denouncing the Memorial of 307 Individuals petitioning for a repeal of the Sabbath observance bill, as "a Public profession of infidelity"—Wrote to Fairbairn thereon

Thursday, October 5, 1837

Dispatched note to Fairbairn rec[d] his reply

[48] J. D. Lushington was an Indian visitor.

Monday, October 9, 1837

Mr Lushington called PPC.

Maggie took back Mr Hough MS. Sermons

Black South Easter

Tuesday, October 10, 1837

Furious South Easter but clear

Wednesday, October 11, 1837

Raging SE wind settling gradually down into perfect calm—all day
Sky superbly clear not a Cloud on hill [a sketch of cloud patterns on
the skyline]

Thursday, October 12, 1837

[Erased] Mr Smith called with a request from Mac

Furious North Gale with unusual phaenomena of Cloud &c See
opp page lasted all night—with drizzling rain from the distant cloud
on T.M.—while all Sky above clear

Friday, October 13, 1837

Mr Smith called with a request from Maclear that I would observe
the Lunar Eclipse tonight Wrote to M. in reply—rec^d^ his Answer &
a Chronometer—& prepared accordingly for the Obs^n^—

Observed the Total Eclipse of the Moon

Sunday, October 15, 1837

Attended Divine Service at Rondebosch—

Mr Maclear called, accompanied us to Church returned & passed
the Evening & night at Feldhausen

(cloudy—all night—a black S Easter)

Monday, October 16, 1837

Black South Easter—Took Horse & rode with Maclear part of way
to Observatory parted behind Col Bell's beyond Truters[49] where I
found an enormous multitude of the White Creamscented Satyrium
with which I loaded myself & returned

Read Shakespeare's (?) Titus Andronicus A mess of horrors which

[49] Olaff John Truter occupied "Welkom," part of the Canigou estate, Rondebosch. A notary and attorney, he was agent for the Dutch "Handelmaatschappy"
(trading company).

must have been written either as a joke or as a bravado as it is perfectly impossible any one could have ever meant it seriously as a Tragedy

Wrote to Lady Smith & Goodenough our Gardener at Slough telling him to give Mrs Rendall notice—as we shall be home in May.

Saturday, October 21, 1837

Rose at 5½ and commenced the hourly Observations for the Month —which I carried on till 4 AM of the 22^d when Marg^t relieved me & took the 2 last hours.

Examined Alexander's Specimen of Native Iron[50] from the Orange river. In 4 [3, erased] grains I detected no perceptible nickel

Mrs Hutchinson left

Miss Pattle dined with us.

Sunday, October 22, 1837

Div Service & Sacrament at Rondebosch Church.

NB. Mrs Smith[51] (Col S) and Mrs Michell (Major M) two catholic ladies, communicated

Stargazing commenced but baffled by clouds

Friday, October 27, 1837

A gloomy hot day (Some Sun but a much greater effect than due to the direct radiation which manifestly comes from the Clouds—

Captn & Mrs Wauchope dined here also Mr Coutts Arbuthnot.

Prepared a new Circumpolar map In Evening worked at Star mapping till 1½ AM when it clouded thick over

Outlined the great Leucospermums

Found all my days work [words erased]

Saturday, October 28, 1837

Worked the whole day at a new skeleton Map of hemisphere N° 2 and at last when quite finished (9 hours good work) found to my dismay I had been doing a duplicate of N° 1 by mistake—N° 2 remaining to do

---

[50] Presumably a meteoritic specimen from this explorer, too (see note 58, August 21, 1836).

[51] This Mrs. Smith is Juanita, afterward Lady Smith, of Spanish birth (see note 108, June 24, 1834).

Sunday, October 29, 1837

At 8ʰ 1/4 AM. our 3ᵈ Son and 6ᵗʰ Child (John) Born[52]

Mr & Mʳˢ Maclear came over & Mʳˢ M. to see Maggie and M for an astron¹ talk

Star Mapping till 2 AM.

Monday, October 30, 1837

Orchard came wʰ a letter about his plot of ground but went away without doing the needful owing to a mistake in one of his drafts (being for £5:4:0) instead of £54:0:0).—

Tuesday, October 31, 1837

Maggie & Baby going on capitally

Wrote to Konigsberg at Stellenbosch to get me bulbs

Wednesday, November 1, 1837

Mʳˢ Hutchinson came out about 1 PM & dined & returned at night

Maggie & Baby going on well

Wrote to Decandolle.[53] letter concluded.

A cloudy night with some Considerable rain

No observing

Thursday, November 2, 1837

Worked at Maps—

Wrote to Mr Harvey.

Rode to Newlands to call on Mʳˢ Wauchope—Went over the house at Newlands wʰ is about to be let for 3 years for a boarding school to a certain Mʳˢ Hull.[54]

Saw reason to modify my bad opinion of its Proprietor Cruywagen[55] who as his wealth increases appears disposed to enlarge his ideas &

---

[52] John Herschel (1837–1921), F.R.S., F.R.A.S., colonel of the Royal Engineers.

[53] Augustin Pyrame de Candolle (1778–1841), Swiss botanist, advocate of the natural system of flower classification as opposed to that of Linnaeus. He held the chair of natural history at Geneva (from 1816), as also did his son Alphonse Louis Pierre Pyrame de Candolle (1806–1893).

[54] Mrs. Hull, a professor of piano, was already running a ladies' seminary at 60 Bree Street, and continued after this date, so the Newlands venture may have been a failure.

[55] J. J. Cruywagen owned "Newlands House" from 1830 to 1852.

act on more liberal principles—This is not the character of avarice as a passion of which his former conduct bore strong marks—I believe a really avaricious man to be rather a rare phaenomenon—though a parsimonious one is common enough.

Mrs W.[auchope] this evening gave a beautiful specimen of that quality of intellect which may be termed pachydermatous—that on which neither talking nor admitted argument & nor experience of facts works any change of opinion—It was in a small matter—but perfectly illustrative—De floribus Capensibus[56] Per tres annos flores Capenses assidue colligit—easque laudat—pretio emit—ac pingendum pictoribus consignat &c—Hodie verum enunciat hasce flores, nec venustate nec odore comparabilis esse cum illis notissimis (Harebell, Cowslip, Primula &c) quas in Scotiâ novit ac dilexit—Odoris respectu (nominative) Flores Capenses penitus esse spernendas—(dictu facile et sentententia maxime vulgaris.)—Illi verum notissimum debet esse opinor [erased] hanc sententiam vulgarem non esse verum, cum flores permultas odore intensissimo ac suavissimo illi saepissime exemplo dedi in manus ac sub nasibus misi—Satyria, Hesperanthos, Gladiolos, Moreas—&c &c

Friday, November 3, 1837

Made out the tables of 6 Sweeps and began their actual reduction.—

---

[56] We are indebted to Dr. William H. Hess, assistant professor of classics at The University of Texas, for aid in transcribing this almost illegible passage, and translating it into English. Sir John seems to have been motivated by an excess of that delicacy which (for different reasons) requires certain passages in the Decameron to be put into a foreign language. Perhaps he was afraid that an exploring child might trumpet his exasperation with Mrs. W. all over Cape Town. The lady might even have been (who knows?) the originator of the famous anti-South African gibe "The birds have no song, the flowers have no scent, the women have no morals." This is what Sir John has to say:
"Concerning the Cape Flowers
    She has been collecting the Cape flowers carefully for three years and praises them—She bought them—and consigned them to a painter to be painted—But today she declares that *these* flowers are not comparable in beauty or smell with those others so well known (Harebell, Cowslip, Primula etc) which she knew and loved in Scotland—in respect of scent—'The Cape flowers are entirely contemptible' (easy to say and an opinion that is especially common).—But it ought to be very well known to her that this common opinion is not true, since I have put as examples in her hand and placed under her nose very many flowers of most intense and pleasant odour—Satyria, Hesperanthos, Gladiolos, Moreas—&c &c"

Outlined Watsonia humilis[57]
Recommenced preparations for polishing—
Maggie got up & sate up all day

Saturday, November 4, 1837

All the Clouds & bad weather swept off as by Magic by a S.
Easter—Sky superb every detail of T M[n] &c &c

Sunday, November 12, 1837

Full Moon Night. Sky cloudless till about 2½ AM. up to which period
no meteors could be seen for [word erased] Moonlight—It then
clouded over steadily from NW & continued cloudy for the next 48
hours and more
Finished copying out my nebulae for the hours 6, 7, 8+

Monday, November 13, 1837

Orchard paid for his Land £93
NB.   A Strange Story he told about Watson
Up to Newlands to finish my Sketch—NB. I never saw the Atmosphere so perfectly transparent—not the smallest detail of the T M[n]
could escape the eye
Uniformly cloudy & an immense sheet of cloud pouring over the
hill from Constantia Kloof to Eckstein's Tower—
Captn Barnard still very ill, no hope can be entertained of his recovery   I don't [erased]   Called on Schonberg to appoint Friday for
him to complete purchase of his allotment of Land
A Cloudy night—not a glimpse of any Star—no Obs[ns] of Meteors

Tuesday, November 14, 1837

Rode over to Observatory   Found the Base line *at last* done
Process of verifying the Transit level in progress—by means of
Spherometer & levers counterbalanced   Somewhat blundered [a
rough sketch of the equipment]   Attempt to work at Eq[1] but cloudy
all but one short interval about [entry ends]

Wednesday, November 15, 1837

Maclear and M[rs] M. came over and dined here.—Admiral Campbell called   Mr. F. Smith   Grandson of W[m] S.   MP for Norwich
called PPC to Sail tomorrow for Swan [end of word illegible] having

[57] *Homoglossum merianellum*, "Flames," a slender member of the Iridaceae
with red and yellowish flowers.

missed Grey & Lushington. Attempted to work at Equatorial but
Clouds would allow no progress

A Letter from Sir John and Lady Herschel, Feldhausen,
to Caroline Lucretia Herschel, Hanover

N⁰ 16

Cape of Good Hope
Nov. 16 1837

[In Lady Herschel's hand]

My very dear Aunt

I hope you have long before this received our last joint letter
giving you the good news of our having now fixed the time of our departure from the Cape, at which, I have my own reasons for being
particularly pleased, without any prejudice to the Cape or its many
powers of pleasing—Herschel, this time leaves me to tell my own
story since the date Sep þᵉ 7ᵗʰ of that last letter & to introduce to you
our third little son who was born on the 29th of October & for whom
I had the name of *John* ready prepared—When he is christened,
Cousin Thomas & his wife & our special favourite Mʳ Whewell* are to
be the Godfathers & godmother—The latter deserves something from
Herschel in return for his excellent work on the "History of the Inductive Sciences" in three volumes which he dedicated to Herschel—
This dear little stranger has been so good hitherto that he gives no
one any trouble, & although he is not three weeks old yet, I have just
come in from a walk round the garden & shrubberies with Herschel
who is very proud of his beautiful Watsonias (called after Sir Wilᵐ
Watson's father) & magnificent Fleur de lis, now adorning the serpentine walks which he had cut out of the wild brushwood with his own
hand when tired of calculations.—We shall leave this pretty place
much prettier than we found it, but it has amply repaid us in enjoyment, & it has given many a beautiful subject for Herschel's sketches,

* Unless on Further consideration Mʳ Jones should be called upon
in that capacity—*NB. JFWH.*

which I hope you may see some day. Table Mountain itself, I fear, may be rejoiced at his departure, for it has sat so often for its likeness that it must be tired by this time, but he has not been guilty of any *Odes* or *Sonnets* to it, for which I am not sorry—Since writing last to you we have engaged our passage in the India ship "*Windsor*" which touched here on its way to Calcutta, & is expected to return hither on the 1ˢᵗ of March next. We know Capᵗ Henning to be a most excellent man & as Lʸ Ryan promised to go with him to England also at this time, we think ourselves very fortunate in having secured a good ship, good Capᵗ, good Cabins & good Company. If good weather be added then we may hope to reach England early in May, & as soon afterwards as possible, Herschel promises to gratify my long most cherished wish, & take me to Hanover to see one who, from my infancy, has had the first place in my admiration & esteem—I might *now* add something more, but it looks weak in the telling—Your dear letter (Nᵒ 8) dated June 11ᵗʰ arrived lately, & gave us as usual on such receipts, a rejoicing day—The departure of the good D. of Cambridge from Hannover must cause a sad loss to you, & your illuminated windows were I am afraid only an *outward* sign of joy at the accession of your new King, but as far as politics are concerned, methinks the whole town must be pledged to preserve *you* from annoyance, & all I demand from the new order of things whatever it may be, [word cut out] clear upon roads to lead English friends to Hannover—Herschel desires me to say that he has already written to Dʳ Heeren acknowledging all the Meteorological Obsⁿˢ which he kindly transmitted to him through you, & he begs that you will now add his thanks for the Spring Observations of 1837 which accompanied your last letter, informing Dʳ Heeren at the same time that Herschel does not intend to observe beyond Decʳ 1837 & therefore Dʳ Heer[en] need not carry on his observations beyond that time either as connected with the Cape of Good Hope. You may be sure I will not forget your message to Mʳˢ Somerville, who I hear is bringing out a *fourth edition* of her book, & Herschel has already sent you acknowledgements to Mʳ Baily & the others.

Pray do not be uneasy about our long voyage, for Herschel & I are famous *packers* in our different departments, & as our three eldest little ones are now really our companions, I only consider that I have *three* in the nursery, & very young children make excellent travellers—Carry & Bella have kept journals for a long time, & I look forward with much enjoyment to their remarks on board ship. Little William is now master of a pony, & if the truth may be told, rides better than Papa, & Alexander promises to be still more manly—but Papa has one decided favourite among them who is little Margaret, for he says she reminds him of what he was himself when young & he seems to recognize her little thoughts, & understand her prattle better than the rest. I hope to find my dear Mama surrounded by her three excellent sons when the delight of meeting is granted to us—& also the dear trio at Anstey Cottage quite well—I wish I could send you something better worth the reading, but in the absence of such power, believe in the most affectionate love of your Herschel, & your attached Niece.   M.B.H.

[On front in Sir John's hand]

Dear Aunt   Last week I finished the reduction of the first nine hours (0^h . . . 9^h) in RA. of my Southern nebulae and double Stars—and before the Thalia Sails I expect to have the first 12 hours ready* to go in charge of Captn Wauchope who will reach England just before us.

I thought you would not be sorry to hear this & so wishing you health till we meet

I remain Your affec^t nephew
JFWH

  * In duplicate of course

The Diary: November 18 to December 31;
Occasional Memoranda; Observations

Saturday, November 18, 1837

A rainy cloudy day—very heavy rain all morning—Worked at θ Orionis Monograph Copying D Stars for the Thalia Repolished mirror N° 3 using the old Colcothar (returned into the general receptacle —only fresh washed over with much care) and putting mirror *at once* in the Polisher without remoulding it—working at first 1 or 2 strokes

& resting 5^m then 100 at a time with 5 or 10^m interval—& lastly as quick as possible 5 or 600

In all 2700 strokes.—*No side motion* came off brilliant & *free of all* bad scratches—so far therefore a successful process & very little trouble

Examined by daylight—It is very scratchy—also it is a bad figure

Night uniformly & densely cloudy

Sunday, November 19, 1837

Cleared late—got some triangles filled in in the Monograph of θ Orionis with new skeletons well prepared. Mirror proves *bad*.

Monday, November 20, 1837

Prepared marks & fixed on the ends of a base line along South Avenue

Went with the Children up to Wynberg Stone & explained to them the nature of a Map, by False Bay, Table Bay & the 2 ranges of Mountains

Took angles (at risk of my neck) from top of Wynberg Stone w^h is a nasty ugly block to *read off* upon.—A great granite Rounded Mass—

Tuesday, November 21, 1837

Rode up to Wynberg Stone & got Androcymbiums—also verified a new Mark on S End of Base line

Ro [entry ends]

Sunday, December 3, 1837

Cloudless & calm from $\odot$ Rise to $\odot$ Set. A most broiling day. Therm 92. Today in Sun Therm 120, under soil at root of a growing fir tree! Wet bulb Therm 63!—Therm in a double glazed frame 218°0!!!

Actinometer Max^m 37.+

Calm & clear at night. Staid at home all day

At night Calm & Clear but Def^n not good.

Monday, December 4, 1837

Expecting another day cloudless from Morning to Night was called at 5 but lo! a dense *fog* over every part of horizon. Its level *here* just cleared the ground but attained the tops of the fir trees, which kept up a pattering of rain!!! which cooled the burning soil beneath.— Therm 64!! This is a Cape Change with a vengeance

As the ⊙ got up the fog dissipated & we have a superb Day with a moderate South Easter, therm 84 & in the wind cool & agreeable

Today I put a fresh Egg into the double glassed frame—merely a tin cup *supported on woollen*—loosely cov^d with a disc of plate glass —the whole in a box buried in the sandy soil & a window pane laid over it on the sand.—

In a couple of hours (for I forgot it) took out the egg which burned þ^e fingers as if fresh from the pot. It was done as hard as a salad Egg and I ate it and gave some to my wife and six small children that they might have it to say they had eaten an egg boiled hard in the Sun in South Africa

To night calm & oppressively hot Therm at 3 AM = 76 It had risen 2° from 76 at midnight to 78 & is *now* 76 as before—not a particle of dew

Tuesday, December 5, 1837

The most burning of all the Cape Days as far as direct radiation goes w^h I have experienced

Max Therm   98 In shade
                        131 in Sun

Actinom^r 40 Steady.

Soil (bulb covered) 159

In double glazed box 248

A retort of water blackened with Ink was maintained all day at boiling temp (under a glass).—

After ⊙ had left the bulb garden Thermom^r 4^in below surface 102°.—

Wednesday, December 6, 1837

Went into Town to Meeting of the Inst^n & to an Exploration Committee

Dr Adamson informed us that last night at 11 PM his thermometer *rose* to 92°— at the Observatory also it was 86 (?) at about the same hour

Thursday, December 7, 1837

Stole a Simile where it was to be *bagged* with tolerable security "Oh that sly night bird Time how he sweeps past us on his silent silky way looking steadily at us while we scarcely see him and stumble on in vain to catch the thief."

Saturday, December 9, 1837
Worked at Equatorial till Daybreak being Superb—but [erased]

Sunday, December 10, 1837
Rose early being roused by Capt[n] Maitland[58] and the Rev[d] Jones
of the Wellesly calling—
Divine Service in Rondebosch Church

Monday, December 11, 1837
Found that the Coachhouse had been broken open last but nothing
Stolen
Sir Frederick & Lady Maitland[59] (the New Admiral on the E. Indian Station) with Miss . . . & certain Middies called
Shewed them the telescopes &c
In Evening Adm[l] Campbell & M[r] Hutchins (of the Thalia . . .)
came & I shewed them the Great Cluster 47 Toucani w[h] in spite of a
full Moon was most superb & the Rose coloured central Mass[60] distinctly seen by both of them (It was most striking & the definition admirably perfect) Also Canopus & Sirius (both so defined as I never
saw them) & the effect of the sieve pasteboard & chairback apertures!!! Most glorious! Also Neb in Orion & the D $*$ ι Orionis
When gone worked at Equatorial till Daylight—reviewing but
bagged hardly anything—

Tuesday, December 12, 1837
Lost part of morning looking for a key. Wind still too strong to take
the angles pour my base[61]
Went into Town with MBH. to call on Sir Frederick & Lady Maitland

---

[58] Captain Thomas Maitland of H.M.S. *Wellesley*, flagship of Sir Frederick
Maitland.
[59] Sir Frederick Lewis Maitland (1777–1839) commanded H.M.S. *Belle-
rophon* in 1815, when Napoleon was taken on board, and, as the sea shanty has
it "Was sent to St. Helena, aboard the Billy Ruffian, Way Hay Ha!"
[60] The remark that the central part of the globular cluster, 47 Tucanae, appeared rose colored in an eighteen-inch reflector, is interesting. The member
stars belong to what is now called Population II, in which the brightest stars are
red.
[61] Sir John, not to be outdone by Maclear, measured a little base line of his
own at Feldhausen and surveyed it into Maclear's reference points, finding, to his
delight, excellent accuracy.

Cooked a Mutton chop & Potatoes in the Sun & ate it with MBH for Lunch It was thoroughly done, and very good

Monday, December 18, 1837
pouring Rain in heavy plumps.—
Distant lightning in all quarters no thunder

Tuesday, December 19, 1837
3¾ AM. Watching a long while for a clear interval between the pouring showers at last got it superbly clear. η Argus[62] is now *hardly inferior* to α Centauri, but it *is* inferior. Very like in colour.

It is much brighter than Rigel. Procyon is left out of all Comparison α is to η like a $*$ 2m to one of 1st

Wednesday, December 20, 1837
1½ AM. η Argus is not so bright as α Centauri but it far exceeds β Orionis now at nearly same Alt. It is much nearer to α Centauri than to β Orionis (Rigel). α Orionis is small and Procyon trifling compared to it

NB α Centauri is low

As to α Crucis, β Centauri &c these are out of all question. Quite clear & Cloudless after a cloudy day & much bad cloud all night up to this time.

Friday, December 22, 1837
Frid^y Night or Sat M^g 2½ AM. η Argus is far Superior to Rigel (as it was also when all clear & of $=$ Altitudes at an earlier hour) and decidedly exceeds α Centauri—but α is lower by far and there is also haze in sky so comparison not fair—yet I *perceived* no haze in or near α.—η is now evidently going on to rival Canopus.—

Saturday, December 23, 1837
Sund^y Morning 3 AM.

Sat night, among cloud and haze so far as I could come to any conclusion by waiting for gleams & clear intervals η Argus is larger than α Centauri, and begins now to approach Canopus—There is not so much diff^e bet^n Canopus & η as between η and α Cent.—

Monday, December 25, 1837
Divine Service at Rondebosch

[62] η Carinae (Argus): Sir John now records the novalike outburst of this object.

Tuesday, December 26, 1837

Repolished Mirror N° 2

Sir G. & Lady Gibbs,[63] Col & Lady C Bell, Col & M[rs] Smith, Capt[n] Williams Mr Parker (Sir G's Mil[y] Secretary) and a M[r] Elliott & a Miss Okes dined with us

Wednesday, December 27, 1837

Skirrow called

A dense Black South Easter, Cloud quite over horizon & far below Blockhouse—not a bit of Stargazing

Thursday, December 28, 1837

Cloudy till 11 PM then cleared—began mapping then clouded again

Occasional Memoranda, November, 1837

Nov 1. In full flower
      Gladiol Blandus
      Satyrium Carneum
      ————Chrysostachyum[64]
      Gladiolus Angustus
      Watsonia Marginata

Going out rapidly—Watsonia humilis
                Gladiolus Patulus
                Ixia Viridiflora & [erased]
                Glad. Natalensis
                Yellow, scarlet & other Ixias
                Babiana Angustifolia

Quite gone or barely   Satyrium piperitum[65]
lingering and seeding   Sweet white Satyrium
                Satyrium curiosore Gnadenthal[66]
                Babiana Spathacea ? Villic (Blue ringens)

---

[63] This should correctly be Sir George Gipps, Governor of New South Wales (see note 85, May 11, 1834).

[64] *Satyrium coriifolium*, Ewwatrewwa (Orchidaceae).

[65] A previously unpublished name, with, apparently, no evidence as to its identity.

[66] A Satyrium, present identity unknown, from the Genadendal Mission station.

Observations, 1837

In the Mass of facts relations exist as Statues exist in Marble—It
is the mind which chisels them out and gives them body—by the in-
strumentality of abstract terms which are the tools and the inward
perception of harmony and beauty which guides them. As there is
but one beauty So there is but one truth—but to recognise it requires
the experience & testimony of whole generations & ages of mankind

So beschriebt mit Figuren der Astronome den Himmel
[Dass in dem ewigen Raum leichter sich finde der Blick]
Knüpft entlegene Sonnen, durch Siriusfernen geschieden,
Uneinander im Schwan und in den Hörnern des Stiers.[67]
&c &c Schiller

[Sir John omits the second line]
Not quite So however—

Observations, 1837

Definition—
Justice is that quality of actions which—*on the great average* in-
flicts the least amount of avoidable evil on the parties concerned in
them—and which therefore on the long run (in which men always
find their true interests) is sure to be acquiesced in as the most reasonable line of conduct—and therefore to afford the greatest satisfaction to all parties on a Calm review in cases where evil must be inflicted in some degree

Hence of all virtues justice is that which goes farthest and commands (ultimately) most applause however unpopular at first—as
the greatest Mountains require to be seen from the greatest distances
to appear in all their Majesty

Questions—How does this apply to Divine Justice—
We know what is justice among men but what is justice with God—
who can say.—With him there are no temporary opinions—no striking of averages—*But* we *may* conceive that the divine mind judges

[67] Lines from Schiller's "Menschliches Wissen": "Thus the astronomer describes the heavens with figures of constellations which make it easier to com-
prehend the view of infinite space; joins together remote Suns at Sirius distances,
in the Swan (Cygnus) and the horns of the Bull (Taurus)."

of the Sum of things while we from our mistaken averages only on an infinitesimal part.

Nevertheless—though the mind of man be incapable of arriving at truth and more incapable of comprehending it if attained it is persuaded that there exists a truth and it is not incapable of perceiving the lines which lead to it—the direction in w^h it lies—and only to crawl a step in that direction is universally felt to be our noblest privilege—Whenever therefore we find those results attributed to Gods justice which the universal mind of man declares to be *contrary* to the results of human justice—we must be wrong—our needle points the other way Either this—or all our faculties are delusive and there IS *no truth* for man

The Same of Divine Benevolence &c &c &c

NB. Justice refers to the distribution of evil (at least more eminently)

i.e. of inevitable evil

Benevolence to that of good.

i.e. of attainable good—*taken in reference to inevitable evil*

Inevitable evil is that which infers a contradiction not to support every *inferior* degree of good to that which is *physically possible* is an evil.—For why should not every being be as happy *as* possible

Every inferior degree of good to the Summum bonum which any existing being has faculties to conceive is an Evil for why should he be less happy than he can believe possible

Great Pain may be a good & a respite—from extreme torture. ∴ Evil must exist where there is inequality of enjoyment—

It is no injustice to reward a man above his merits otherwise than as such reward operates as a discouragement to others by abstracting a portion of the "Matière de Bien" from a limited stock. But infinite power may reward as largely as it will without injustice for who can complain that they are happy and retain his title to be regarded as benevolent

Observations, 1837

Novr 17/37 One trifle I must note—observed almost every night since our arrival here

In England it is difficult (though not impossible—with luck) to *blow a* candle *in* after blowing it out—

At the Cape on þe contrary it is hardly possible *not* to blow it *in* again if the blast is continued for ever so short a time after the first puff—I speak of the common mould candles of commerce—

What is odd is that the same facility of relighting pertains to certain *white wax* taper which I brought with me from England though I do not remember ever to have noticed that property at home

Observations, 1837

Nov. 22 1837.—Latin. participle in *ax*
Indicative of capacity to *do* or tendency to *do* i.e. Action without time

Capax.—Capio Loquax Mendax
Fugax   Audax   Rapax
Edax   tenax

All modifications of verbs in a high degree *active*

Cape Noises
    Woodpigeons[68]—Paparrrrra
                    paparrr-a
    Day and night
    Blasop.[69]   Ki-ki-krrrrrrr . . .
    Goatsucker[70] [Figure 9]
    D° ? Quack—quack quack . . . just like a duck[71] only louder and quicker

Death Watch . . . .
    .   .   .   .
    .   .   .

Buckmakeiri[72]

---

[68] The Cape Turtle-Dove (*Streptopelia capicola*), the call usually rendered as "Dear father, dear father . . ."

[69] Something which blows itself up; often used for a fish, but in this case, probably refers to the fact that the Cape Coloured people have long given the name "Blaasop" to a small, ground-dwelling frog, *Breviceps*. Its call, which could be rendered as shown, is very insistent in this area in spring and early summer.

[70] See note 88, Occasional Memoranda, March, 1836.

[71] Possibly the Dikkop, Stone Curlew (*Burhinus capensis*), which calls at night; alternatively (see note 88, Occasional Memoranda, March, 1836), the European Nightjar has a "chuck-chuck-chuck" call.

[72] The Bokmakierie (*Telophorus zeylonus*), a shrike with a remarkable variety of calls. Mrs. Rowan remarks: "The bird is one of several members of the shrike

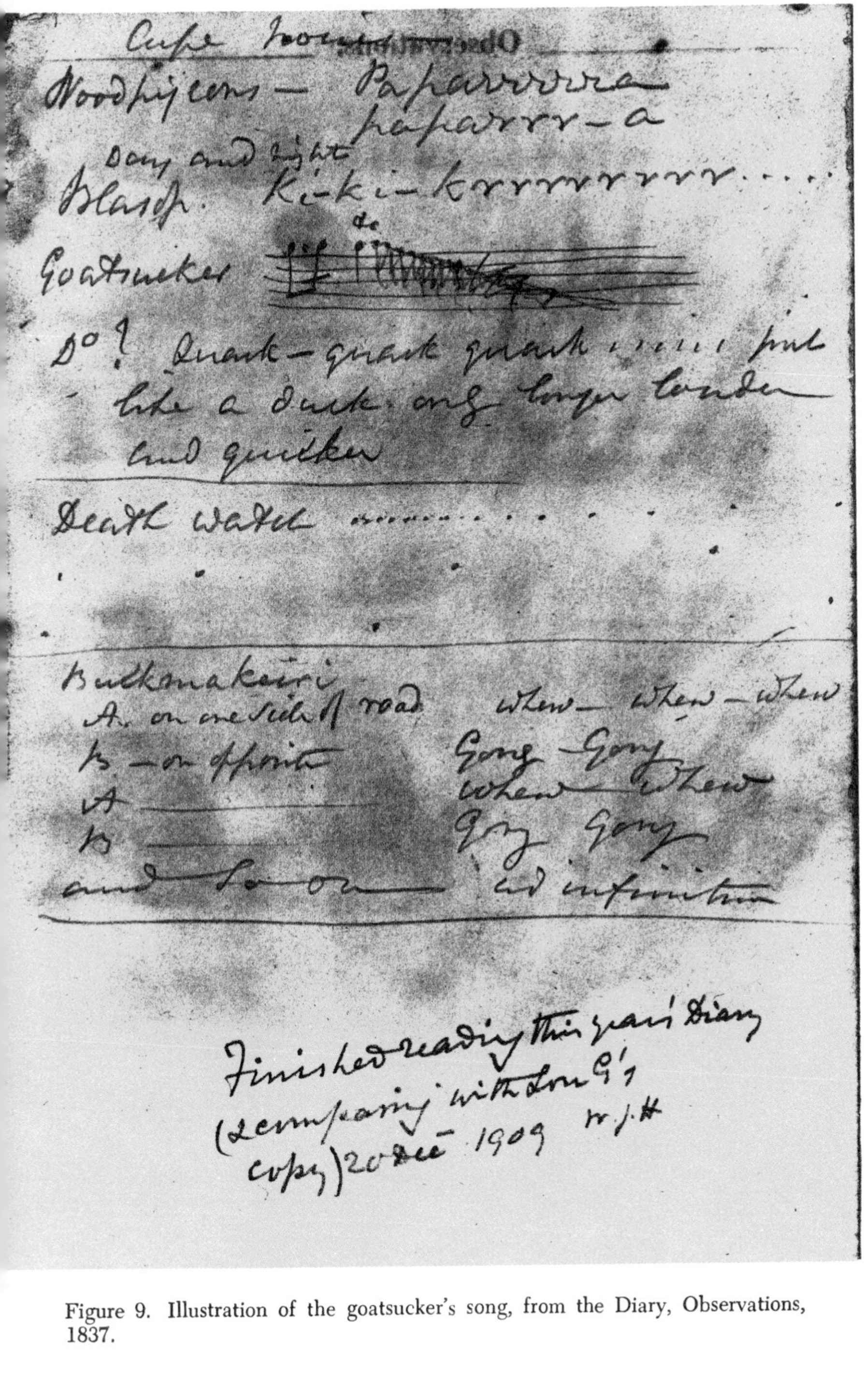

Figure 9. Illustration of the goatsucker's song, from the Diary, Observations, 1837.

A. on one side of road   whew—whew—whew
B. —on opposite   Gong—gong
A ———   Whew—whew
B ———   Gong   Gong
and so on ad infinitum

[In another hand]
Finished reading this year's Diary
& comparing with Lou G's copy)
    20 Dec 1909   W.J.H.

---

family which are notable for their habit of duetting, male answering female, or vice versa. Sound spectrograms reveal that their reaction times are far superior to anything achieved by *Homo sapiens.*"

1838

Wednesday, January 31, 1838
Drove into Town & attended Gen¹ Napier's¹ first Levee—thence
with Maclear & Lieut. Williams to M^rs De Wetts² to reconnoitre La
Caille's Station

Thursday, February 1, 1838
Col & Lady C Bell, Mr Chas Bell and M^r & Mrs Hough Dined here
also Capt & Mrs Wauchope. In Ev[en]ing came His Ex.^cy Genl Napier
& 2 Misses Napier to see stars &c with Capt Napier³ Mr. Craigie,³
Mr Banbury &c &c also a Mrs. ———
Viewed Canopus with apertures, η Argus and the Milky Way & one
of the clusters near η Argus

Tuesday, February 6, 1838
Hard at work all morning copying D Stars to go by Thalia   Captn
and Mrs. Wauchope & Wm W. came to lunch in way to Simon's Town.
& took away with them Margt. with Baby-John, and Willie to make a
farewell visit to Simons Town & Take leave of the Campbells.—
Capt Wauchope took charge of a Tin case addressed to myself care

¹ Sir George Thomas Napier (1784–1855), Governor of the Cape in succession
to Sir Benjamin D'Urban. He served in the Peninsular War where he lost his right
arm. CB, 1815, Major General, 1837, Governor at the Cape from October 4, 1837
to December, 1843, he enforced the abolition of slavery. He led a military expedition to Port Natal (Durban) and secured Natal for the British. After retirement
in 1844 he resided at Nice and died at Geneva.
² The house of the Widow De Witt in Strand Street, Cape Town. It was established by the researches of Maclear and others as the site of Lacaille's observatory (1751–1753) and the southernmost point of his geodetic triangulation. It is
now marked by a memorial plaque.
³ Sir George Napier's son Lieutenant George Napier was his A.D.C., and his
son Lieutenant John Moore Napier, extra A.D.C. Mr. J. M. Craig (not Craigie)
was Sir George's private secretary, being probably a relative of his first wife, who
was a Miss Craig.

of P. Stewart N° 57 York Terrace, Regents Park L^n containing Reduced obs^{ns} of

$$\left.\begin{array}{l} \text{1232 Nebulae} \\ \text{1194 Double Stars} \end{array}\right\}\quad \begin{array}{l} \text{obs}^d \text{ with} \\ \text{20 feet Refl} \end{array}$$

as also 1071 sets of measures of 407 Double Stars observed with the Equatorial. They were hardly gone when it began blowing hard from NW

Wednesday, February 7, 1838

Dull day. Rode into Town via Observatory to meet Maclear at Mrs De Witts. Met him—conferred about the Sector. It appears to me that there is flexure in the tube w^h M. attributes to the counterweights & expects to take off by equalizing their strain but I think it is in the tube along its whole length Rode with M & Lieut Williams up to the "Mill" & the "Platte Klip"[4] on the face of the Table M^n to select a spot for erection of the Sector[5] to get attraction of the hill

---

[4] Literally "flat stone," the name of the deep gorge in the front face of Table Mountain, affording a laborious means of ascent to the summit.

[5] Refreshing one's memory by reference to the notes on Lacaille (note 188, December 28, 1835), on surveying (note 15, March 20, 1837), and on the Widow De Witt's house (note 2, January 31, 1838), one has the picture as it presented itself to Maclear and Sir John. This was, that Lacaille had come to measure the figure of the Earth in the southern hemisphere, by measuring over the ground the length corresponding to a north-south line stretching through one degree of latitude. If the Earth is oblate, this length increases from the equator toward higher latitudes, because one thus moves into regions of lower surface curvature. However, Lacaille's result at the Cape indicated the opposite, and was attributed to the attraction of Table Mountain at his southern station causing a deviation of the vertical. To clear up this whole matter had been one of the prime tasks assigned officially to Maclear. Airy suggested that one way of investigating this deviation of the vertical was to use the zenith sector of Bradley, and this they were now about to do. James Bradley (1693–1762) was third Astronomer Royal (1742–1761), and he had discovered the phenomenon of the aberration of light using his zenith sector. This was an instrument for measuring positions of stars close to the zenith, where other effects, such as refraction by the atmosphere of the Earth, would have little or no effect. Aberration is, of course, an affair of very small angles, being the deviation in apparent direction of incoming starlight due to the, relative to the velocity of light, small velocity of the Earth through space. The zenith deviation was expected to be of the same sort of size. The zenith sector, now in the old Royal Observatory at Greenwich, which has been converted into a museum, was a hefty instrument, requiring a tent seventeen feet high to protect it from the wind. Maclear was now to lug it half way up the slopes of Table Mountain, and in the years to

Hence (on the spot selected or a little below it) took sketch of Cape Town (N° 964) w^h is very pretty from this point [Plate 19].— [In another hand] Genl no 192 (also 193)

Dr Adamson walked up & met us here at a small house in the deep ravine by the waterfall from the Platte Klip   observed granite veins penetrating the Schistose black rock which here protrudes under the sandstone & above the granite. Noticed also a cavern where there are Dikes but it came on to rain hard & obliged to leave them unexplored

Maggie came back to Dinner from Simon's town with Johnnie & Willy all well   Heavy rain at night

Thursday, February 8, 1838

A showery day and cloudy at night with exception of about 2 hours shortly after full moon rose.—Resumed obs^ns on α Hydrae   Occupied all day in packing and in revising Books for packing—D° Chemicals Dismounting of the 20-feet in progress

Monday, February 12, 1838

Dined at Gov^t House   present only Genl Napier & family After tea Miss Napier sang several very beautiful songs in a very beautiful style a regular musical treat.

Tuesday, February 13, 1838

Pulled down the Roof of the Equatorial and dismounted & pulled to pieces the Instrument ready for packing

Wednesday, March 7, 1838

The Windsor came in last night, so all now put in activity for Embarkation[6]

Sunday, March 11, 1838

Rose at ¼ before 1 AM and with great exertion completed final packing & all arrangements for leaving Feldhausen. Dr Liesching & Mr Smith (Surgeon of the Windsor) reported Willy fit for going on board, so putting him in Blankets with Mamma & Mrs. Hutchinson in

---

come would transport it hundreds of miles by ox wagon all over the roadless and difficult country of the western Cape, even up to the Orange River, where a peak still rejoices, no doubt to the mystification of the local inhabitants, in the name "Sector Berg."

6 The Honourable East India Company's Ship *Windsor* left Calcutta on January 16, and arrived at the Cape on March 4.

the Carriage and the children & Nurses (Macqueen & M⁣ʳˢ Humphries Leah & Catherine) into Baron Ludwig's Waggon, packing & dispatching the cart full of Baggage with Stone & Mʳˢ Hutchinson's Covered Cart also loaded with James as Driver, mounted "Old Harry" and after Brief adieus Galloped away Down the glorious avenue of Feldhausen—never turning a head to look back!—

Called at Govᵗ House *just* in time before Genˡ Napier went to Church, with Fairbairn's letter about Stockenstrom⁷ in hand which N. very properly refused to hear read, giving reasons quite unanswerable, but adding afterwards (in private convⁿ) that shewed any such letter to be quite unnecessary. viz: that with execution of his Office he should support S. to the utmost—consider any mark of disrespect shewn to him as shewn to himself, and to the further effect that he considers S's character to be quite untouched by the result of this trial. Took leave of His Excʸ and Miss Napier—also C. Banbury whom I met in street From Saunders's⁸ addressed a note to Fairbairn (which I gave to Maclear in þᵉ boat) stating result (in so far as þᵉ Governor's *public* conduct & opinions went) Rode to the jetty found MBH & all ready & just embarking in Captⁿ Bance's Launch (Captⁿ B having obligingly placed it at our disposal & coming on board himself with us) Willie Blanketted up in a Sedan chair in the Boat and—about Noon, were fairly aboard the Windsor which immediately weighed anchor and was off with a moderate, but perfectly fair, SE. Wind and in less than half an hour was rounding Green Point   at 2½ PM fairly off the land with the noblest view of the T. Mountain &c imaginable   Sketched it—and passed rest of the day in getting into some degree of *endurable disorder*

Monday, March 12, 1838

Passed an uncomfortable night, no undressing and little sleep and today set to work getting into sailing order, but all goes slow & nothing gets done, it seems as if every motion required to be planned

---

⁷ Captain Duncan Campbell, civil commissioner for Albany in the eastern Cape, claimed that Stockenström had deliberately and brutally shot a Kaffir boy in 1813. Stockenström brought an action for libel against Campbell in November, 1837, and on March 1 next year judgment was given in favor of Campbell on the facts; however, on May 21 a Commission of Enquiry exonerated Stockenström of any brutality.

⁸ John Saunders kept a confectioner's shop in the Heerengracht.

beforehand and every now & then there comes a still-stand—Sea-sickness would almost be better than this—but M. (who is suffering under *that* evil) says decidedly No.

Wind steady & almost right aft & scudding through a moderate swell in gallant Style with every bit of canvas spread—*Place at Noon — 32°30′, — 15°16′*

The sky throughout the day grew gradually more & more overcast & at length it grew quite cloudy—at night more so—no moon— Many Albatrosses Sea grown very blue Begun tonight to take temperatures NB Captn H. observes Barom &c regularly.

Rigged up a net as a hammock in the awning cabin—but it proved too narrow & the diagonal meshes bothered the thing so between dozing and starting up to rearrange, I passed a very bad night—Willy decidedly off the sick list & beginning to long to get out.—M still very ill.—Carry as lively as possible

Tuesday, March 13, 1838

Rose at 6 more tired than refreshed with a bad night

8 AM γ^9 in air not obs^d In water = 68.0

No events—Cloudy sky with a few trifling patches of dull blue now & then but wind as good as possible—

Found my cabin boxes &c terribly mashed by the horrid Cape Waggons in their journey to Cape Town & shipping

8 PM; γ in air 65.2 In water 69.0

Wednesday, March 14, 1838

8 AM. Therm. γ. in Air = 66.3, in Water 68.3 Sky uniformly cloudy a perfectly even stratum of Cumulostratus clouds evidently of very moderate thickness. Wind right aft (SSE)

Unpacked and planted in Earth all my Satyrium Carneum,[10] Chrysostachyum[10] & Disa graminifolia—Examined Satyrium Roseum[11]

---

[9] Denotes a particular thermometer which he used to take the temperatures.

[10] According to a note published in November, 1838, both these plants survived the journey, but *S. chrysostachyum* (= *S. corrifolium*) did not flower in the first season in England (*Botanical Register*, 1838, Misc. 84).

[11] A previously unpublished name, identity unknown. The reference to "Harvey's Herschelia" is puzzling, as it was John Lindley, not the Dublin botanist, Harvey, who first published the genus Herschelia. Perhaps it was intended that Harvey should describe it, but it appears that the plant died (see diary for April 12, 1838).

(Harvey's Herschelia) found them much disturbed by the carting to Cape Town. So replanted them   No events—began reducing the Magellan Cloud-map Observations but with little progress

Thursday, March 15, 1838

The whole day has been uniformly cloudy—a stratum of dense black cumulo-stratus clouds of ? 2000 feet high & ? 500 feet thick covering every part of horizon

At night numerous fire-masses were seen in the ship's wake. They seemed to be of the Pyrosoma kind, concentrated glowing bodies of considerable size

Friday, March 16, 1838

Rigged up a net of willow-wove work on the principle of a long cone to give the water means of escape. At night tried it & caught a few luminous shrimp-like (very small) insects & some little lumps of transparent jelly about as large as peas. But the net proved too large & difficult of management. The hoop was 10 inches diam^r which at a rate of 5 or 6 knots an hour is almost as much as it is possible to haul in & is quite useless with so fine a net as is necessary. The insects are all beaten to pieces—

Saturday, March 17, 1838

Adapted the neck of a bottle to my net thus [a sketch][12] to which attached a bag of bunting to receive the insects.— In so doing, I learnt from one of the Crew a very neat mode of *cutting a bottle* clean in half as if with a knife at any given place. Thus [another sketch][12] Take a strong *new* twine or whipcord   tie it to a hook A and let it make one turn round the bottle B at the place where it is to be cut. Stretch it very tight by holding & pulling at C & then run the bottle swiftly back & forward in the loop. The excessive friction generates heat so suddenly that the bottle presently cracks asunder with a neat clean cut all round or if not, a dash of cold water determines the fissure.—An Argand lamp[13] burner was cut in the same way

---

[12] In the first sketch the bottom of the bottle is seen cut off, the top inserted neck down in the open point of a conical net, and a bag tied around the bottle neck. This would be trawled through the water, the large end of the cone first. The second sketch is hardly necessary, as the description is excellent.

[13] The first scientifically designed oil burner, by Aimé Argand, of Geneva, in 1784. The cylindrical wick was confined between two concentric tubes, with a

The luminous masses however had disappeared & nothing more was caught. No events, same sailing

Sunday, March 18, 1838
Divine Service performed by Captn Henning—on Deck attended by all the Crew & 40 invalided soldiers of the . . . with Captn Chadwick in Command a fine veteran looking set of men & very orderly and well behaved on board—cum wivibus and childibus. About the most pitch dark night I ever saw

Wednesday, March 21, 1838
A superb starlight ev[en]ing.
α Hydrae $<$ δ Canis, $>$ δ Argus, $>$ γ Leonis[14]

Thursday, March 22, 1838
St Helena in sight at day break and nearing and developing till noon when it was on the lee beam & rounding rapidly—as we turned the NE corner of it to get to anchor off James Town. Constantly engaged in sketching it. It is a brown barren precipitous-coasted mass crowned by a broken & jagged summit[15]

Thursday, March 29, 1838
The progress of the Ship being slower, proceeded to fish with the fine bobbin-net fishing net & caught several sorts of very odd creatures[16]

1.st   a most decided Sea-butterfly, A.
2^d   a shrimp shaped Insect—B
3^d   another D° D° (perhaps a young one of B)—C
4th   Angular masses of Jelly with 3 cornered yellow hearts D
5th   Globular D° D° with globular hearts E
6 . . . A very minute fish, with disproportionately large Eye F

---

current of air playing on its inner surface. A glass chimney increased the draft and improved the illumination.

[14] Herschel is arranging various bright stars in order of brightness, presumably by simple estimation.

[15] Sir John must have landed, for he made a sketch of Napoleon's place of exile, "Longwood House."

[16] Professor Day's identifications are (1) Pteropod, possibly Cavolina, (2) Amphipod, and (3) probably young Amphipod, or another species, ? Hyperia. Remainder uncertain.

Saturday, March 31, 1838

Just across the line   Almost becalmed the whole day. At 10 AM another shark caught a *blue* shark nose pointed, large Eyes, about 8 feet long. Out of it were taken 24 *live* young sharks, which although the umbilical cord was still attached, yet lived and were very active in a pail of sea water[17] Put three of them in Brandy. This shark was attended by 2 Pilot fishes[18] 1 about 18^{in} long the other about 1 foot. They baffled all attempts to spear them. Also 3 sucking fishes[19] (a brown fish about 6^{in} long with an oval sucking apparatus close under the mouth These swim about the shark freely & attach & detach themselves ad libitum. Bottled one of them. Dissected this sharks Eye. The vitreous humour nearly all resolved by stirring into clear liquid slightly ropy.

Fished with large old willow net & caught only 3 of the purple shell fish[20] (A. Mar 29) Got a new Bobbin net rigged & fished. At 9 PM had caught nothing but 1 insect [A drawing labeled G] Fished again at night and caught (H) a very remarkable flat transparent insect[21] about 2 inches long with 10 legs, oval body, broad angular shield &c   also divers shrimps and a flat blue disk[22] [Drawings labeled H and I; Figure 10]

Sunday, April 1, 1838

On Examining net found it stuffed with straw & damaged—However there was in it another of the flat insects H in a much better state & larger of w^{h} I made a careful drawing of twice the size by measuring every joint &c. The two balls at the tail are *Eyes*, regular polyscopic eyes & the black center is the Retina, when magnified the Hexagonal divisions are well seen—

Tuesday, April 3, 1838

Lat 3 22 N   Lon 19 49 W

Caught a quantity of non-luminous Jelly much bruised—but one

---

[17] Not an umbilical cord, but a yolk sac, absorbed just before birth.

[18] *Naucrates ductor.*

[19] Remora—the sucking disc is not under the mouth, but on the back; it is a modified dorsal fin.

[20] *Janthina* sp. These are planktonic, and produce frothy bubbles as an egg raft, and are found in all warm seas.

[21] Phyllosoma larva of a rock lobster; Sir John's specimen was a large one.

[22] Porpita—a coelenterate related to the "bluebottle," Physalia. It has a vane above, which acts like a sail.

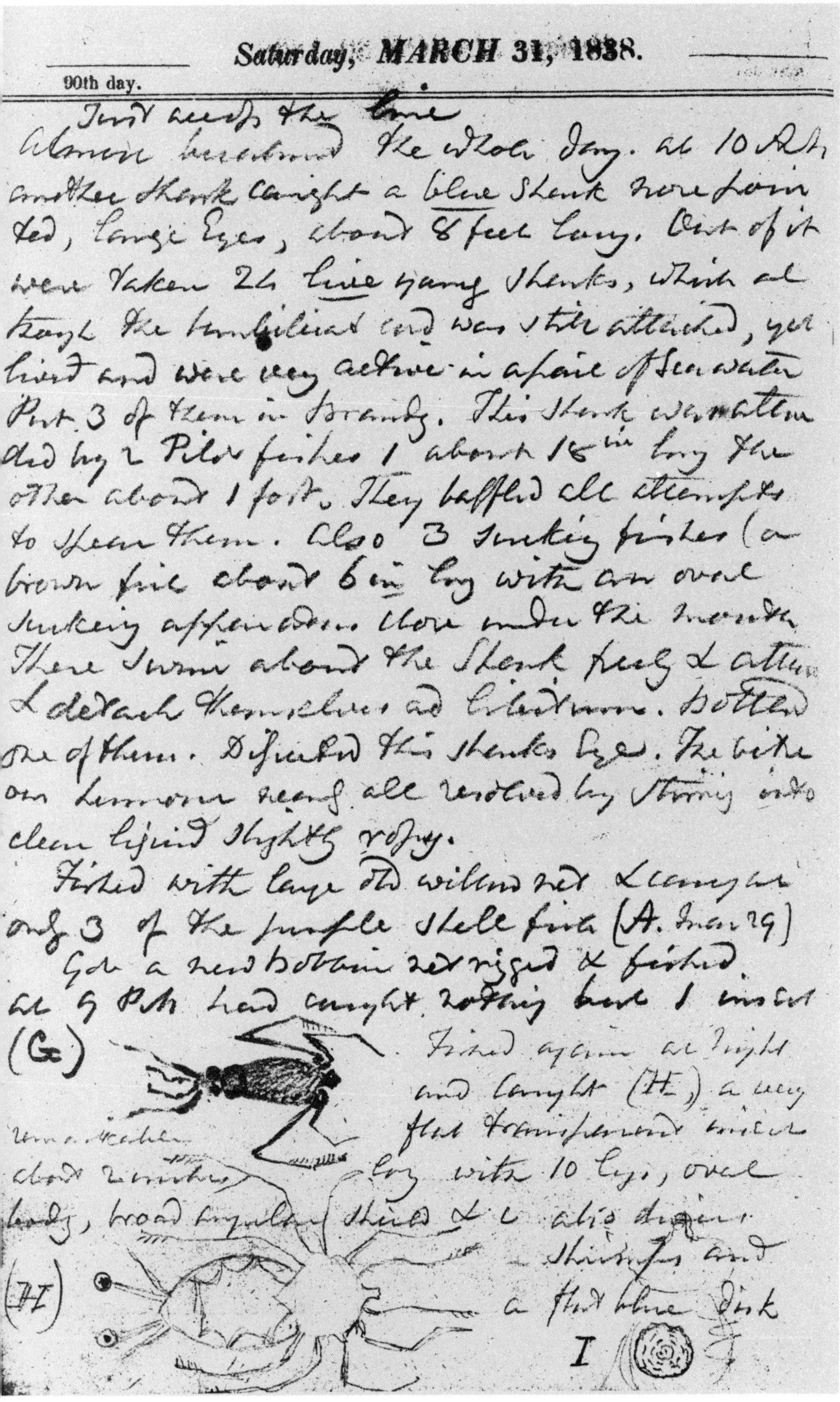

almost becalmed the whole day. At 10 A.M.
another shark caught a blue shark more pointed, large eyes, about 8 feet long. Out of it
were taken 24 live young sharks, which al-
though the umbilical cord was still attached, yet
lived and were very active in a pail of sea water.
Put 3 of them in brandy. This shark was attended by 2 Pilot fishes 1 about 18 in long the
other about 1 foot. They baffled all attempts
to spear them. Also 3 sucking fishes (a
brown fish about 6 in long with an oval
sucking appendan above under the mouth
These swam about the shark freely & attach
& detach themselves ad libitum. Bottled
one of them. Dissected this sharks eye. The bile
our humour nearly all resolved by staring into
clean liquid slightly rosy.

Fished with large old willow net & caught
only 3 of the purple stell fish (A. Mar 29)
Got a new bottom net rigged & fished
at 5 P.M. had caught nothing but 1 insect
(G)   Fished again at night
                                         and caught (H) a very
                                         flat transparent insect
                                         long with 10 legs, oval
                                         body, broad angular shield & c also different
                                                         shrimps and
                                                         a flat blue disk
(H)
                                         I

Figure 10.   Sketches of insects, from the Diary, March 31, 1838.

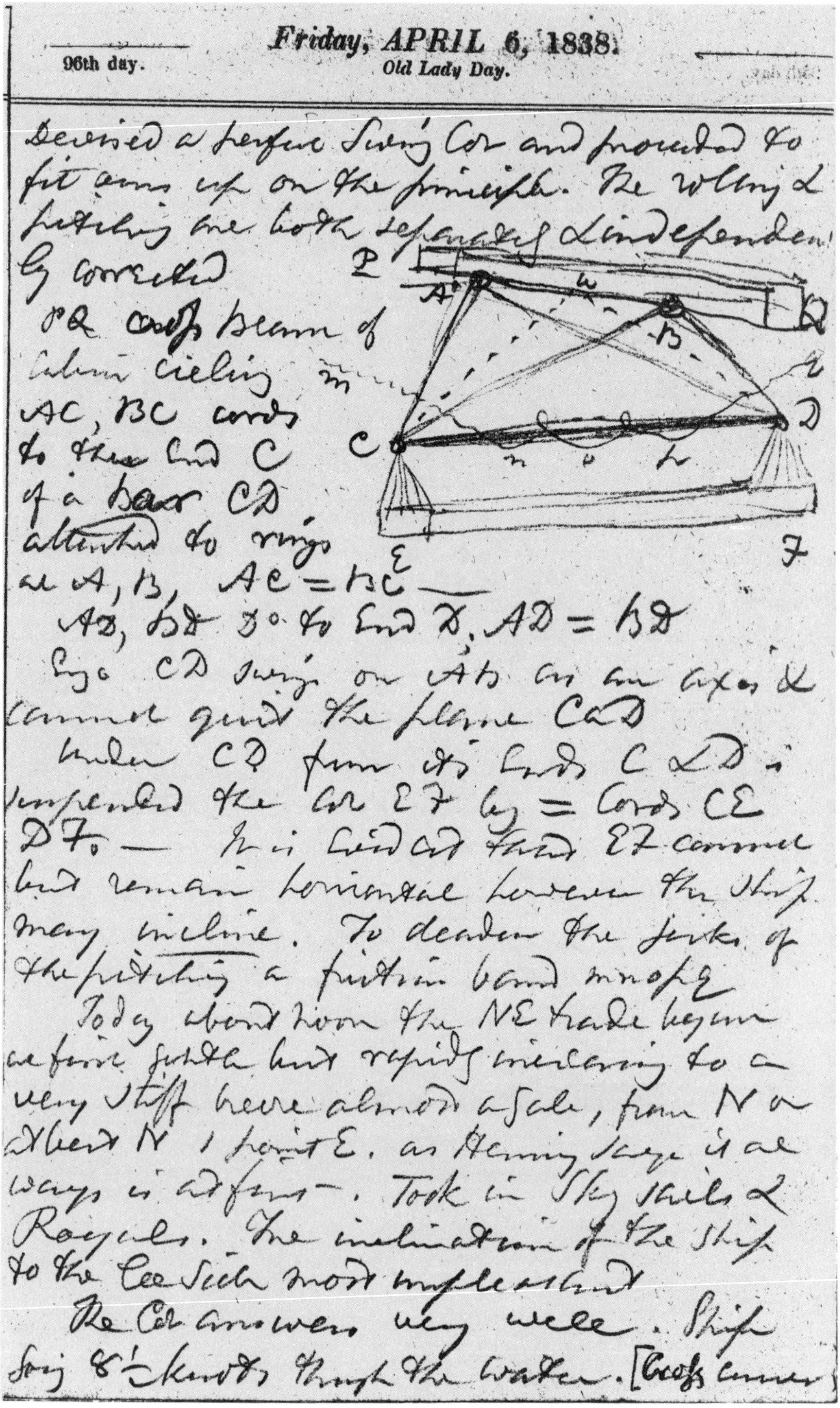

Devised a perfect Swing Cot and proceeded to fit arms up on the principle. The rolling & pitching are both separately & independently corrected

PQ cross beam of cabin cieling. In AC, BC cords to the end C of a bar CD attached to rings at A, B, $AC = BC$

AD, BD do. to end D, $AD = BD$

So CD swings on its on an axis & cannot quit the plane CaD

Under CD from its ends C & D is suspended the Cot EF by = cords CE DF. — It is found that EF cannot but remain horizontal however the ship may incline. To deaden the jerks of the pitching a friction band mnopq

Today about noon the NE trade began we first gentle but rapidly increasing to a very stiff breeze almost a gale, from N or about N 1 point E. as Herring says is always in adfins. Took in Sky sails & Royals. The inclination of the ship to the lee side most unpleasant

The Cot answers very well. Ship going 8½ knots through the water. [Cross anns

Figure 11.　Sketch of Sir John's swing cot, from the Diary, April 6, 1838.

mass well preserved & forming an evidently organised structure like a bag or pouch nearly spherical with a mouth round w^h are 5 or 6 rounded protruberances like this figure J Nothing visible inside of it Very clear & transparent [A drawing labeled J]

At Dinner the ship got stern-way & the net got entangled in the Sternpost so I was obliged to cut it adrift & desist from my fishing till another could be got

Thursday, April 5, 1838

A list-less morning occupied with trials to improve the suspension of a Cot   After Dinner various trials of strength &c. 1st I lay down & being told to "inhale" drew a full breath when 4 persons also "inhaling" lifted me off þe Deck each using one hand horizontal from elbow joint Tried same on "the Doctor" but he *proved too heavy* &, inhaling as much as we pleased we (4) could hardly get him off the ground. I see nothing extraordinary in this trick though it has been made a wonder of.

Next we (Captn Boswell, Mr Ramsay "the Doctor"—self & young Shore a lad of 11 or 12 stood holding out both arms horizontally. We all held out 10 minutes & then *struck work* but Shore who went on to 14½ when the Doctor pulled his arms down.

Dead Calm or next to no wind. It is surmised that the N & S Trades have formed a Trades Union & struck work

Friday, April 6, 1838

Devised a perfect Swing Cot [Figure 11] and proceeded to fit ours up on the principle. The rolling & pitching are both separately & independently corrected. P Q Cross Beam of Cabin cieling AC, BC cords to the end C of a Bar CD attached to rings at A, B

$$AC = BC— \qquad AD, BD\ D° \text{ to end } D. \quad AD = BD$$

Ergo   CD swings on AB as an axis & cannot quit the plane CaD Under CD from its Ends C & D is suspended the Cot EF by = [equal] Cords   CE   DF.—It is Evident that EF cannot but remain horizontal however the ship may *incline*. To deaden the jerks of the pitching a friction band mnopq

Today about noon the NE trade began at first gentle but rapidly increasing to a very stiff breeze almost a gale, from N or at best N 1 point E.   as Henning says it always is at first.—Took in Sky sails & Royals. The inclination of the ship to the lee side most unpleasant

The Cot answers very well. Ship going at 8½ knots through the water. (Cross currents)

Saturday, April 7, 1838

Friday Contin.—2 Strata of Cloud 1 low drifts parallel to the surface wind. 1 high & mackarelly at first *exactly* opposite & [illegible] by degrees more to the West of South so as to make $\angle$ of 150° $\pm$

Tuesday, April 10, 1838

Lat 10 49 N Log 27 34 W

Henning Related that in one of his voyages in þe Duke of Buccleugh that ship had gone 6153 miles in 30 successive days in *one* of which the run was only 97 miles, I take down the numbers directly after the Conversation.—Of course *sea* miles[23] (minutes of a great circle) are spoken of

Lunar [erased]

Got a few imperfect obsᵘˢ of Star-comparisons but Cirro-Cumulae Clouds very troublesome & at length put Stop to all obsⁿ

Wednesday, April 11, 1838

Lat 12°31′ N   Lon 29° 15′ W

A very fine day   Therm. 75. Wind very Moderate   Sea smooth comfortable Sailing

Many "Portugueze Men of War"[24] passed ship   none caught. Are like semitransparent pink bladders.—Rigged out a new Net the last having been lost (entangled in Rudder) & at night caught 1ˢᵗ 2 very small fishes heads & fins lost—2ᵈ 3 shells, clear, colourless transparent evidently a 2ᵈ species of the shells A See March 29 wʰ I call K They differ in 2 respects 1ˢᵗ They have no projecting beak at the orifice 2ᵈʸ they have a thorny process at the tail[25] [sketches of fish labeled K]

---

[23] A sea mile consists of 6,080 feet, as against 5,280 for a statute mile.

[24] Any of several large siphonophores (genus *Physalia*) having a large bladderlike sac or cyst with a broad crest on the upper side, by means of which they float at the surface. The creatures sail with the wind and their long fine hairlike appendages can inflict severe damage to the human skin on contact.

[25] The shells A: Pteropods. Professor Day has the following identifications:

K: Pteropod shell	R: Mollusc, ? Heleropod
L, M, N: Probably fish eggs	T: Copepod, possibly Sapphirina
O: Pteropod, ? Cavolina	U: Pteropod; the "wings" are parts
Q: Ctenophoran, ? Beroe	of the foot

There was also some kind of bony apparatus with jelly & stringy fibres which I presumed to be the ruins of a small Portugueze man of war (They say the latter sting *fearfully*)

Observed some star Comparisons.

NB. The Satyriums exposed all day & left out at night   i e The 2 growing boxes   Sat[yrium] Hersc[heliae] & Disa Gramin[ifolia]— The former begins to droop [scored through] look very sickly

Finished Scotts Pirate[26] also began Boswell's[27] tour to the Hebrides Got a new net ready

Thursday, April 12, 1838

Lat 14 8 N   Long 30 33 W

The Satyrium Hersc^{ae} looks worse The plants are quite yellow & apparently dying.—Fished & brought up several *spherical* bodies of transparent Jelly which when examined appear to be of 2 or perhaps 3 distinct sorts. L. M. N. [Figure 12]

In each of L there is a small curved white body c and a dim portion d.

M is set all over with delicate short spines & has in it a Curved body extending from what seems to be an aperture b at one end to D^o at the other.—N is set all over close with whitish papillae & resembles very much on a minute scale the "Pyrosoma Atlantica" which I caught in the outward voyage   It is a little larger than the others. The small body c in L is a little white cartilaginous or bony disc bent into double curvature thus & is evidently some vital organ.

Friday, April 13, 1838

Lat 15 48 N   Long 31 59 W

Saturday, April 14, 1838

Lat 17 12 N   Long 32 42 W

Wind Dying & at about 10 AM dead calm. While þ^e ship had head-way fished but caught nothing   Calm & serene till 11 PM though "in the strength of þ^e NE trade w^{h} the sailors think very extraordinary. Obs^d Actinometer. NB   Venus seen by J. Stone with naked Eye and by him shewn to us all—79° ± Alt[itude] very conspicuous & fine. Observed before Moonrise a set of Star-Comparisons & Fished again

---

[26] Sir Walter Scott, Bart (1771–1832), published *The Pirate* in 1822.

[27] James Boswell (1740–1795), biographer of Samuel Johnson, published an account of a journey in company in *Journal of a Tour to the Hebrides* in 1785.

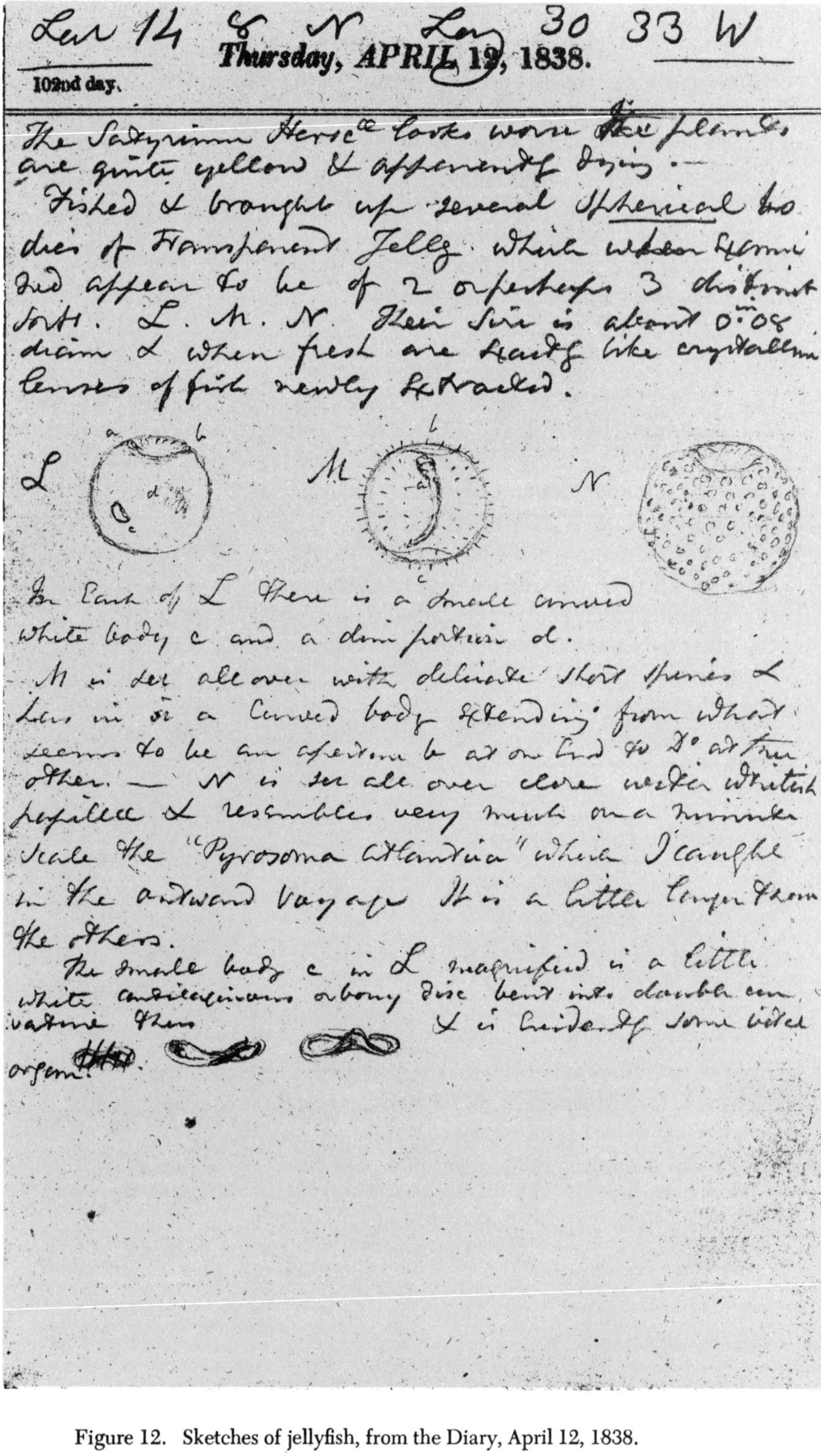

Lat 14 S N Long 30 33 W

Thursday, APRIL 12, 1838.

102nd day.

The Satyrium Herse^ce looks worse the plants are quite yellow & apparently dying.—
Fished & brought up several Spherical bodies of Transparent Jelly which when examined appear to be of 2 or perhaps 3 distinct sorts. L. M. N. Their size is about 0in.08 diam & when fresh are exactly like crystalline lenses of fish newly extracted.

In Case of L there is a small curved white body c and a dim portion d.
M is set all over with delicate short spines & has in it a curved body extending from what seems to be an aperture b at one end to do. at the other. — N is set all over close with whitish papillae & resembles very much on a minute scale the "Pyrosoma Atlantica" which I caught in the outward voyage. It is a little larger than the others.
The small body c in L magnified is a little white cartilaginous or bony disc bent into double concavities thus & is evidently some vital organ.

Figure 12. Sketches of jellyfish, from the Diary, April 12, 1838.

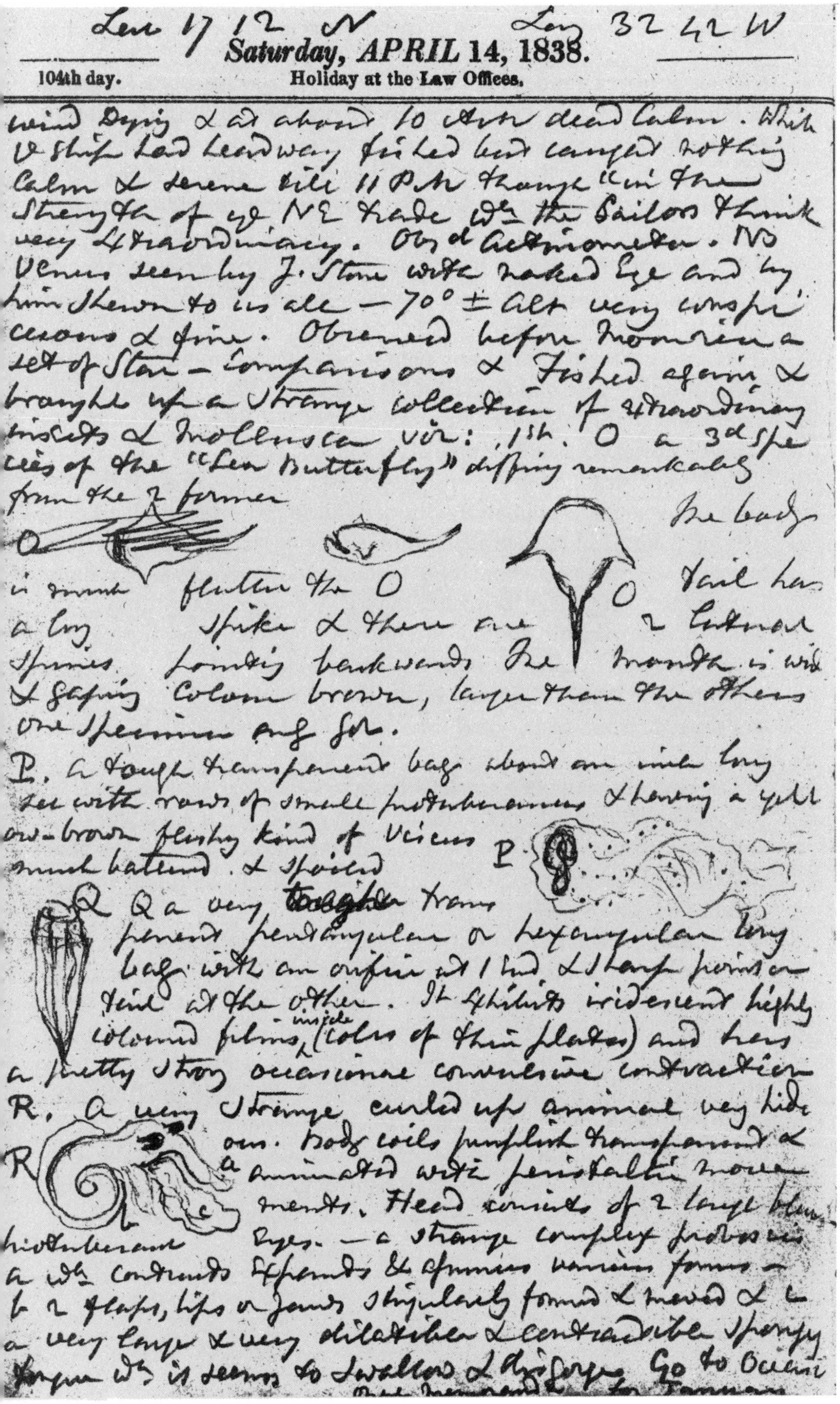

Lat 17 12 N   Long 32 42 W

Saturday, APRIL 14, 1838.

104th day.   Holiday at the Law Offices.

wind Dying & at about 10 A.m. dead Calm. While
& ship had headway fished but caught nothing
Calm & serene till 11 P.m. change "in the
Strength of the N.E. trade wh the Sailors think
very extraordinary. Obs.d Actinometer. No
Venus seen by J. Stone with naked eye and by
him shewn to us all — 70° ± Alt very conspicuous & fine. Observed before noon ire a
set of Star — comparisons & Fished again &
brought up a strange collection of extraordinary
insects & Mollusca viz: 1st. ○ a 3d species of the "Sea Butterfly" differing remarkably
from the 2 former

The body
is small flutter the ○ tail has
a long spike & there are 2 external
spines. Pointing backwards. The mouth is wide
& gaping. Colour brown, larger than the others
one specimen only got.

P. A tough transparent bag about an inch long
See with rows of small protuberances & having a yellow-brown fleshy kind of Viscus
much battered & spoiled P

Q a very tough trans-
parent pentangular or hexangular long
bag with an orifice at 1 end & sharp point or
tail at the other. It exhibits iridescent highly
coloured films, (colors of thin plates) and has
a pretty strong occasional convulsive contraction

R. A very strange curled up animal very hideous. Body coils purplish transparent &
animated with peristaltic movements. Head consists of 2 large blue
protuberances eyes. — a strange complex pro
which contracts expands & assumes various forms
& 2 flaps, lips or jaws singularly formed & moved &
a very large & very dilatable & contractible spongy
tongue with is seems to swallow & dis Go to Ocean

Figure 13. Sketches of various insects and Mollusca, from the Diary, April 14, 1838.

& brought up a strange collection of extraordinary insects & Mollusca [Figure 13] viz: 1ˢᵗ O a 3ᵈ species of the "Sea Butterfly" differing remarkably from the 2 former

The body is much flatter the tail has a long spike & there are 2 lateral spines pointing backwards The mouth is wide & gaping colour brown, larger than the others one specimen only got.

P. A tough transparent bag about one inch long set with rows of small protruberances & having a yellow-brown fleshy kind of viscus much battered & spoiled.

Q a very tough transparent pentangular or hexangular long bag with an orifice at 1 end & sharp point on tail at the other. It exhibits irridescent highly coloured films inside (colʳˢ of thin plates) and has a pretty strong occasional convulsive contraction.

R. A very strange curled up animal very hideous. Body coils purplish transparent & animated with peristaltic movements. Head consists of 2 large blue & protruberant eyes.—a strange complex proboscis *a* wʰ contracts expands & assumes various forms—*b* 2 flaps, lips or jaws singularly formed & moved & *c* a very large & very dilatible & contractible spongy tongue wʰ it seems to swallow & disgorge. Go to Occasional memoranda for January.

[Entry for April 14 continued in this part of the diary and now transferred from Occasional Memoranda, January, to its proper place]

S. A kind of shrimp of a scarlet colour

T. A very singular insect so far as I can perceive having no legs but moving with great suddenness & rapidity by fibres in his tail *a*. The parts *b c d* are fringed with hairs or most likely infinitely minute paddles the size is about 0.15 inch long

U a thin transparent but strong shell 0.36 in. in length in wʰ lies coiled a brown animal. In repose it only shows a kind of gauzy film or double tongue (a) projecting beyond its *flattened* orifice (the Chamber is cylindrical or at least round) But when water comes to it it gradually projects these tongues which then appear to be long thin *wings* or flappers which it works actively like the "Sea Butterfly" of wʰ it is perhaps another *genus*. Between them also comes out its head a very extraordinary piece of organisation, a sort of double cartilaginous plate undulated at edges, brown, flexible into various forms & attached to the body in the chamber by a long brown neck

Sunday, April 15, 1838
Lat 18 25 N   Long 33 14 W

Monday, April 16, 1838
Lat 19 43 N   Long 34 33 W
[An entry of eight lines erased]

Tuesday, April 17, 1838
Lat 21 54 N   Long 36 7 W
A few patches of the Gulf weed first seen today

Wednesday, April 18, 1838
Lat 23 30 N   Long 37 42 W
A good deal of the Gulf weed seen here & there in small round patches Read the abominable History of the Thugs.[28] Every thing one reads or hears about India & Indian affairs seems more hateful than the last—

Thursday, April 19, 1838
Lat 24 33 N   Long 38 30 W
The Gulf weed now floats past in considerable abundance, it lies in *lines* of *insulated* masses *each a single plant* When I say abundant I dont mean that it occupies any appreciable aliquot part of the surface—One plant to a square of a hundred yards would be *very large* allowance   often 5 minutes or more would elapse without seeing a bit (ship going 5 knots) All sorts of contrivances at work to catch it—Capt[n] Boswell made a tool of a wooden stick set with nails thus [a sketch] Mine was a set of hooks & crooked nails tied round a stick to fling at it [a sketch] Thus by hook & by crook a great deal was caught. Found on it 1st Live Crabs—2[d] Very small turbinated Shells (Buceinum?) 3[d] Pretty large yellow slugs or Mollusca—The weed itself is yellow inclining to green & evidently fresh & growing. It ramifies from a Center spherically & shews no signs of being detached from any rock or larger mass of weed   the branches thicken to a

[28] The Thugs were a well-organized confederacy of professional assassins in India, who insinuated themselves into the confidence of travelers, then strangled them with a handkerchief or noose, robbed them and buried them, all according to a strict religious ritual. The power of the Thugs was effectively broken by Lord William Bentinck during his time as Governor General of India (see note 100, May 23, 1835).

certain max^m & then decrease   It is covered with hollow spheres as
large as peas w^h seem to be the fructification & w^h. act as floats. The
leaves are long & serrated, thus [two sketches of the weed]

Friday, April 20, 1838

Lat 25 54 N   Lon 36 55 W

Took Drawing of the Gulf weed slug (which the Sailors call a
squid)   Read "Yes or No" Vol I[29]

Poor little Bella sea sick.

Caroline learnt 2 or 3 new songs very readily and prettily & sang
them to her Mamma's second in a way which promises that if she
takes pains she will one day sing very well. "Make a Cake," "One a
penny" "High diddle diddle" &c. Towards evening squally and after
sunset it increased to a gale so as to render it necessary to take in all
sail but what sufficed to keep the Ship's head to the wind. Also Rainy
& very dark. Sea at first kept down by the wind but as night advanced
growing heavy. Much pitching & rolling & everything very uncomfortable By daylight found the wind moderated but the sea high &
course very far from advantageous being at one time to the South of
East.

Saturday, April 21, 1838

Headaches and utter discomposure from the Ships movements.—
To live on board ship is equivalent to living in a perpetual earthquake on land with the prospect of being drowned instead of being
crushed, as a catastrophe. Lay about on sofas listless & nauseated,
reading "Yes or No" Vol 2.

Meanwhile it has fallen calm or what wind there is is nullified by
the "Head Swell" and horrible cross sea.

Tuesday, April 24, 1838

Observed the zodiacal light[30] very bright involving Mercury, Ple-

---

[29] *Yes or No: A Tale of the Day*, 1827, by Constantine Henry Phipps, First
Marquis of Normanby.

[30] Illuminated area of sky on either side of the Sun, elongated in the plane of
the orbit of the Earth. It is best seen in low latitudes just before dawn or just
after sunset, when the Sun is below the horizon. It then appears as part of an
oval patch of light extending upward from the horizon, centered on the (invisible) Sun. Produced by scattering of sunlight by particles of various kinds
in interplanetary space, it is basically an enormous extension of the solar corona.

iades, passing through Gemini South of Castor & Pollux & attaining Praesepe[31] Cancri

Wednesday, April 25, 1838

My net (the 4th) gone—The String having been *cut* off close to the Ship—having no more proper Net (Bobbin-net) gave up fishing. The Zodiacal light observed tonight very bright involving Mercury [scored through]

Thursday, April 26, 1838

Ship rolling so as to make it hardly possible to stand.
α Hydrae < Castor, > (very little) than γ Leonis

Sunday, April 29, 1838

Today at daybreak, Pico & St. Georges[32] were seen (not by me) but the Peak of the great Volcano of Pico was quite invisible—towards noon Terceira[32] was seen It is a long Island thus [a sketch] and after dinner a very distant but sure glimpse of St. Michael's.[32] One high Conical hill was well seen.—The wind w[h] had all day been freshening now got up to a heavy gale with lightning in the night and severe squalls. As we ran before the Sea, we carried on with 2 reefs in Main Topsail—no sail aft—&c.
A vile comfortless squally rainy night as dark as pitch.

Monday, April 30, 1838

The gale abated but wind still strong and the ship lays over so on the larboard tack that the sea horizon appeared through the Cuddy skylight as we sate at the end of the table—thus—[a sketch] not by fits & momentary jerks, but steadily

Wednesday, May 2, 1838

Drawing in Morning—Reading Hadji Baba[33] & Music in Evening

Thursday, May 3, 1838

Drawing the whole day.

---

[31] A cluster of stars.

[32] Pico, St. George's, Terceira, and St. Michael's are islands of the central group of the Azores.

[33] *Adventures of Hadji Baba of Ispahan,* by James Justinian Morier (c. 1780–1849). This was a remarkable Persian romance of great popularity.

Dreadfully cold a dry cutting cold wind (54°) and the wind growing contrary & veering to NE w^h is precisely the wrong quarter.

Being so cold took in my Satyriums & Disas.

Friday, May 4, 1838

Another miserably cold day & NE wind, making no progress.—

A man disorderly, & violent, had struck one of the officers so the Captn put him in irons & on bread & water till the ship shall be in dock & then assembled the Ship's company at the Capstan and addressed them on the Subject of Subordination & temperance & discipline on w^h it is to be hoped they "trembled" as he "reasoned."

Tacked Ship, wind being Easterly and seeming to become more so

Saturday, May 5, 1838

Observed α Hydrae[34]—It is visibly less than last night—It is now < or at the utmost barely = γ Leonis—but only once so estimated while many est^{ns} made it <γ.—It is < Polaris and << β Ursae Minoris. The season is now growing too late for it in this latitude.

Monday, May 7, 1838

3 Ships in Sight—one supposed the Wellington w^h left St. Helena the day before us.

Green Water almost bottle Green

No Bottom found with 120 fathoms

This morning Captn Boswell saw a *very large* whale *close* under the ship's stern, back above water.

Tonight obs^d γ Leonis >> α Hydrae (γ 50° α 15°)

β Aurigae > α Hydrae (β 20° α 15°)

Polaris >> α Hydrae

β Ursae Min >>> α Hydrae

Though low, yet it is now decidedly an insignificant star. It is very obviously much diminished since Saty night ☽ and low alt[itude] both unfavorable for the obs^n but it leaves no doubt on my mind of the Minimum being nearly attained

Tuesday, May 8, 1838

Full Moon. Superbly clear

As soon as dark enough began Star comp^{ns}—α Hydrae still high

---

[34] In modern literature the variability of α Hydrae is regarded as doubtful.

(30° ±) and quite free of all cloud & haze in a very fine blue sky (fairly above the N East haze) is very decidedly inferior to γ Leonis. —α Hyd << γ Leonis α Hyd, *little* < Polaris, <<< β Ur[sae] Min[oris] α Hyd is now *nearly* = Polaris

NB. Tonight (Great Bear near Zenith)

α Ursae >> ε Ursae, there can be no comparison

—Former obs^{ns} made them nearly or quite = but then Ursa was low.—However NB. this must be looked to—NB NB See obs of Apl 15

I incline to place the minimum of α last night as α is tonight > β Aurigae at certain intervals of its twinkling. On the whole however the ✳ is higher tonight & circumstances far more favorable to good comparisons.

On the whole after many comparisons α Hydrae is rather inferior to β Aurigae

Wednesday, May 9, 1838

Packing & arranging for landing if we ever *are* to land but this obstinate NE wind making þ^e prospect very poor—

Full moon.—Cleared but soon clouded

In good intervals

α Hydrae < γ Leonis but not much < very evidently on the rapid increase

α Hyd > β Aurigae, [following words erased]

The minimum is certainly fairly passed & the ✳ is rapidly regaining its light—;

α Ursae > ε

However the cloud cover makes this [scored out]

Thursday, May 10, 1838

Obs^d α Hydrae much < γ Leonis

  not > but rather < β Aurigae

NB.   Alt of α Hyd = that of Capella at time of Obs^n

It is still about its Minimum

Friday, May 11, 1838

Superb sky

α Hydrae tonight obs^d at alt = ½ that of Regulus, hour angle = about 3^h W

α Hydrae > β Aurigae no doubt

  β much *higher*

α Hyd nearly $=$ but a little $<$ γ Leonis.

($=$ When it flashes by scintillation)

α Hyd $<<$ Polaris

NB.   Lyra $>$ Capella though at less altitude but the Diff^e is not striking

α Ursae Maj decidedly $>>$ ε Ursae

Saturday, May 12, 1838

α Ursae ⎫

ϵ Ursae   Obs^d while in strong twilight, α Hydrae more than

γ Leonis   ½ height of Regulus—at 8½ PM.—NB. The diff^e

α Hydrae   of α and ε Ursae tho' material is yet not near so

  Castor   striking as some nights ago (May 8), ϵ $>>$ Castor

β Aurigae ⎭

I .˙. violently suspect ϵ Ursae of variation

Reobs^d when darker

α Ursae $>$ It would appear thus that α H

ϵ Ursae $<$ increases more slowly from the

γ Leonis $>$ minimum than it dim^{shes} to it

Castor   ⎫

           nearly $=$

α Hydrae ⎭

β Aurigae much less, quite left behind.

Lyra $>$ Capella very decided

Arcturus $>$ Lyra when the latter 45° high and quite fairly clear of all haze. A Superb Night. haze bank $=$ 5° high   very definite   Wind changed from E to N.

Tuesday, May 15, 1838

[In another hand]   Landed at Tower Stairs 8 P.M.

See M B H to Baldwins.

A Letter from Sir John, Feldhausen,
and on board the Windsor
to James Calder Stewart, London

Feldhausen March 6/38

Dearest Jamie,
          We are so near day of Sailing, the Windsor being come in Strict
to her appointed day that we are all taken aback or rather *un-pack*
for all is at 6'es and 7's and there is no end of Tin cases lying about
and crying "Pack me pack me" and I write this just on a chance of
getting it off by the True Briton w^h is a faster sailing ship & starts
before the Windsor just to thank you in the first place for your many
dated favours carrying as they do on them the impress of such va-
rieties of place circumstance and sentiment. As Shakespeare's lover
could find countries in his Mistresses features so I could find mile-
stones in your letters as they drop in from all quarters of the globe
and mark what my good friend Sir W^m R Hamilton would call a
"thought-paven" highway along this world of ours.—For your sake
& my own I would I could soar with the Lark or tie myself on to
the tail of a Skyrocket (if better could not be) that I might keep pace
with you in some of your excursive flights—But let that pass—I want
more particularly to reply to those parts of your last w^h relate to
merely mundane matters.
          You speak of 500 £ a year rent for certain houses in Regents Park—
and to be explicit, I won't pay it. We must cut our coats to our cloth
and in fact I cannot very well decide till we get home what to do. My
present impression is to pass 3 days in London Leaving then Maggie
in a Lodging taken for one Month as near Grandmamma and Ah—
[illegible name] as convenient as may be gotten visit Slough myself
(alone) for 3 or 4 days or a week more & get my *Bulbs* disposed of
(for ridiculous as it may seem they won't brook delay)—return pass
a fortnight in Town and clear all my matters at the Custom Houses
& arrange for their temporary disposal—then carry down such of the

children as may seem capable of enduring English country air to Slough[35] and thence take Maggie & the rest off to Hanover always supposing a guillotine not to be erected in the great square of that devoted city in which event I shall go alone & try and bring away my poor old Aunt out of harm's way.

The upshot of all this is that somewhere about the 2^d week in May if you can have your eye on any lodgings (furnished) where we can be received for a month without the necessity of hiring additional servants—such will be infinitely more in accordance with the general uncertainty of our movements than taking a house for any longer term. Therefore you need not bother yourself or Peter about any purchase or hire of Furniture or other matters of that sort, and I only regret that we have led you into any field of enquiry of that nature

May 6. 1838   On Board the Windsor

Here we are on the parallel of Nantes, becalmed. I would I could borrow the wings of a sea gull just seen skimming that way & perhaps I might catch a sight of Johnnie in his glory. However we can't be much longer now before we all meet & as Maggie said this morning when every body else was grumbly she was so happy to be nearer England than she had been for the last 5 years. She could not join in the complaints—

This will of course go into Port at our first Landing which *may* be at Portsmouth if circumstances favour in w^h case we shall stop a day at Anstey in our way to London   However I shall leave it open to the last & so for the present good bye

[In pencil] P.S. Here we are off Poole—Probably shall land there but winds & circumstances render it doubtful

---

[35] Slough, one should perhaps remind oneself, is only twenty miles from the center of London. Considering where Sir John had been willing to take his children, this change of scale of his thoughts strikes oddly.

The Diary: Selected Entries

Thursday, November 22, 1838
[Reference to Cape Education scheme]
2 PM. Colonial Office (waited ¾ hour) and saw Sir G Grey[36] in lieu of Ld Glenelg[37] who had app^d about the Cape Educ^n scheme w^h is to be carried into effect by þ^e app^ting of 12 School masters at salaries from 150 to 300 and a Superintendent at 500

Friday, November 23, 1838
[Extract]
NB. In this conver^s (with Lord Minto[38]) attended purposely to the Cape Obs^y w^h drew from Ld Minto incidentally the remark "that all that I had recommended" (in the Letter to Beaufort) "for that Establishment was to be carried into effect but had not yet been put on the Estimates" or words to that effect.

Sunday, November 25, 1838
[Extract]
Wrote to *Captn Beaufort* returning Hurricane papers about the 'Thunder' and pressing Maclears Road from Observatory to Cape Town

Saturday, December 1, 1838
[Extract]
At 2^h 5^m went back to Spring Gardens where met by appointment ... Hutton Esqr MP for Dublin who has a relative going to the Cape as Attorney General in place of Oliphant. Walked with him back to Somerset House found the Commiss^n still sitting. Messrs. Haggart & Sharpe in process of examining.
Called on Stockenstrom at 73 Pall Mall Not at home

Wednesday, December 12, 1838
Slough At 3 PM. M^r & Mrs Lyell, John Stewart, Captn Stocken-

---

[36] Under Secretary for the Colonies.
[37] Colonial Secretary.
[38] First Lord of the Admiralty.

strom came to Dinner by the 2 o'clock train. Taking it very quietly Dinner over at 6   Coffee & the Lyells returned by 7 o'clock Train. S. staid to see Windsor Castle tomorrow with M.   Read S's dispatches from þᵉ Col office and *his* to Lᵈ Glenelg & Gnl Napiers letters in clearance of S's character from the Caffer-shooting charge. All as satisfactory as possible

Thursday, December 13, 1838

Slough. Stockenstrom[39] breakfasted & Jⁿ Sᵗ They went to see Castle & I at 12ʰ 15 m—to Paddington per train in 34m. Tried to see Lord Glenelg—he at Windsor.—Appointed 2 on Saturday [entry continues]

Saturday, December 15, 1838

[Extract]

Called (by my own appointment) at Colonial office. Handed in a paper on Schools at Cape (B). Communicated Sir G. Napier's message (or impression or whatnot) respecting Col Somerset's Command as Military Commandant of Caffraria—

NB.   This the first proper occasion to do so   the General subject of uniting the Civil & Military Commands on the East frontier being mooted by Sir G. N. himself.

Took care to compromise nobody—Then had some Conversation about Stockenstrom of which it will be curious to see if anything results. Down to Slough by train at 4 PM.—40 m.

Wednesday, December 19, 1838

[Extract]

Wrote to Mʳ Cook. About Cape Schools—

Sunday, December 23, 1838

Wrote to Lord Northampton about the Deputation of the RS.—See Copy—

---

[39] Stockenström left for England on the conclusion of the enquiry (see note 7, March 11, 1838), and resigned the Lieutenant-Governorship on arrival. He withdrew his resignation on Lord Glenelg's expressing full confidence in him. Glenelg, however, resigned in February, 1839, and his successor, Lord Normanby, decided not to reappoint Stockenström in view of his unpopularity in the Eastern Province of the Cape. Stockenström was made a baronet and given a pension of £700 per annum.

—To Capt. Stockenstrom congratulating on his reappointment & extracting passages from Sir G. Napier's letter praising him (S)

—Messrs. Drummond arranging remittances &c

—Captn Beaufort enclosing Maclear's letter and urging pressing & pushing the Cape Obs^r

BIBLIOGRAPHY

Airy, Wilfred (ed.). *Autobiography of Sir George Biddell Airy.* Cambridge, England: University Press, 1896.

Alexander, Sir James Edward. *An Expedition of Discovery into the Interior of Africa, through the hitherto undescribed countries of the Great Namaquas, Boschmans, and Hill Damaras &c &c.* Two vols. London, 1838.

Armitage, Angus. *William Herschel.* London: Thomas Nelson & Sons, 1962.

Bradlow, Frank R. *Baron von Ludwig and the Ludwig's-burg Garden.* Cape Town: Balkema, 1965.

——. *Thomas Bowler, His Life and Work.* Cape Town: Balkema, 1967.

——, and Edna Bradlow. *Thomas Bowler of the Cape of Good Hope: His Life and Works, with a Catalogue of Extant Paintings; with a Commentary on the Bowler Prints by A. Gordon Brown.* Cape Town: Balkema, 1955.

Brown, Harrison (ed.). *A Bibliography on Meteorites.* Chicago: University of Chicago Press, 1953.

Buttmann, Günther. *John Herschel, Lebensbild eines Naturforschers.* Stuttgart: Wissenschaftliche Verlagsgesellschaft M.B.H., 1966.

Cameron, Roderick. *The Golden Haze: With Captain Cook in the South Pacific.* London: Weidenfield and Nicolson, 1964.

Campbell, the Reverend John. *Travels in South Africa, undertaken at the Request of the Missionary Society.* London: Black, Parry & Co., and T. Hamilton, 1815.

——. *Travels in South Africa, undertaken at the Request of the Missionary Society, being a narrative of a second journey in the interior of that country.* Two vols. London: Francis Westley, 1822.

Chapman, Sydney. "Edmond Halley as Physical Geographer, and the Story of His Charts," *Occasional Notes of the Royal Astronomical Society,* 1 (1941), 122.

Clerke, Agnes M. *Dictionary of National Biography.* Lives of Caroline Lucretia Herschel, Sir John Frederick William Herschel, and Sir William Herschel.

Darwin, Charles Robert. *The Zoology of the Voyage of H.M.S. Beagle under the command of Captain Fitzroy R. N. during the years 1832 to 1836.* Five parts. London: Smith, Elder & Co., 1840.

Evans, David S. "The Astronomical Work of Sir John Herschel at the Cape," *Quarterly Bulletin of the South African Public Library,* 12 (1957), 44.

——. "Dashing and Dutiful," *Science,* 127 (1958), 935.

——. "La Caille: 10,000 Stars in Two Years," *Discovery,* 12 (1951), 315.

Gill, Sir David. *A History and Description of the Royal Observatory, Cape of Good Hope.* London: H. M. Stationery Office, 1913.

Gill, E. Leonard. *A First Guide to South African Birds.* Cape Town: Maskew Miller, Ltd., 1950.

Herschel, Sir John F. W. *Results of Astronomical Observations made during the years 1834, 5, 6, 7, 8 at the Cape of Good Hope, being a completion of a telescopic survey of the whole surface of the visible heavens commenced in 1825.* London: Smith, Elder & Co., 1847.

——. [Catalogue of His Sketches in the South African Public Library (with a Note on the Camera Lucida)], *Quarterly Bulletin of the South African Public Library,* 12 (1957), 72.

Herschel, Mrs. John. *Memoir and Correspondence of Caroline Herschel.* New York: Appleton & Co., 1876.

Holman, James. *A Voyage round the World including Travels in Africa, Asia, Australasia, America &c from MDCCCXXVII to MDCCCXXXII.* Four vols. London: Smith, Elder & Co., 1834.

Huxley, Leonard. *The House of Smith Elder.* London: Printed for private circulation, 1923.

Kirby, Percival R. *Sir Andrew Smith M.D., K.C.B.: His Life and Works.* Cape Town: Balkema, 1965.

Linscott, Robert N. (ed.). *Complete Poems and Selected Letters of Michelangelo,* translated by Creighton Gilbert. New York: Modern Library, 1965.

Lubbock, Constance A. (ed.). *The Herschel Chronicle.* Cambridge, England: University Press, 1933.

Luyten, Willem J. "The Grootfontein Meteorite," *South African Journal of Science,* 26 (1929), 19.

McIntyre, Donald G. "The Herschel Obelisk," *Quarterly Bulletin of the South African Library,* 8 (1954), 87.

Maclear, Sir Thomas. *Verification and Extension of La Caille's Arc of*

Meridian at the Cape of Good Hope. Two vols. London: Lords Commissioners of the Admiralty, 1866.

——. [Unsigned Obituary of Maclear], *Monthly Notices of the Royal
Astronomical Society,* 40 (1880), 200.

MacPike, Eugene Fairfield (ed.). *Correspondence and Papers of Edmond
Halley.* Oxford: Clarendon Press, 1932.

Pells, E. G. "Sir John Herschel's Contribution to Educational Developments at the Cape of Good Hope," *Quarterly Bulletin of the South
African Library,* 12 (1957), 58.

P(ritchard), C(harles). "Obituary of Sir John Herschel," *Monthly Notices
of the Royal Astronomical Society,* 32 (1872), 122.

Rae, Isobel. *The Strange Story of Dr. James Barry, Army Surgeon, Inspector General of Hospitals, Discovered on Death To Be a Woman.*
London: Longmans, 1958.

Roberts, A. A. *South African Legal Bibliography.* Pretoria: A. A. Roberts,
1942.

Roberts, Austin. *Birds of South Africa,* revised by G. R. McLachlan and R.
Liversidge. Cape Town: Central News Agency, 1958.

Robinson, A. M. Lewin. "The Cape in the 1830's," *Quarterly Bulletin of
the South African Library,* 12 (1957), 65.

Robinson, H. B. (ed.). *Narrative of Voyages to Explore the Shores of
Africa, Arabia and Madagascar, performed in H.M. Ships* Leven *and*
Barracouta *under Captain W.F.W. Owen.* London, 1833.

Sidgwick, J. B. *William Herschel, Explorer of the Heavens.* London:
Faber and Faber, 1953.

Sillery, A. *The Bechuanaland Protectorate.* Oxford: Oxford University
Press, 1952.

Smith, Sir Andrew. *Illustrations of the Zoology of South Africa; consisting
chiefly of figures and descriptions of the objects of natural history collected during an expedition into the interior of South Africa, in the years
1834, 1835 and 1836; fitted out by 'The Cape of Good Hope Association for Exploring Central Africa', Published under the Authority of the
Lords Commissioners of Her Majesty's Treasury.* No pagination; vari-
ously, three, four, or five vols. London: Smith, Elder & Co., 1849.

Smyth, Admiral William Henry. *A Cycle of Celestial Objects: Vol. I, Pro-
legomena; Vol. II, The Bedford Catalogue.* London: John W. Parker,
1844.

———. *Speculum Hartwellianum: The Cycle of Celestial Objects continued at the Hartwell Observatory to 1859*. London: Printed for private circulation at Dr. Lee's expense, 1860.

Stoy, Richard H. "Sir John Herschel (1792–1871)," *Quarterly Bulletin of the South African Library*, 12 (1957), 41.

Varley, D. H. "Sir John Herschel at the Cape," *Quarterly Bulletin of the South African Library*, 12 (1957), 39.

Wales, William, and William Bayly. *The Original Astronomical Observations, made in the course of a Voyage towards the South Pole, and round the World, in His Majesty's Ships the* Resolution *and* Adventure, *in the Years MDCCLXXII, MDCCLXXIII, MDCCLXXIV, and MDCCLXXV*. London: Published by Order of the Board of Longitude, 1777.

Walker, Eric A. *A History of Southern Africa*. London: Longmans, Green & Co., 1957.

INDEX

aardwolf: 80 n.
Abel, Clarke: 296
Abercorn (ship): 294
aberration: 111 n., 342 n.
Aborigines Committee: 311
Abraham's Cottage: 310
Acacia: Australian, 110 n.; *lophantha*,
  161
Achmet: 210, 223
actine: 275
actinometer: 16, 17, 32, 33, 34, 43,
  136, 152, 153, 168, 247, 248, 267,
  273, 275, 276, 281, 283, 294, 295,
  296, 307, 311, 329, 330, 353
Adamson, Rev. Dr. James: biographical note on, 69 n.; mentioned, 41 n.,
  69, 85 n., 90, 108, 141, 175, 176,
  177, 179, 203, 238, 239, 245, 265,
  267, 330, 343
Adare, Lord: 132
Adventures of Hadji Baba of Ispahan:
  359
Africaner (Griqua leader): 92 n.
Afrotis afra afra: 101 n.
Agama atra: 94 n.
Agapanthus: general, 206, 270; *africanus*, 270
Agar, Mr.: 99
Airy, Sir George Biddell: biographical
  note on, 113 n.; mentioned, 113,
  116, 125, 142, 158 n., 186, 342 n.
albatross: 34, 35, 345
Albermarle, Duke of (General Monk):
  31
Albizia distachya: 161 n.
Albuca: *viridiflora*, 142; *spiralis*, 142
Alcibiades: 92
alcohol, distillation of: 84, 186
Alexander, Capt. Sir James Edward:

biographical note on, 209 n.; expe-
  dition by, 249, 252; meteorite speci-
  men from, 322; mentioned, 209, 242,
  243, 251
Algoa Bay: 42 n.
Allan, Mrs.: 113, 118
Aloe: 157 n., 178
Altenstaedt, Alida Maria (wife of
  Baron Ludwig): 38 n.
Altona, Germany: 145
Amaryllis belladonna: 140 n., 141, 258
Amblyomma: 317 n.
Amherst, Lord: 296 n.
Amici, Giovanni Battista: 198 n.
Amphipods: 21 n., 347 n.
Analytical Society, the: xxii, 45 n.
Andreas (servant): xxiv, 204
Androcymbium: 329
Anne Robertson (ship): 164
Anstey Cottage: 328
Antares: 144
antheap, cottages built of: 223
Anthericum: general, 58, 86, 91; *fal-
  catum*, 142 n.
Antholyza: *ringens*, 157, 180, 221, 249,
  269; *praealta*, 221, 269
Aphateia: 197
Aponogeton distachyon: 120, 124
Aquarius: ʒ Aquarii, 88
Aquila: 1 Aquilae, 214
Arago, Dominique François Jean:
  biographical note on, 57 n.; men-
  tioned, 57, 58, 61, 114, 294, 296
Arbuthnot, Mr. Coutts: 322
Arcturus: 362
Argand, Aimé: 346 n.
Argo: η Argus, 48, 58, 148, 171, 222,
  228, 229, 236, 238, 239, 240, 242,
  253, 286, 296, 304, 305, 307; out-